Hyperspectral Remote Sensing in Urban Environments

This book is intended to provide a detailed perspective on techniques and challenges in detecting urban materials using hyperspectral data including a systematic perspective on the spectral properties of the materials and methods. It adopts a process chain approach in describing the topic and explains image processing steps from reflectance calibration to final insights. The objective of the book is to provide in-depth information on hyperspectral remote sensing of urban materials covering global case studies as applicable.

Features:

- Covers the complete processing chain of hyperspectral data specifically in urban environments;
- Gives more information about the mapping and classification of urban scenes;
- Includes information from basic imaging spectroscopy to advanced methods such as deep learning for imaging spectroscopy;
- Reviews detailed spectral characteristics of urban materials commonly found in world cities;
- Discusses advanced supervised methods such as deep learning with a due focus on hyperspectral data analysis.

This book is aimed at professionals and graduate students in Hyperspectral Imaging, Urban Remote Sensing, and Hyperspectral Image Processing.

Hyperspectral Remote Sensing in Urban Environments

Shailesh Shankar Deshpande and Arun B. Inamdar

CRC Press
Taylor & Francis Group
Boca Raton London New York

CRC Press is an imprint of the
Taylor & Francis Group, an **informa** business

Designed cover image: Shailesh Shankar Deshpande

First edition published 2024
by CRC Press
6000 Broken Sound Parkway NW, Suite 300, Boca Raton, FL 33487-2742

and by CRC Press
4 Park Square, Milton Park, Abingdon, Oxon, OX14 4RN

CRC Press is an imprint of Taylor & Francis Group, LLC

© 2024 Shailesh Shankar Deshpande and Arun B. Inamdar

Reasonable efforts have been made to publish reliable data and information, but the author and publisher cannot assume responsibility for the validity of all materials or the consequences of their use. The authors and publishers have attempted to trace the copyright holders of all material reproduced in this publication and apologize to copyright holders if permission to publish in this form has not been obtained. If any copyright material has not been acknowledged please write and let us know so we may rectify in any future reprint.

Except as permitted under U.S. Copyright Law, no part of this book may be reprinted, reproduced, transmitted, or utilized in any form by any electronic, mechanical, or other means, now known or hereafter invented, including photocopying, microfilming, and recording, or in any information storage or retrieval system, without written permission from the publishers.

For permission to photocopy or use material electronically from this work, access www.copyright.com or contact the Copyright Clearance Center, Inc. (CCC), 222 Rosewood Drive, Danvers, MA 01923, 978-750-8400. For works that are not available on CCC please contact mpkbookspermissions@tandf.co.uk

Trademark notice: Product or corporate names may be trademarks or registered trademarks and are used only for identification and explanation without intent to infringe.

ISBN: 9781032359106 (hbk)
ISBN: 9781032359113 (pbk)
ISBN: 9781003329312 (ebk)

DOI: 10.1201/9781003329312

Typeset in Times
by Newgen Publishing UK

Dedicated to

My parents:
Sulabha Deshpande, Shankar Deshpande,
and my family:
Wife Priya Deshpande, and son Abhiraj Deshpande

Contents

Preface xiii
Acknowledgements xv
About the Authors xvii

Chapter 1   Cities Means to Menace? 1

     1.1   Urban Enterprise: Chaos of the People, by the
People, for the People! 1
         1.1.1   Impact of Urbanization on Water 3
         1.1.2   Impact of Urbanization on Air 7
     1.2   Urban Sustainability and Systemic View of an
Urban Enterprise 10
         1.2.1   Urban Systems 11
         1.2.2   Measuring Sustainability – Sustainable
Development Goals Indicator Framework 12
     1.3   Urban Metabolism Model as an Instrument to Reduce
Complexity 15
         1.3.1   Urban Metabolism Framework Using Remote
Sensing 17
         1.3.2   Modules of Urban Metabolism Framework 20
         1.3.3   Example Remote Sensing Models in Urban
Metabolism Framework 22
     1.4   Remote Sensing Opportunities 32
     1.5   Summary 34

Chapter 2   Introduction to Hyperspectral Imaging 41

     2.1   Hyperspectral Data 41
         2.1.1   Definition and Characteristics 41
         2.1.2   Some Useful Terms 43
         2.1.3   Quantum Mechanical Explanation of Radiation
Absorption 46
     2.2   Multispectral, Superspectral, Hyperspectral, Ultraspectral 47
     2.3   Challenges 49
         2.3.1   Coarse Spatial Resolution 49
         2.3.2   Increased Data Size 51
         2.3.3   Small Number of Samples and High Dimensions 51
         2.3.4   Lack of Unique Spectral Signature 52
     2.4   Sensors and Platforms 52
         2.4.1   Spaceborne Platforms 52

		2.4.2	Airborne Sensors	56
		2.4.3	Standard Data Sets for Research	59
	2.5	Visual Interpretation of Hyperspectral Images		59
		2.5.1	Vegetation	62
		2.5.2	Soil	63
		2.5.3	Impervious Surfaces (Built-up)	64
	2.6	Common Workflow for Hyperspectral Image Processing		65
		2.6.1	Convert the Digital Numbers (DN) to Radiances	65
		2.6.2	Solar Elevation Angle Correction	66
		2.6.3	Convert Radiance to Reflectance	66
		2.6.4	Extract the Features	66
		2.6.5	Use Supervised or Unsupervised Learning Approach	67
		2.6.6	Dissemination	67
	2.7	Summary		69
Chapter 3	Reflectance Calibration			73
	3.1	Reflectance Calibration		73
		3.1.1	Atmospheric Effects	74
	3.2	Physics-based Methods		76
		3.2.1	EO-1 Hyperion Equation for Clear Atmosphere	76
		3.2.2	Second Simulation of a Satellite Signal in the Solar Spectrum-Vector (6SV)	77
		3.2.3	MODerate Resolution Atmospheric TRANsmission (MODTRAN®)	81
		3.2.4	Discrete Ordinated Radiative Transfer Code (DISORT)	81
	3.3	Image-based Methods		82
		3.3.1	Dark Object Subtraction	82
		3.3.2	Improved Dark Object Subtraction	82
		3.3.3	Improved Dark Object with Overlap Correction	87
		3.3.4	IAR and FAR	95
	3.4	Empirical Line Method		98
	3.5	Simplifying Assumptions and the Methods		99
	3.6	Working Example		100
		3.6.1	Data	100
		3.6.2	Method	105
		3.6.3	Experimental Set up for Classification Experiments	107
		3.6.4	Results of Image-based Reflectance Calibration Experiments	107
		3.6.5	Calibration Performance for IAR and FAR	109

		3.6.6	Overall Performance of Physics-based Methods 113
		3.6.7	Optimal Flat Field Assessment 113
		3.6.8	Summary of the Case Study for Image-based Methods 114
	3.7	Summary 119

Chapter 4   Spectral Resources 125

	4.1	Significance of Spectral Resources 125
	4.2	Spectral Libraries 127
		4.2.1	USGS Digital Spectral Library – Version 7.0 127
		4.2.2	ECOSTRESS Spectral Library – Version 1.0 130
		4.2.3	JPL Library 131
		4.2.4	Tarang – Spectral Library of Indian Urban Materials 132
	4.3	Analysis Methods 141
		4.3.1	Basic Statistics 141
		4.3.2	The Bhattacharyya Distance Analysis for Urban Classes 144
		4.3.3	Most Significant Wavelength Span: Fine Granularity 145
		4.3.4	Most Significant Span: Coarse Granularity (VNIR vs SWIR) 146
		4.3.5	Chromatic Properties of the Urban Materials 147
	4.4	Spectral Analysis 148
		4.4.1	Spectral Properties of Various Urban Materials 148
		4.4.2	The Bhattacharyya Distance Analysis for Urban Classes 151
		4.4.3	B-distances for Broader Urban Classes 151
		4.4.4	Bhattacharyya Distances for Cement Composites and Metal Roofs 153
		4.4.5	Most Significant Wavelengths for Discrimination of Urban Classes 154
		4.4.6	Most Optimal Span and Important Wavelengths for Urban Material at Intra-class Level for a Few Important Classes Such as Cement Composites and Metal Roofs 155
		4.4.7	Chromatic Properties 157
	4.5	Tarang – The Spectral Library and Web Services 161
		4.5.1	Tarang Components 162
		4.5.2	Using a Spectral Library 165

4.6	Spectral Characteristics of Urban Materials	166
	4.6.1 Park City and Santa Barbara/Goleta Observations	167
	4.6.2 Tarang Observations	168
	4.6.3 Significant Bands for Urban Materials	171
4.7	Summary	171

Chapter 5 Classification of Urban Land Use and Land Cover 175

- 5.1 Introduction 175
  - 5.1.1 Classification 177
  - 5.1.2 Hyperspectral Image Classification 179
  - 5.1.3 Mathematical Preliminaries 180
  - 5.1.4 Features for Hyperspectral Image Classification 181
  - 5.1.5 Urban Land Cover and Land Use 182
- 5.2 Classification with Urban Indices 188
  - 5.2.1 Useful Properties of Urban Materials for Developing Indices 190
  - 5.2.2 Various Urban Indices 190
- 5.3 Nearest Neighbour Classification 193
  - 5.3.1 Spectral Angel or Cosine Similarity 198
  - 5.3.2 Normalized Spectral Similarity Score (NS3) 199
  - 5.3.3 Hamming Distance 200
  - 5.3.4 Cross Correlogram Spectral Matching (CCSM) 201
- 5.4 Maximum Likelihood Classifier 202
- 5.5 Economic Zone Classification 207
  - 5.5.1 Study Area 209
  - 5.5.2 Hyperspectral Data 209
  - 5.5.3 Semi-supervised Learning Background 210
  - 5.5.4 Experimental Setup 216
  - 5.5.5 Results of Land Use Analysis Using Proxy Features 218
- 5.6 Advanced Machine Learning Methods Such as Deep Learning 229
  - 5.6.1 Understanding 1D, 2D, and 3D Convolutions 232
  - 5.6.2 Deep Learning for Hyperspectral Images 236
  - 5.6.3 Spectral, Spatial, and Spectral-Spatial Neural Networks 238
- 5.7 Performance Evaluation of Various Classifiers on Benchmark Datasets 240
- 5.8 Explaining Convolutional Spectral Features 243
  - 5.8.1 Capsule Net 245
- 5.9 Summary 250

Chapter 6	Appendix		259
	6.1	Summary of Important Work in Hyperspectral Remote Sensing of Urban Areas	259
	6.2	Hyperion Top of the Atmosphere Irradiance (Wavelength in nm; Irradiance in W m^{-2} μm^{-1})	266
	6.3	The Book Code Help	268
		6.3.1 Code Components	269
		6.3.2 Main Calibration Script	271
		6.3.3 Deep Learning Models	272

Index 275

Preface

Nearly half of humanity lives in urban conglomerates. Though they are economic prime movers of society, they are becoming the cause of environmental concerns too. Thus, monitoring the positive and the negative effects of urban areas is becoming critical day by day, for making cities more sustainable, energetic places to live in. Land use and land covers are the two important agents of these favourable or adverse effects. Of these two, land cover is much easier to monitor as it is directly amenable to remote sensing observations. Various composite pavements and roof materials such as concrete and asphalt dominate the urban landscape. Precision monitoring of urban land cover changes is essential for establishing their causal relation with environmental adverse effects. This book is about that – identifying urban materials as precisely as possible. The book is intended to provide a detailed perspective on the techniques and challenges in detecting urban materials using hyperspectral data. It provides a systematic perspective on the spectral properties of the materials and methods to exploit them. The book adopts a process chain approach in describing the topics and takes readers through image processing steps from reflectance calibration to final insights. Remote sensing researchers, engineers and environmental scientists all should find this book helpful in their endeavour as it attempts to strike a balance between introductory and advanced material on the topic.

Hyperspectral data provides many advantages over conventional multispectral data for impervious surface detection in an urban environment. The minute details available in the spectral signature of a particular material are lost in broad band data collection. In contrast, the signatures provided by hyperspectral data are unique, most of the time, enabling discrimination of a large variety of natural and manmade materials. Direct identification of a particular material by comparing the target spectrum with the laboratory references has thus become a viable option. Despite the advantages, hyperspectral data has not been used for urban studies as extensively as it is used for minerals, vegetation, and environmental monitoring studies. This is because the local spectral resources required are not easily available. The proposed book attempts to fill this gap. It is intended to provide end-to-end information for hyperspectral remote sensing of urban materials, which is difficult to find in one place, especially image-based calibration and spectral library and its analysis.

The objective of the book is to provide in-depth information to the reader on hyperspectral remote sensing of urban materials – covering international and Indian case studies as applicable. Additional emphasis is given to providing materials with diverse characteristics originating from various parts of the world. The book intends to provide a good introduction to hyperspectral remote sensing and should be suitable for teaching a course. It covers the entire processing chain for hyperspectral data – from reflectance calibration to land use land cover classification of urban areas. Furthermore, it presents to the reader detailed spectral characteristics of urban materials commonly found in world cities.

Other books available provide a broader perspective on urban remote sensing addressing the breadth of the topic. They commonly explore all the modalities and methods for urban remote sensing addressing common urban issues. This book focuses on the spectral properties of urban materials and describes how these urban materials can be detected using hyperspectral data. Thus, it focuses on the methods that leverage the spectral properties of urban materials. Furthermore, it provides an end-to-end process chain for the hyperspectral remote sensing of urban data, which is especially useful for students and practitioners. This book would provide a wide range of material from basic imaging spectroscopy to advanced methods such as deep learning for imaging spectroscopy.

The first two chapters of the book cover the background of both urban settings and hyperspectral data. The first chapter covers the motivation for precision monitoring of urban areas in the context of the urbanization impact on the local and global environment. It discusses a few important examples of the adverse effects of urbanization such as urban flooding and urban heat island. It further provides glimpses of urban impact assessment models and modelling approaches and makes a case for remote sensing modelling of urban metabolism. The next chapter on the introduction to hyperspectral remote sensing provides a general introduction to the topic; it introduces the common vocabulary used in the context of hyperspectral remote sensing. It also provides important principles of spectroscopy in a crisp manner so that the motivation of hyperspectral data analysis methods is well understood by the reader. With this background, the next chapters, one by one, cover each processing step in hyperspectral data processing.

The chapter on reflectance calibration explains image-based and physics-based methods for atmospheric corrections – it covers how to convert radiances to reflectance values of the spectrum. The next important chapter takes a detailed look at the spectral characteristics of urban materials. It establishes some of the fundamental properties of urban material helpful in detection. After establishing the basic image processing procedures and the characteristics of the data, the book provides basic and advanced methods of supervised and unsupervised machine learning for exploiting spectral characteristics. Advanced supervised methods such as deep learning are covered with a due focus on hyperspectral data analysis. Appropriate international and Indian case studies are presented throughout this book, as applicable.

The hyperspectral data through spaceborne platforms (such as PRISMA, SHALOM) will be available to common users soon. Furthermore, many private/public enterprises (HySpecIQ, Orbital Sidekick) have launched or are planning to launch hyperspectral spaceborne platforms, including a satellite launched by Indian Indian private company Pixxel. Overall, a large amount of hyperspectral data will be available for analysis. Generating meaningful insights from the same for social and commercial benefits will be essential. In this context, the code used for some of the case studies and analysis results in the book is made available to the readers. This will be helpful support to the text description of the topic. Readers are encouraged to use it as applicable. Thus, the book will be a useful resource to all researchers and engineers who are eager to use the spectral data for urban application, and maybe more!

Shailesh Shankar Deshpande
Arun Inamdar

Acknowledgements

Work of this nature and involvement is not possible without sound mentors, well-wishers, and friends. I would like to thank all of them who supported and helped me during the work.

I would like to thank both of my mentors Prof. Arun Inamdar and Prof. Harrick Vin for providing constant encouragement and guidance. Constructive feedback is an important part of research. I would like to thank Prof. Krishna Mohan Buddhiraju and Prof. Shirish Gedam for their useful feedback and suggestions for improvements. I would like to thank Dr Balamurlaidhar and Dr Arpan Pal, for their constant support throughout the project.

Big thanks to Dr Daniel Schläpfer for a valuable discussion on spectral deconvolution, and Dr David Jupp for providing useful links for Hyperion spectral response functions. I appreciate their earnest help. I would like to thank Robin Wilson for his support for the Py6S interface. I would like to thank Prof. Martin Herold for providing me with the necessary data to reproduce the spectral curves of a few urban materials. I also thank principal investigators for their effort in establishing and maintaining the AERONET Pune site. Special thanks to two of my associates Chaman Banolia and Karan Owalekar – it was because of their help that I was able to complete the project in a timely manner. Thanks to Chaman for preparing data in some of the required cases. Big thanks to Karan. He painstakingly verified all the Python scripts and maintained the *splibtarang* site as required.

I would like to thank Priya Deshpande, Piyush Yadav, and Sachin Gupte for their assistance during the fieldwork. It was good to have such friends around, always eager to help. The staff at CSRE were always helpful. I would like to thank Mr Nikam, who kindly arranged spectrometers for the field data collection. Special thanks to Manoj Apte, my friend at TRDDC, for meticulously going through the figures and pointing out the errors. I would like to thank Sachin Pawar for his helpful suggestions on one of the most important chapters in the book on image classification. Some of the fieldwork required for the spectral resources was done at IIT-Bombay. I would like to thank Prof. Bakul Rao for her support during my entire stay at IIT. She gave me a home away from home. Special thanks to Priya Sathe for sharing her feedback on the cover page.

Last but not least I would like to thank the Indian Society of Remote Sensing (ISRS), Dr Gagandeep Singh, and Aditi for their support during the work on this book. Their earnest professional help was comforting.

Finally, to all my friends – thanks for standing by my side when I needed you most.

About the Authors

Shailesh Shankar Deshpande is currently working as Principal Scientist at Tata Research Development and Design Centre, TCS research, Pune, India. He is a part of Robotics and Autonomous Systems (Remote sensing and space-tech division) of TCS research. He has been one of the lead contributors in important artificial intelligence (AI), machine learning (ML), and remote sensing research and solutions TCS research has developed over the years.

He has an M-tech and PhD in Hyperspectral Remote Sensing from IIT-Bombay, Mumbai. His main research interest includes the processing of hyperspectral and multispectral remotely sensed data. He is particularly interested in using the spectral and spatial properties of earth observations to generate meaningful insights.

With research and development experience of nearly 25 years in multidisciplinary areas such as remote sensing and machine learning, he has published over 50 research papers in refereed journals and conferences of international repute. He has authored 5 book chapters and has more than 10 national and international granted patents. A couple of his papers at international conferences have received Best Paper awards. He serves as a reviewer for major remote sensing journals. He has chaired sessions at the international IEEE-WHISPERS 2018 to 2023 conferences, and he is a member of the review committee of IEEE-IGARSS. He has supervised a few B-tech projects and several internships.

Arun B. Inamdar was at the Centre of Studies in Resources Engineering, IIT-Bombay, Mumbai 400076, India. He has an MSc and PhD in Geology from IIT-Bombay, Mumbai. His research interests include the use of remotely sensed data in various land and ocean applications especially in assessing/monitoring various water quality parameters of the coastal ocean; coastal land use and vulnerability studies; wetlands monitoring; integrated coastal management. He has published over 70 papers and guided 10 doctoral and 20 master's students, besides executing more than 50 research and consultancy projects in various multidisciplinary areas. He has been a reviewer for various national and international journals, including *ISRS Journal* and *Current Science & Environmental Monitoring and Assessment*. He was a member of multiple expert committees advising state and national governments on various marine and urban environmental issues.

1 Cities Means to Menace?

This chapter will discuss:

- Complexity of the urban environment
  - Impact of urbanization on its surroundings mediated by impervious surfaces
  - A few case studies of urban impact for urban catchment changes, and urban heat island
- Urban sustainability and systemic view
  - Model for urban interaction with its surroundings
  - Burden, benefit, and vulnerability perspective
- Urban metabolism framework for understanding urban impact and reducing the complexity of sustainability assessment
  - Modules of the framework
  - A few example models in the framework
- Hyperspectral remote sensing opportunities for urban area

1.1 URBAN ENTERPRISE: CHAOS OF THE PEOPLE, BY THE PEOPLE, FOR THE PEOPLE!

Fifty per cent (3482 million urban vs. 3387 rural) of the world population lives in cities even though cities occupy just ~4% of the area on the earth. It is estimated that they consume 75% of the world's resources (The World Bank Group, 2016). By 2030 the urban dwelling population will increase to 60% (United Nations, 2002). The number of towns in India, the second most populous country in the world, is 3894 and these towns are home to 31.16% (377 million) of India's population (1210 million) (Office of the Registrar General & Census Commissioner, India, 2011-2). The Indian urban population is expected to grow to 35% by 2021.[1] Out of these 3894 towns, fifty-three are million-plus cities and they account for 42.6% (160.7 million) of the urban population. The urban population has increased by 31.6% (Office of the Registrar General & Census Commissioner, India, 2011-1) since 2001. India just covers 2.4% (135.79 sq. km) of the world's surface but it supports 17.5% of the world population (Office of the Registrar General & Census Commissioner, India, 2011-2). Maharashtra's urban population stands at 46%, with a 22% increase from 2001 to 2014 (Office of the Registrar General & Census Commissioner, India, 2011-1).

Urban areas are important conglomerates of socio-economic activities and are regarded as prime movers of any state. However, at the same time, because of the population density, they act as critical agents in changing their surroundings – on a substantial spatial and temporal scale. Urban areas act as sinks of a large number of societies' resources and major sources of environmental change. These land use and land cover (LU/LC) changes caused by urbanization affect urban surroundings adversely. Increased run off, ground water depletion, change in ambient temperatures (Heat Island Effect), and increase in air and water pollution are a few examples of these adverse effects. Some of the changes affect local surroundings whereas some of them affect the global environment as well.

For example, local changes in the hydrological cycle. Impervious surface reduces the quantity of ground water recharge. As a result, ground water depletion is an inevitable consequence if the mitigation steps are not taken. Furthermore, the quantity of urban runoff is increased, and peak runoff is achieved in a brief time. Urban flash floods have become a routine phenomenon now in many metros. The urban runoff naturally adds to the non-point source pollution, which finally reaches the natural sinks such as rivers, coastal water and so on. Non-point source pollution is among the largest pollutants in the United States of America (USA) (Arnold & Gibbons, 1996).

Whereas urban impervious surfaces have a more pronounced local effect, other artefacts of urbanization, say air pollution by cities, have a global effect. Such emissions by cities are becoming a global issue, in addition to local issues. CO_2 emissions, for example, are a leading cause of global warming and climate change. The particulate matter, under certain conditions, shows a potential to change the amount of precipitation. The effects of particulate matter on precipitation are not very well understood yet (Khain et al., 2008). Because of these adverse effects, cities are becoming unsustainable day by day.

Figure 1.1 shows a common sequence of urbanization events that mediate the changes in the urban surroundings. The thickness of the connecting arrows in the diagram approximately indicates the lag between the cause and effect. Thick line indicates direct impact or immediate effect whereas thin line indicates longer duration between cause and effect. The changes in air, water/hydrological cycle, and soil may lead to changes in land use patterns after a long time. Land use changes are associated with land cover changes. The vegetation or soil covers are changed to concrete or other urban composite materials along with the changes in land use. The new land covers have different physical/chemical responses to solar radiation, rain etc. compared to the land cover it replaced. The heat capacity of soil and concrete is different, as is their permeability. The effect on the surroundings is a natural consequence of these changed properties and the new activities according to the land use. In case the effect is more than the surroundings can dissipate safely or adapt to, the adverse effect of them on the urban area is inevitable.

In the next sections, we discuss two such major adverse effects mediated by land cover changes: changes in the hydrological cycle, and changes in the temperature regime of a city. The severity of these two problems is experienced by the local population. Hence these are the major problems faced by most of the cities in the world.

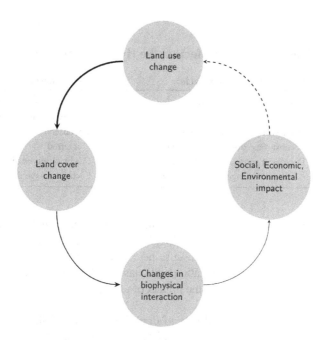

FIGURE 1.1 Sequence of urban development events.

1.1.1 Impact of Urbanization on Water

The effects of urbanization on the hydrological cycle were recognized early. However, the urgency of mitigation measures was not realized until recently. This is due to increased urban floods and logging incidents in the recent past, accompanied by heavy rains. Developing countries are especially more susceptible to urban flood incidents (National Institute of Urban Affairs, 2016). The effect of urbanization on the hydrological cycle within the city has multiple effects. To list a few:

a. It replaces water-pervious surfaces such as soil with impervious surfaces (concrete, bitumen etc.) and this increases the amount of runoff,
b. It affects the natural drainage carrying capacity of the urban catchment area; lost drainage or decreased carrying capacity are two main effects if the natural drainage in the urban catchment is ignored,
c. In the case of poor waste management in the city, it results in the blocking of the storm sewers (artificial drainage system), worsening the situation further (National Institute of Urban Affairs, 2016).

Table 1.1 shows the common effects of land cover changes on the hydrological cycle in the urban catchment. It also shows which of the changes are amenable to remote observations. Multiple modalities of data acquisition are useful indeed. However, hyperspectral data provides more possibilities because of its ability to resolve a target material or its chemical properties more precisely (more about this in Chapters 2 and 3).

TABLE 1.1
Land Cover Changes and Their Effect on the Hydrological Cycle

Changes	Effect on hydrological cycle	Remote observation
Clearing of vegetation	Increased flow rate, less percolation, stability of slopes	Can be observed directly, land cover change detection
Soil compaction	Decreased permeability, and subsurface flow, subsurface storage, increased runoff	Not observed directly, can be inferred
New pavements, buildings	Decreased permeability, increased runoff, blocked natural drainage or reduced capacity	Can be observed directly, land cover change detection

As can be seen, a simultaneous increase in the runoff and a decrease in carrying capacity enhances the chances of urban logging or flooding. Extreme rainfall events aggravate the situation further. We discuss here a few examples for understanding the causal relation between urbanization and changes in the hydrological cycle. We present two case studies as well on the impact of urbanization on natural drainage.

Ogden et al. (2011) studied the effects of drainage density and the width function in addition to the effect of impervious surfaces in the catchment. Drainage density is defined as a ratio of the total length of channels and the total area of watershed. Thus, drainage density has a unit of length per area. Width function is defined as a number of channel segments at a given distance from the basin outlet. Width function indicates the drainage density at a particular distance from the basin outlet. They found interesting relations between drainage density, flood peaks, and impervious surface area. The impervious surface area did not affect the flood peak if the drainage density was far from the outlet. Contrary to belief, width function was observed to have a more pronounced effect on runoff volumes and efficiencies than imperviousness.

Oudin et al. (2018) studied various metrics of urbanization and their relationship with the hydrological impact of urbanization. The focus was to study whether the impervious surface area alone explains the hydrological impact or if its location, spatial distribution and so on are required too. The authors did not calculate the lost natural drainage in the urban catchment. Instead, three metrics that indicate the proximity of the impervious surfaces to the natural drainage were considered. They were: RDIST.NET, RDIST.OUT, IMP.100. RDIST.NET – is a ratio of the mean distance of urban pixels to the hydrological structure to the mean distance of the catchment pixels to the hydrological structure. Higher values indicate that the hydrological network is far from the city. RDIST.OUT is a similar ratio for catchment outlets. Higher distance indicates the catchment outlet is far from the city. IMP.100 is the mean imperviousness of the land use within the 100 m corridor of the large natural stream. Higher value indicates that the buffer zone is having more impervious area. The metrics can be treated as proxies for the dysfunctional natural drainage, assuming that the proximity of urbanization decreases the natural drainage capacity.

GAGES-II database was used for the study. 142 catchments from GAGES-II were selected for further studies. Most of them were situated on the east coast of the United States of America (USA). The catchments with large urban areas and the availability of flow measurements were the factors for selecting only a few areas. The study underlined the importance of impervious surfaces as a generic indicator for the hydrological impact of urbanization. The hydrological network was away from the main urban areas for the selected catchments and hence the proximity of impervious surfaces and hydrological impact relation was inconclusive.

Urbanization impacts the natural drainage system adversely. A drainage system includes the land cover such as soil-grass or mixture, subsurface material, natural stream channels, subsurface storage, and stream channels. An artificial drainage system includes all the structures built for carrying surface runoff to the nearest natural drainage such as streams, rivers, or oceans. This includes pipes, canals, culverts and so on. Booth (1991) considers the urban land covers such as impervious surfaces as the part of drainage system too. The water from the drainage systems overflows and floods urban areas. Floods may cause further sediment deposits or erosion of the areas which otherwise would not have been exposed to the floods. This is predominantly because of the water channel cross-sectional changes, outright blockages of the channel at some times and places, increased runoff because of impervious surfaces and so on. According to Booth (1991), the data to show the effectiveness of the corrective changes in drainage in reducing urban floods is missing.

The effects on drainage pattern changes are cascaded to both surface flow and subsurface flow as well. Usually, the effect of urbanization is an increase in peak runoff and changes in the peak time of the runoff. However, the net impact will depend upon the original regime before urbanization. For example, if the regime was predominantly the surface flow regime and the vegetation cover is lost because of urbanization; though the runoff quantity may not change, the peak runoff may be achieved much more quickly than before. In cases where the land cover is changed to concrete or asphalt pavements and the regime was subsurface flow; then the runoff will increase multi-fold as all the water which was part of subsurface flow is now run off. Thus, urbanization affects the subsurface flow regime more (Booth, 1991). The impact of urbanization is not limited to surfaces; it is in a vertical direction above and below the surface as well. More systematic studies are required to understand the impact of urbanization on the subsurface flow regime. Measuring the subsurface impact is much more challenging than monitoring above-ground changes because of urbanization.

1.1.1.1 Mashhad Case Study

The effect of urbanization on the watersheds in the city of Mashhad, Iran, was studied by Hosseinzadeh (2005). The focus was on anticipating increased runoff and taking preventive measures against recurring flash floods and urban floods. The study focused on establishing causal relations between urban development and floods. Mashhad is the second largest city in Iran and is the capital of Khorssan province. It is situated in the northeastern part of Iran in the Kashaf Rood basin. Geomorphologically, the city is located in the flood plains and extends towards the mountains in the south or southeast of the city. The study area bounded by $36°-13°$ to $36°-25°$ N and $56°-28°$ to $56°-44°$ E was chosen.

The area was divided into nineteen smaller catchments. The amount of flood water entering the city from these catchments was determined by the study of historical records including aerial photographs, field observations and so on. The past boundary of the city was determined by the 1966 photograph. The growth of the city from the 1966 baseline was determined using satellite images. Changes in drainage patterns, which was one of the main objectives of the study, were determined using the following steps: First, the natural drainage streams were marked on the base map. Then, the natural water channels affected by urban expansion were recorded using aerial photographs and field observations.

The main reasons for changes in the natural water channel were excavation for stones for construction from mines within the urban catchments, dumping of debris in the channel, and extraction of sand from the riverbeds for construction activities. The main channels of ~140 km (277 nos.) in length were blocked or severely affected because of urbanization. Secondary channels of ~83 km (43 nos.) were lost because of urbanization. The city area was increased ~100 times (from 0.975 to 104.4 km^2 in alluvial fans and ~4 times (from 28.075 to 109.7 km^2) in flood plain regions of the city basin. All of these changes resulted in doubling the frequency of urban floods.

1.1.1.2 Pune Case Study

Pune is a cultural, educational, and industrial hub of western Maharashtra, India. The city is located beside the foothills of the Sahyadri mountains and on the confluence of the rivers Mula and Mutha. The city has seen tremendous growth in recent years because of flourishing information technology and the auto sector. Many software technology business centres have come up on the fringes of the city and this has resulted in growth in infrastructure development projects in and around the city.

Yadav and Deshpande (2015) studied the growth of the city over a decade from 2001 to 2014 using remotely sensed data. The focus of the study was to assess the increase in impervious areas in the city over the years and to study the impact of growth on natural drainage patterns. Pune has started facing urban flooding and water logging issues in the rainy season and the frequency of the same has increased noticeably. This was the motivation for these studies.

Landsat data (USGS, 2015) between 2001 and 2014 was used for the analysis of the land use and the land cover changes. The images from the months of April and May were chosen for the study as they provide a decent quality of cloud-free images. This is a 1500 x 1500 pixels seven-band multispectral image with a 30 m spatial resolution. The images were corrected for atmospheric effects and scan line corrector (SLC) errors. SLC^2 errors were corrected using a moving window approach. The image was convolved with a mode filter to remove the scan line drops. The Digital Elevation Model (DEM) Images were collected from the CARTOSAT-1 satellite of the Indian Space Research Organization (ISRO). The satellite uses PAN (2.5 m) stereo data sensor with a spatial resolution of one arc sec (BHUVAN, 2015). The natural drainage network analysis was done using the Digital Elevation Model (DEM). The DEM image of the target area was converted to slope using QGIS DEM Terrain models (GDALDEM, 2015). The network of streams was identified using the watershed model of GRASS module which takes DEM as an input to extract the river streams and their basin areas (Watershed, 2015). Some of the drainage streams were manually corrected by overlapping them on google earth images. The channel loss was identified by overlaying corrected stream networks over the satellite images.

Analysis findings suggested that the Pune municipal corporation (PMC) area and the Pimpari-Chichwad municipal corporation (PCMC) area together have observed a total loss of 4.14% as of 2014 from the base year of 2002–04. There was a total loss of 5.30% of the channel network in the areas that had maximum urban penetration, whereas the loss of only 0.51% of streams was observed in the areas with low urban penetration. Thus, the association between urbanization and loss of natural drainage was evident. Further studies (Yadav & Deshpande, 2016) divided the Pune basin into seven urban catchments for detailed analysis of increases in impervious areas according to the catchments. The focus of this study was to anticipate the increase in runoff for different urban catchments because of increases in impervious surfaces. The study showed that there was tremendous growth in the built-up area, affecting the overall catchment of various rivers. The overall impervious area in the PMC and the PCMC corporation has almost doubled since 2001 (to 2014). The increase in impervious surface has led to an increase in runoff as well. The comparison of 2001 and 2014 runoff from the PMC-PPCMC indicated an increase of 87.8% just because of increases in impervious surfaces. One of the urban catchments named Hadapsar was found to be the most urbanized and most affected urban catchment with 62.3% impervious area, and runoff was increased to 168.89% (Yadav & Deshpande, 2016).

1.1.2 Impact of Urbanization on Air

The urban heat island is a well-known phenomenon. The urban area is marked by higher ambient air temperatures than its surroundings, which is referred to as an Urban Heat Island (UHI). The Urban Heat Island effect is observed at two levels; surface temperature as well as ambient air temperatures (Phelan et al., 2015). The reasons for urban heat islands are well understood. However, it is difficult to model UHI because of the complex interaction between multiple urban entities (Phelan et al., 2015). The higher temperature of the urban area than its surroundings is because of human activities in the urban area and the changes in the land cover. The heat exchange response of soil and vegetation is entirely different compared to concrete and other pavements. The radiated heat by urban structures, heat created by internal combustion engines, and heat by power plants are a few of the major causes of the increased temperatures. The increased temperatures, in addition to discomfort, increase the cooling needs of the city. The elevated temperatures have shown a positive correlation with mortality (Basu & Samet, 2002). India experienced ~17000 deaths in the last ~50 years (1970–2019). The increase in mortality because of heatwaves in India during 2000–2019 was ~62% compared with the duration 1980–1999 (Ray et al., 2021).

The difficulties in the observation of temperatures or any other urban meteorological attribute in a city make it more difficult to build predictive models (Oke, 1982). Remote sensing is one of the useful tools for modelling urban heat island (Streutker, 2003). Large-scale spatial and temporal measurements of temperatures are provided by various remote sensing platforms (Mirzaei, 2015). Earlier remote sensing studies were focused on plotting the cross-sectional surface temperature profiles of the city. These studies were essential for establishing site-specific deviations, if any, from the general observed phenomenon of urban heat island. One of the early remote sensing studies of urban heat island by Streutker of Houston, Texas, USA, had two objectives: the primary objective was to quantify the urban surface heat island, and the secondary objective was to study the correlation between the rural temperature and the magnitude of urban heat island (Streutker, 2003). Recent remote sensing studies try to understand the effect of land use and land cover on surface temperatures, in addition to surface temperature studies. Chen et al. (2006) used Landsat Enhanced Thematic Mapper (ETM) data between the years 1990 to 2000 in the Pearl River Delta of Guangdong Province, southern China, to understand the land use land cover changes and associated surface temperature changes. The temperature changes were explained using urban indices including the newly developed Normalized Difference Bareness Index (NDBaI). The surface temperature was found to be positively correlated with the built-up index. More et al. (2015) studied the urban heat island effect for Pune in an equivalent manner. However, the urban indices were not used. In addition to conventional analysis methods, Hilbert Huang Transform (HHT) was applied to the series of temperatures over time. Hilbert Huang Transform is useful in identifying long trends in the time series, if any. Hilbert Huang Transform represents the data as an additive combination of the basic waves which are not sine or cosine waves. The months of May and April were used for the study. Urban temperatures were found to be higher by 2 degrees on average than their immediate surroundings.

As compared to the drainage network impact assessment, UHI studies are straightforward. The thermal bands of satellite platforms are used to retrieve the surface temperatures. The surface temperatures are plotted against the time and space intersection of the target area. The statistics of the data are interpreted in the context to understand the intensity of the UHI. We discuss two case studies in more detail: Houston and Pune.

1.1.2.1 Houston Case Study

According to the 1996 census, Houston was the fourth largest city in the United States of America (USA). It remains the fourth largest city; the difference is that the population has increased to ~2.28 million from 1.6 in 1996 (Unites States Census Bureau, 2023). AVHRR level 1b High Resolution Picture Transmission (HRPT) data between April 1998 to January 2000 was used for the study. A total of 48 temperature maps were prepared for the duration. A $2° \times 2°$ box centred around the city was considered as a study area. The radiance values from the images were corrected using calibration constants and then converted to the brightness temperatures using the inverse of Plank's equation. Water and cloud pixels were removed from the image for further analysis. The background rural temperatures were subtracted from the remaining land pixels. A Gaussian profile was fitted to the urban temperature component. In addition to the Gaussian model, a best-fit profile to the natural logarithm of the temperatures was obtained using linear regression.

The Gaussian model underestimated the magnitude of the urban heat island effect as some of the pixels had a greater difference than suggested by the Gaussian model. The intensity of the urban heat island and the spatial extent were found to be correlated. The magnitude of the urban heat island and the rural temperatures were found to be negatively correlated. That is, in the situation of a high rural temperature, it was difficult to detect the urban heat island. This was a significant conclusion.

1.1.2.2 Pune Case Study

Twenty-six L7-ETM+ images from the years 2002 to 2015 were used for understanding the urban heat island effect, if any, for the city of Pune, Maharashtra, India (More et al., 2015). The data from the months of May and April was considered as these two months provide a decent quality of cloud-free images. L7-ETM+ path 147 and row 47 cover Pune city. The revisited period of the Landsat 7 is 16 days. The thermal bands of the Landsat 7 images have a spatial resolution of 120 m. The images were cropped as per the target area and dropped lines were replaced using the sliding window approach (Yadav & Deshpande, 2015). The digital numbers of the images were converted to radiance values and then the United States Geological Survey (USGS) thermal band algorithm (NASA, 2011) was used to retrieve the surface temperature. Equations are as follows:

$$T = K2/ln(K1/L + 1) \qquad \text{Equation 1.1}$$

$$LST = T/(1 + (\lambda.T/\rho).ln\varepsilon) \qquad \text{Equation 1.2}$$

where,
L = spectral radiance in watts.meter^{-2}.ster^{-1}μm^{-1}
K1 = calibration constant 1 in watts.meter^{-2}.ster^{-1}μm^{-1} = 666.09 (Landsat 7)
K2 = calibration constant 2 in Kelvin = 1282.71 (Landsat 7)
T = effective at-satellite surface temperature in Kelvin
ε = emissivity = 0.004.PV + 0.986
P_V = ((NDVI-NDVI$_{min}$)/(NDVI$_{max}$-NDVI$_{min}$))^2
Normalized Difference Vegetation Index (NDVI) = (Band4-Band3)/(Band4+Band3)

Hilbert Huang Transform (HHT) was used to identify long-term trends, if any. In addition, land use land cover classification of the images was performed using Support Vector Machines (SVM). This was for understanding the temperature and land cover correlation, if any, especially in a predictive sense. The two cross sections of the city were considered for plotting the temperatures.

The urban heat island effect was not as pronounced as it was expected to be. Three temperature regimes were clearly identified in both cross sections: the city core with higher temperatures, a marginally cooler immediate surrounding, and then the mixed open areas outside the city extent with higher temperatures again. The intensity of UHI for the later years did not show in increase over the previous years. An almost constant higher temperature of ~2 degrees was observed for the entire period. Table 1.2 shows the fraction of each land use land cover class in the total study area and the average temperature of pixels in the corresponding region. Thus, surface temperatures recorded at ~10.30 AM in the given geographical and urban structure were not sufficient to understand the UHI completely.

These are just a few examples of the impact of urbanization on the local and global scale. The purpose of the discussion was to motivate the need of urban impact assessment for sustainable development. Cities have played a significant role in changing the world's climate. If there is a case for reversing human-mediated climate change for our harmonious co-existence with nature, cities need to play a critical role.

TABLE 1.2
Land Use Wise Temperature Distribution

Area	Percentage cover	Average temperature (in Celsius)
Deciduous	3.06%	46.82
Forest	0.51%	40.57
Agriculture	9.95%	41.57
Burnt grass	25.00%	46.09
Grass	3.31%	43.43
Residential	28.87%	46.32
Industrial	0.38%	45.45
Bright soil	1.11%	46.06
Common open area	26.54%	48.11
Water	1.09%	32.65

1.2 URBAN SUSTAINABILITY AND SYSTEMIC VIEW OF AN URBAN ENTERPRISE

Sustainability is defined as *"an ability to meet the present needs without compromising future ..."* (United Nations, 1987). The goal is for the entire humanity and each individual part of it. Whereas sustainable energy determines the sustainability of the system to a considerable extent, a sustainable system is not possible without sustainable practices and sustainable parts. Cities are one of them. Cities provide an important socio-ecological function of wealth generation for individuals and society. While policies play a key role in regulating social resources, cities' burden on social resources because of their activities is inevitable. Cities are viable only if they remain relevant socio-ecologically. In addition to the United Nations (UN) definition, sustainability can be considered simply as *a likelihood of successfully continuation of the socio-ecological functions, for a sufficiently long time*. This depends upon many factors; availability of energy and other resources at a sustainable rate, sustainable operations, and sustainable socio-ecological function are three underlying parameters of the sustainability of an enterprise such as a city. The following list elaborates on conditions under which an enterprise like a city may remain sustainable. These factors follow an input-process-output model, that is, each factor is derived from input space, process space, and output space of an enterprise. For example, an enterprise consumes some resources (energy, time, materials) and produces products (soft or hard) by some process (men, machines, methods). The sign of the list item indicates the factors leading to the positive or negative effect on social, economic, environmental/ecological (SEE) benefits (United Nations, 1987).

   a. [-] Rate of consumption of raw materials is less than rate of production of raw materials (the mass balance relates to the sustainability; it is environmental performance indicator as well).[3]

b. [+] Physical infrastructure, men and materials are not threatened or are resilient under any external calamity resulting from natural or anthropogenic causes. This is the ability to function under internally and externally adverse conditions.
c. [+] Rate of economic and social service delivery is above or equal to the rate of demand (consumers).
d. [-] It produces by-products which are manageable and do not threaten all above 3 (this is a direct measure of the environmental performance).

1.2.1 Urban Systems

The complexity of an urban system is very well recognized. The effects of urban activities on its surroundings are far-reaching. Abstracting those activities in some high-level categories in a holistic manner is helpful in reducing the complexity. The abstraction is required for understanding the impact of urbanization and for designing mitigation measures for sustainable cities. There are several ways to do it; we discuss one of them.

A natural or anthropogenic process of any kind requires some resources including energy for its maintenance and growth. The process produces some outcomes that may be useful, or harmful. The time and space extent of the adverse effects will depend upon many conditions; for example the potency of the outcome, the adaptive capacity of the surrounding system, and so on. The same model, that is, input-process-output is extremely helpful in explaining the impact of any enterprise, small or large, like a city. In its most generic sense, an enterprise is an entity or a group of people with some infrastructure and processes bringing social, economic, and environmental change. Hence it could be as small as a retail shop, or it could be as large as an industrial setup like a power plant, or even a metro city, as in this case. The word "enterprise" refers to a city most of the time in this chapter, if not stated. However, it must be remembered that the discussion applies to any small or large enterprise. The size determines the complexity of the impact assessment. The larger the enterprise the more complex it is to audit all the burdens, benefits, and vulnerabilities. We attempt to understand the impact of a city on the society and environment in the context of sustainability.

The activities of an enterprise, however small or big, impact its surroundings. The enterprise consumes society's resources and provides services to the individual members of society. Each of these activities exhibits positive or negative effects on society, the economy, and the environment (SEE) (United Nations, 1987). Figure 1.2 illustrates the categories of each such activity at the broader level. The employment and services provided by an enterprise affect society positively. On the other hand, the waste produced by an enterprise pollutes the environment. Consumption of the resources adds to the environmental burden. An enterprise and its activities are exposed to natural hazards as well. Natural hazards not only affect enterprise activities adversely but also natural resources and society, in general. Resilience provides protective cover against natural hazards. Thus, any of the enterprise activities can be broadly categorized into benefit, burden to society, economy, and the environment

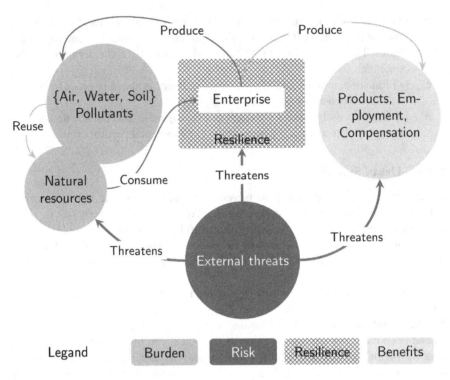

FIGURE 1.2 Systemic view of a city.

(SEE) easily. This can be summarized into the following approximate mathematical form:

Sustainability $\cong$ SEE compensation^4 (COM) + SEE benefits (SB) −
SEE burden (EB) − Vulnerabilities (V);   *Equation 1.3*

where,
COM = SEE compensation,
SB = SEE benefits,
EB = SEE burden,
V = Vulnerability,
V $\cong$ [resilience (sensitivity + adaptivity + strength) + exposure (likelihood + potency].

1.2.2 Measuring Sustainability – Sustainable Development Goals Indicator Framework

The United Nations (UN) Sustainable Development Goals (SDG) framework (United Nations, 2022) is an umbrella framework for measuring the SDG performance of a large enterprise such as a country. The 17 SDG goals are divided into sub-goals. The sub-goal description is a high-level achievement target or objective activities geared to achieve a specific SDG. Each sub-goal has indicators associated with it and they are metrics that indicate some measure of achievement. The SDG goals are for humanity in general and addressed to governments or administrative units. Table 1.3

TABLE 1.3
Sustainable Development Goals and Cities

Goal description	Administrative scope	Impact
1: No Poverty	Managed through policy instruments	Positive
2: Zero Hunger	Managed through policy instruments and social welfare schemes	Positive
3: Good Health and Well-being	Directly responsible for providing healthy living conditions through good waste management, and health services for the society	Positive
4: Quality Education	Managed through policy instruments and social welfare schemes	Positive
5: Gender Equality	Managed through policy instruments and social welfare schemes	Positive
6: Clean Water and Sanitation	City administrative departments are solely responsible for supplying clean water and sanitation	Positive
7: Affordable and Clean Energy	City administrative unit is responsible for this	Negative
8: Decent Work and Economic Growth	Managed through policy instruments and social welfare schemes	Positive
9: Industry, Innovation and Infrastructure	Managed through policy instruments and social welfare schemes	Positive
10: Reduced Inequality	Managed through policy instruments and social welfare schemes	Positive
11: Sustainable Cities and Communities	This goal is about cities	Negative
12: Responsible Consumption and Production	Managed through policy instruments and social welfare schemes	Negative
13: Climate Action	Cities are responsible for this as they are the primary source of carbon dioxide emissions	Negative
14: Life Below Water	Managed through policy instruments and social welfare schemes	Negative
15: Life on Land	Managed through policy instruments and social welfare schemes	Negative
16: Peace and Justice Strong Institutions	Managed through policy instruments	Neutral
17: Partnerships To Achieve The Goal	Managed through policy instruments	Neutral

shows the main goals and their relevance to the cities. The relevance of the goal to the city administrative department and its activities is discussed along with the goals. The relevance is discussed with respect to the city's administrative scope. The "impact" column suggests the general perception of the impact of urban living on the goal. For example, if one had a better chance of receiving a good education because of urban living or otherwise. The comments are qualitative statements.

How to achieve the goal or how to measure the parameters of the process for achieving the goal is not described in the framework. At times, as expected, the monitoring of an indicator on a continuous basis is the only way to measure the goal. In that sense, the metrics are designed for a conventional auditing outlook (wherein the output quantity is taken as evidence of the correct process). For example, 1.5 – build resilience for the poor goal has an indicator 1.5.2 – direct loss attributed to disasters. That is, resilience is believed to be achieved if the economic loss is reduced over the years – this may not be useful other than for reporting. A similar outlook exists, for example, 2.4 – 2.4.1; 9.1 – 9.1.1–9.1.2 (United Nations, 2022). Furthermore, the ideal metric values or some sort of benchmarks are not provided; those would have been helpful to assess the progress on each goal.

Building a resilient system is one of the important objectives for achieving the SDGs (United Nations, 2022). However, resilience and vulnerability are not defined or measured directly. It is believed that reduced economic loss indicates that we have built a resilient infrastructure and we have identified vulnerabilities. It is indirectly defined at 9.1.1, 1.2; 11.2.1 (United Nations, 2022). However, more comprehensive treatment is not reflected in the metrics for resilience. The goals and the metrics are based on audit or reporting requirements. They do not give enough thought to measuring process parameters, especially for an important parameter like resilience. Thus, at a broader level, there is greater scope for designing metrics for the process of achieving the goal; for example, resilience.

The Global Reporting Initiative (GRI) guidelines enable sustainability reporting for a business enterprise. While enterprises across the globe are submitting GRI reports, it is a challenging ask still. The goals are intertwined with each other, and these guidelines do not attempt to reduce the data complexity by providing a simplified conceptual framework. The typical reports run into hundreds of pages without providing meaningful insights. The SEE sense is lost in reporting numbers.

Some of the attempts in literature do not provide a holistic view and meagrely map the SDGs to the identified pillars for sustainability (Institute of Chemical Engineers, n.d.; Industrial Environmental Performance Metrics: Challenges and Opportunities, 1999). There are some past studies, which mathematically formulate the environmental burden or environmental performance of a company. Mathematical formulation is certainly helpful. However, it assumes that the data is available and collected by an enterprise on a regular basis. Furthermore, it does not provide any guidelines on measurements. The index calculation assumes the point source data for most of the environmental factors. These isolated efforts need to be integrated for an enterprise, adapting an eco-services-based approach for sustainability.

There is a need to organize and interpret the SDGs for an enterprise for its sustainability and its contribution towards a sustainable earth. The environmental performance of an enterprise is seen as one of the key differentiators for its socio-economic sustainability. An enterprise does not operate in isolation. It interacts with its social, economic, and ecological/environmental surroundings. The SDGs provide the broader framework for any entity of SEE change for remaining relevant to the socio-ecology for a longer time.[5]

One of the alternatives for the sustainability performance assessment of a large enterprise such as a city is to design a separate assessment module for each ecosystem (for example, air, water, and soil). Remote sensing is extremely helpful in such situations, especially for monitoring large ecosystems such as oceans, forests, the atmosphere, and so on. For example, the number of water bodies with good water quality around the city can be easily monitored using remote sensing. The specific indicators related to the measures are 6.3.2, 6.4.2, and 6.6.1 (United Nations, 2022). This perspective would be helpful for goals directed toward an environmental impact. However, assessing the indirect contributions of a city's activities to other sustainability factors is still complicated. Instead, building a model with the activities of a city that affects the environment (burden/environmental performance) and the activities in the surroundings that affect the enterprise (vulnerability to external threats) would be operationally easy (see Equation 1.3, and Figure 1.3). In other words, a simplified view of what a city does to the environment and what the environment does to a city provides operational ease. Such a unified view would be further helpful in slicing and dicing the information for calculating the burden/benefits it brings to society and hence sustainability.

1.3 URBAN METABOLISM MODEL AS AN INSTRUMENT TO REDUCE COMPLEXITY

Let us consider the basic stages of urbanization that will help us in conceptualizing a model for the impact assessment of a large enterprise such as a city on its surroundings. Let us also assume that the area is entirely devoid of any urban development and that we are at the very beginning of urban development. If we consider different phases of urbanization the sequence would be as follows: human activities change land use, changed land use entails changes in land cover, changes in land use and land cover affect the environment (local and global) because of the changes in the bio-physical properties. The ecological stresses, if not managed, threaten a city's health and well-being in a social, economic, and environmental sense (SEE). These activities of a "living" city can be considered as metabolic processes of a city "cell" or a city "organism". It is analogous to a biotic entity consuming natural resources for its survival and generating metabolic waste. Similarly, a city requires energy – food, petroleum, and other similar materials – and while maintaining its state or growth it produces waste products of the metabolism like air pollutants, water pollutants, solid waste, and so on.

The idea of considering the city as a biotic entity is not new. We review some of the initial concepts and methods. This is not a comprehensive survey but a concise survey citing important works as examples. The roots of the idea were in the neighbourhoods or micro units imagined by Perry, Reichow (Elke, 2003) and many other pioneers. These units were treated as a cell – a unit of urban development. Like a biotic entity, they can be born, grow, mature and so on. Thus, the urban agglomerate is nothing but a collection of these neighbourhood cells.

Wolman (1965), among the early pioneers of the concept, highlighted the importance of treating waste generated by a city. He compared the waste generated by a city with the metabolic waste and insists on its removal without any harmful effect on the city. Though he explained the concept of urban metabolism with respect to water demand and CO_2 emissions, most of the discussion is based on the reported data by the city. There was no model involved for the prediction or estimation of these quantities; it was a study for understanding the inflows and outflows of a few important parameters of the living city.

Another notable study was by Duvigneaud and Denaeyer of Brussels city (Duvigneaud & Denaeyer-De Smet, 1977). It was one of the most comprehensive urban metabolism studies. The city activities were modelled based on ecological principles. It created a detailed account of energy flow, material flow, water, and pollutant fluxes. The focus again was on the accounting and analysis of the various parameters for city planning. Though it was a comprehensive study, the focus was not on predicting or estimating the flows, it was an accounting based on the available data.

Decker et al. (2000) study twenty-five megacities for energy, water, and material flow. The focus was on Mexico City because of the availability of the data. They indicated that there are limited studies on the impact of cities on biogeochemical cycles and ecological processes. They suggested comprehensive research for modelling the metabolism for predicting material flow for a given spatio-temporal cross-section easily.

A helpful review of some of the early and latest concepts can be found in the article by Kennedy et al. (2010). They point out that most of the urban metabolism studies are accounting studies, that is, they provide an audit of inflows and outflows on an as-is basis. They also discussed some of the recent efforts for modelling urban metabolism – for example, STAN. However, STAN assures consistency of inflows and outflows by using mass balance equations and data checks in the process for calculating the flows. It is based on the inventory of the data still. The focus of this software is more on the graphical representation of the flows and correcting inconsistencies in the data.

As can be seen from earlier work on this topic, most of it is for accounting metabolism based on the available inventory data. The data collection itself is a

Burden	Risk	Resilience	Benefit
CO_2, SO_x, NO_x, BOD, COD, TSS, Nutrients, Oil, Other soil contaminants, Water used, air used, Other eco-services, Minerals, and fossil fuel … **SDG:** 2.4; 6.3, 6.4; 6.6; 7.2, 8.4, 9.4, 11, 12.4, 12.5; 14.1 to 14.3 (coastal cities); 15	Earthquake, Extreme weather episodes, Rain, Flood, Pollution episodes, Resource crunch …	Road network, Open spaces, Natural drainage, Good ventilation, No. of good quality water bodies, Vegetation cover, Economic well being, Education, Proximity to rescue and recovery stations or large portal … **SDG:** 1.5; 2.4; 9.1, 9.2, 9.3, 9.4; 13.1	GDP contribution, Contribution to the circular economy, Employment generated, Other direct or indirect socially beneficial activities (CSR), for example, research on climate change … **SDG:** 1.5; 2.4; 2.5; 3; 8.1; 9.3, 11, 12 a; 14.5, 14.a.1; 15.3

FIGURE 1.3 Burdens and benefits of a city and its SDG indicator mapping.

Cities Means to Menace?

challenging exercise. The maturity, consistency, standard methods for collection and reporting, and completeness of the inventory data are some of the critical issues in such methods. Maintaining the inventory and addressing all these issues is a mammoth task. Such studies can be used for predicting the future metabolism of a city. For example, city water demands can be predicted using time series data for population and per capita demands. However, the time series methods ignore spatial distribution as the inventory may or may not use the spatial attributes of the data. Furthermore, in some cases, the outflows of the materials estimated/calculated by inventory-based methods require verification by additional methods. For example, the Intergovernmental Panel on Climate Change (IPCC) survey for city carbon emissions seeks information about the verification method, if any, for reported CO_2 emissions by the city. The reported data in such surveys goes unverified because of a lack of a verification method. To summarize, the drawbacks of the existing urban metabolism models are:

a. At present, there is no single comprehensive model for the estimation of metabolic parameters without data inventory. There are some isolated efforts in the direction of estimating the metabolic quantity by remote sensing or other methods without data inventory. They provide point solutions and do not provide a holistic view required to solve the problem.
b. For many urban metabolism parameters, the future projections are based on time series data. Geospatial distribution of the metabolic parameter is not considered in estimation, or prediction.
c. Existing models do not generate "what if" scenarios according to the urban development plans easily. This is because the existing models do not use the land cover land use classes to estimate or predict the metabolic parameters. The levers for changing the quantity of metabolic parameters are zoning strategies for land use land cover patterns in the city. For example, the goal of reduction in CO_2, say by 2030, would require changes in land use patterns and other policy measures such as the use of clean technologies and so on. Furthermore, some of the emission rates especially for carbon are available for land use changes but not for the activities within the given land use.

1.3.1 Urban Metabolism Framework Using Remote Sensing

With this background, we present a possible remote sensing framework for estimating urban metabolism. (The model is a more detailed view of the "produce" arrow in the systemic view (Figure 1.2)). The model simplifies the assessment of the complex interaction between an urban area and the surroundings and hence is useful in the SDG assessment of any city as well. The framework is illustrated in Figure 1.4 and Figure 1.5. Each rectangular block indicates a module in the system. The types of models are labelled with the names, for example, "LULC Models". The input and output of each model are stated below the model name. Important modules are numbered for citing them easily in the discussion. The framework has two basic layers:

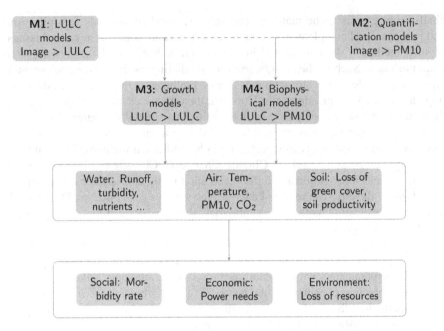

FIGURE 1.4 A framework for urban metabolism using remotely sensed data.

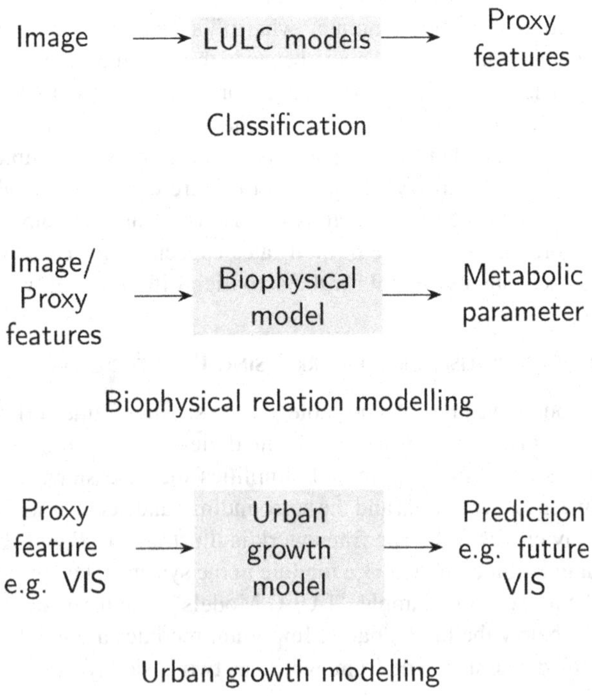

Image ⟶ LULC models ⟶ Proxy features

Classification

Image/Proxy features ⟶ Biophysical model ⟶ Metabolic parameter

Biophysical relation modelling

Proxy feature e.g. VIS ⟶ Urban growth model ⟶ Prediction e.g. future VIS

Urban growth modelling

FIGURE 1.5 Models in urban metabolism framework.

1. Remote sensing and urban growth model aided by remotely sensed data for predicting the metabolic parameters such as resource consumption and emissions from a city given the remotely sensed data. Source layer.
2. Socio-economic impact models for predicting the socio-economic impact (SEE) given the quantities of biological, physical, and chemical parameters of a city. Sink Layer.

We will focus on the first layer of the framework and discuss it in detail. The basic workflow for generating an urban metabolism scenario using the framework is as follows:

1. Identify features from the image and then develop/use relations between the image feature and the urban metabolic parameter. For example, run off model (module 4) using impervious surfaces extracted using module 1, radiant heat using basic land covers such as vegetation, impervious surface, soil (VIS). The image features may be direct features or proxies for the metabolic parameter. For example, run off is related to the land covers, whereas emissions such as CO_2 are related to fossil fuel burned. Remote sensing cannot observe that directly. However, it can observe road length or built-up area which can be correlated with CO_2. Such indirect features are called proxy features. Use of proxies is common in remote sensing, as remote sensing observes land covers only (more later). Extraction of urban features is performed by image classification models.
2. Develop/use an urban growth model which predicts the spatial extent of the given proxy feature. For example, given the spatial distribution of impervious surfaces as the input, develop/use the model for its prediction in the future (module 3). This is an important class of models as they help in generating different scenarios for a city. The type of input and output of these models is equally important. These models predicting the same proxy variable state used by the metabolic parameter model (module 4) are more useful, for example, VIS is used for CO_2 estimation and the urban growth model produces the VIS output as well.
3. Use the model in module 4 for predicting the urban metabolic parameter which takes the output of models in module 3 as an input. The output of module 3 is subjected to different conditions. The urban growth scenario for each condition is different. By changing the conditions of the model, different growth scenarios are played out. For each such scenario, an urban metabolic parameter is predicted by taking the output of 3 for each scenario as an input.

Each of the modules is described in more detail in the next section. Each module can be imagined as a set of functions of a library performing similar tasks. It can be seen as a library of functions that can be used by any other system as well. The other software architectural details are not provided now.

1.3.2 Modules of Urban Metabolism Framework

Module 1

This is the module for land use and land cover models. Land use and land cover models (module 1) are the fundamental methods for identifying different land use and land cover regions in a city scene. The models in this module take any satellite image as an input and identify the LU/LC as a city scene. There are many supervised and unsupervised machine learning techniques available for the same. Any one of them can be used in this module. However, it is to be noted that the development of models giving the right proxies as outputs is more useful. The outputs should be such that they are useful proxies as inputs to the models in module 4 for metabolic parameter models. Thus, the development of module 4 models may require a lot of experimentation for the right proxy land cover or land use for a given metabolic parameter. For example, if road length is a proxy for CO_2, then the model should correctly retrieve the road length given in a city scene.

Module 2

This is the module for quantification models. Models in this module often take a satellite image as an input and produce a quantitative map of a particular parameter of the region, for example, CO_2. This module is a supplementary module in cases where the metabolic parameter is not available for the entire city, or its spatial distribution is not available with the required resolution. This is also useful in cases where only a few local measurements, recorded by the measuring instrument, are available. For example, the CO_2 concentration measurements are available only for a few total stations (say 20) in the city. Either using optical satellite data or CO_2 measurements data (which is available at a much coarser level – a few km by km), we can create models to predict the CO_2 concentration for the city at a much finer resolution, say 30 m by 30 m. These are relations between remotely sensed data and the metabolic parameter. It would require one model for each metabolic parameter and for each parameter, a different image feature/s is used.

Module 3

This is the module for urban growth models. Urban growth models (module 3) are key ingredients of the present system. Especially their form: they take impervious surfaces or different land covers and their spatial distribution as an input and predict the spatial distribution of the same over a city. This cannot be achieved by simple time series models. Markov models or cellular automata models are required. For example, VIS distribution over a city is to be predicted considering detailed transition probabilities for the VIS classes at a coarse or fine level. Thus, given historical land use and land cover data of required granularity, the model learns the transition probabilities of the changes and using such probabilities it predicts the given land cover for the future. The land cover historical data is extracted using remotely sensed data. Furthermore, other driver parameters of the growth are incorporated as well. This is helpful in playing out different growth scenarios for different socio-economic conditions.

Module 4

This is the module for biophysical relation models or metabolic models. The models in this module (module 4) take image features as an input and predict a required urban metabolic parameter. For example, given impervious surfaces, runoff from the area is predicted. The features such as impervious surfaces are extracted using satellite imagery. In some conditions, the factors that are directly correlated with the metabolic parameter may not be observable. In such conditions, a proxy feature from the image is extracted and correlated with the metabolic parameter. For example, fossil fuel burned in a city is correlated with impervious surfaces. Many such proxies are used. Some of the examples are listed here in Table 1.4.

These models are straightforward in cases where the emission rates are directly available for the extracted image features. Simple multiplication of the extracted image

TABLE 1.4
Proxy Features and Metabolic Parameter

City features	Proxy for	Metabolic parameter
Impervious surfaces	For most of the metabolic processes of the city	CO_2, SO_x, NO_x, particulate matter, solid waste, wastewater, temperature, energy requirement, electricity usage and on
Vegetation, impervious surfaces, soil	Same as above with fine granularity	Same as above
Building count	Energy usage, fossil fuel burned, solid waste, air, and water emissions, and so on	Same as above
Building count type wise (residential, commercial, industrial)	Same as above with fine granularity	Same as above
Proportions of or area of {building types or zones residential upmarket, slums, mid-economy res., small-scale industries, large-scale industries, malls, theatres}	Fuel burned for cooking heating cooling, transportation, other inflows, and outflows	Same as above
Road length, road area, area of different road surfaces such as tar, soil etc.	Transport volume, average trip length based on spatial distribution of residential and commercial zones	CO_2, SO_x, NO_x,
Population/density, building volume	Same as row one	CO_2, CO, solid waste, fossil fuel burned, wastewater, temperature energy requirement, electricity usage and so on
Temperature, humidity, wind	Heating needs	Energy for heating and cooling, comfort index

feature for a target area and the emission factor will lead to the total emissions for the target area. Emission rates are calculated by a variety of means: first-principle, empirical studies or plot studies, and remote sensing itself. First-principle methods will use chemical reactions to work out the quantity; for example, one litre of petrol will generate x amount of CO_2. Other complex phenomena may require plot studies or physical models of the process, for example, vegetation emits x amount of CO_2 annually. Another method is to develop a correlation between the image feature and the city feature based on historical data. This is useful when the emission factors are not available or not easy to calculate but the quantities from inventory methods are available. For example, emissions because of fossil fuel burning – including cooking, transportation, etc.

1.3.3 Example Remote Sensing Models in Urban Metabolism Framework

We discuss below a few examples of the models required by an urban metabolism framework. We focus on urban growth models and the models for biophysical parameters, say CO_2. This is just to showcase the possibility of building such a framework for urban analysis. There are many other models required for a complete urban metabolism framework as envisaged.

The cellular automata (CA) based models such as SLEUTH (Clarke et al., 1997) are especially suitable as the urban growth is to be expressed as the changes in land cover in each time step (growth cycle). The changes in land covers during a development cycle are encapsulated in CA or CA-like rules. For example, if the undeveloped region is in proximity to the main road, then in the next cycle (say 5 years), the undeveloped cell will be developed, if the vegetation pixel is surrounded by impervious surfaces on all sides, then in the next cycle the vegetation changes to an impervious surface. The chances of changes in the land cover are determined by historical data, called a calibration step. These model parameters control the behaviour of CA rules. For example, the CA rule may suggest that the cell next to the road would be developed in the next cycle, however, the chances of that would be determined by historical records of the changes in that city. Historical data for urban development is often available through remote sensing. The chances are that it may be the only source in many cases. The CA rules, in principle, can be defined over two classes, such as built-up vs non-built-up, or multiple classes, such as low-level land cover and land use. TerrSet is a useful tool too. TerrSet's Land Change Modeller (transition potential modelling) calculates the transition probabilities (called transition potential) of the classes and then it uses them for predicting the future scenario (Eastman, 2016). This functionality is especially useful because transition probabilities by considering additional driver variables can be generated and plugged into the module for prediction (Yadav et al., 2019).

1.3.3.1 Urban Growth

1.3.3.1.1 Johannesburg

This is a case study for Johannesburg. Verma et al. (2016) analysed patterns of land use land cover changes in the city of Johannesburg, South Africa. The focus was to study the increase in the impervious surface area and loss of green cover if any. The changes in various classes such as industrial, forest, shrubs, water bodies, and open areas were evaluated using Landsat imagery from 2000 to 2016. The Support Vector Machines (SVM) classifier was used. We present this case study in detail.

With a population of 3.8 million people (year 2010) Johannesburg is the largest city in the Gauteng province of South Africa. It is almost 30% of the population of the Gauteng province. The population has increased by 16% as compared with the 2001 population. It is one of the fastest-growing cities. Johannesburg shows a characteristic dual urban structure: the urban zones with drastically different amenities coexist. This is the legacy of the colonial era. Economic opportunities and the availability of other means of livelihood have made Johannesburg a most favoured destination for workers from all over the continent. The unanticipated urbanization has resulted in extensive pressure on the natural and economic resources of the city (Schäffler & Swilling, 2013). Urbanization has resulted in some of the common problems faced by the cities in developing countries, such as urban sprawl, loss of agricultural land and so on. Thus, the city provides an interesting case, like Pune in many ways, for studying urban growth. The population of the two cities is close to each other. The urban structure with mixed financial status and mixed amenities exists in Pune as well. The drivers of urban growth remain the same, being a city in a developing country.

Figure 1.6 illustrates the data-processing steps adapted for urban growth studies. The data processing includes two main stages: data pre-processing for correcting the images, and the extraction of the required parameters from them using classification or other data-processing methods. The details of the steps are explained below.

The Landsat data was used for classifying the land use and land cover of the city. The stripped lines in each image, because of the sensor problem in the old Landsat platform, were filled using a sliding window approach (Yadav & Deshpande, 2015). Then, the digital numbers of each image were converted to radiance values by applying the sensor calibration constants. Next, radiance values were converted to top of the atmosphere reflectance values. Solar elevation angle correction was

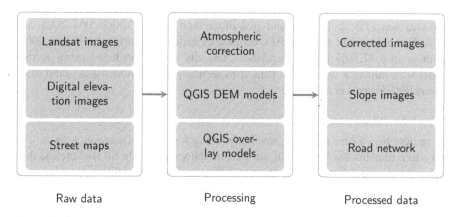

FIGURE 1.6 Data processing steps for urban growth studies.

applied too. After all the pre-processing steps were completed, each image was classified into three basic classes, namely vegetation, impervious surface, and soil (VIS). The classification was performed in two steps: first, the image was classified into classes namely soil-grass (mixture), soil, water, road, residential, industrial, vegetation, forest, golf area. A total of fourteen classes were considered by considering subclasses as dominant classes, such as residential and industrial, and they showed a wide range of signatures. They were combined into VIS after classification into fourteen classes. This process, that is, identifying fine granular classes first and then combining them into coarse VIS classes, was adapted for more accurate classification. The overall accuracy of classification was around 84% to 91%. The remaining errors in the classification were corrected manually using high-resolution imagery.

In addition, the digital elevation model (DEM) data was acquired from CARTOSAT-1 satellite, operated by the Indian Space Research Organization (ISRO). The DEM images were converted to slope using QGIS DEM Terrain models (GDAL_PROXIMITY, 2015). The road network for the present work was acquired from Open Street Maps (OPENSTREETMAP, 2015). Only primary motorway roads were considered for assessment due to their significant impact on the economy. The obtained network was converted to raster distances with the use of QGIS proximity raster distance (GDALDEM, 2015). Both the parameters such as road network and slope are important drivers of urban growth. A higher degree slope, generally speaking, decreases the chances of urban development, whereas a good road network increases it. Road networks and urban growth share a complex relation with each other. Highly developed areas can lead to efficient road networks, and good road networks can also contribute towards development. These two maps, slope, and road are the input to the growth model in addition to the transition probabilities. The transition probabilities of VIS changes were calculated and used for predicting the city's growth by 2016 and 2020.

VIS analysis of the city for the past 16 years, from 2000 to 2016 revealed rapid urbanization. More pronounced development was observed on the fringes of the city, maybe because of the availability of land for public and private infrastructure. The impervious surface showed an increase during the study period. The impervious surface area was 17%, 22%, 24%, and 26% of the total area for 2000, 2004, 2008, and 2016. This amounts to almost a 50% increase in impervious surface area since 2000, resulting in a 47.80% loss of vegetation. The city growth predicted using TerrSet model was validated using 2016 data. The accuracy of the model was found to be around 80% (Verma et al., 2016).

1.3.3.1.2 Pune

We present here a case study of growth modelling for Pune, Maharashtra, India. The procedure remains the same as in the Johannesburg case study till the satellite data is processed. The difference is in the way VIS transition probabilities are calculated. VIS transition probabilities are calculated by a hidden Markov model (HMM). The working hypothesis is that VIS transitions are affected by the economic status of the country during the development cycle, for example, liquidity affects the rate of urban development. This is intuitive as the land cover transitions are interconnected with the economic variables of the country and should be considered in calculating the transition probabilities. Thus, the mechanism provides finer control over the transition probabilities and the values are more realistic too. The argument is elaborated on in the next section.

Often urban growth models consider the influence of growth factors on land use and land cover for predicting the changes. These factors belong to three main categories: temporal, spatial, or spatio-temporal. The temporal factors predominantly change over time and their spatial variation can be ignored or they are true of sufficiently large areas, for example, Gross Domestic Product (GDP). The spatial factors change over space mainly and their variation, if any, over space can be ignored, for example, the digital elevation of a target area. The spatio-temporal factors are the factors which change over space and time both in a considerable manner, for example, proximity to the road. This is because the road network is very dynamic and new roads or changes in the importance of a road keep on changing during the urban development cycle. These are the direct factors and a link between them and urban growth is easily established. The temporal factors can be further divided into supply-side factors and demand-side factors. Examples of supply-side factors are the availability of workforce, materials, liquidity and so on. Examples of demand-side factors are the availability of jobs, and the standards of living, education, and health. The probability of converting a land cover into another, say vegetation to urban, not only depends upon spatial or spatio-temporal factors such as road networks and so on but also on the temporal factor such as liquidity. The link between liquidity and infrastructure development is well understood. There are other temporal factors too; for example GDP, the interest rate cycle and so on.

We need a different model to consider the influence of these factors, in addition to the land cover class transitions. We can consider three basic VIS classes as three different states a city cell can take, and we can imagine a state transition model for transitions between each VIS class at a broader level. If we consider the transitions to be conditionally dependent on temporal factors such as liquidity, then we can define a complete Hidden Markov Model (HMM). The temporal variables as observable states and VIS states as hidden states completely define the HMM for urban growth. It is formally expressed as:

$$a\ set\ of\ N\ hidden\ states\ (S)\ and\ a\ set\ of\ parameters\ \theta = \{\pi, A, B\}$$ Equation 1.4

where,

π = vector of prior probabilities with $\pi_i = P(q_1 = s_i), i = 1...N$, i.e., probability that s_i is the first state (q_1) of a state sequence.

A = matrix of state transition probabilities Each $a_{ij} = P(q_{n+1} = s_j | q_n = s_i)$ represents the probability of transitioning from n^{th} state s_i to $(n+1)^{th}$ state s_j.

B = matrix of emission probabilities that characterize the likelihood of seeing an observation X_n.

Pune data between the years 2001 to 2014 was considered for the study (Yadav et al., 2019). The economic variables considered were Gross Domestic Product (GDP), interest rate cycle, consumer price index inflation, gross fixed capital formation, urban population growth, per capita electricity, and road length added. Table 1.5 provides the list of driver parameters and their rationale.

The HMM and Logistic Regression (LR) model was built to predict the urban scenario for 2014 (so that the models can be validated). The results of HMM-LR were compared with the standard Markov Chain-Logistic Regression (MC-LR) model to assess the improvement by HMM-LR, if any. Visual analysis of the prediction by 2014 showed significant improvement over MC-LR model prediction. MC-LR prediction made a gross error in predicting growth in local urban pockets by either showing very low or very high urban growth. Table 1.6 shows the VIS transition probabilities calculated with different methods. The VIS transition probabilities calculated using HMM are more realistic as compared to the Markov chain. For example, inertia of impervious surface changes, that is, impervious surface (I) remaining impervious surface (I) is captured well by HMM. Overall, the HMM output was well-balanced and resembled the actual output better. HMM-LR resulted in an 11% increase in the precision of the persistence of impervious surfaces. Similarly, the precision of the soil class jumped up by 26%. However, there was a drop in the precision of the vegetation class marginally by 6% (Yadav et al., 2019).

TABLE 1.5
Economic Drivers of Urban Growth

Growth Factors	Definition
Gross Domestic Product	Gross Domestic Produce (GDP) growth rate of a country of local region as applicable
Interest Rate Cycle	A tight monetary policy affects the overall investment policy which leads to slowdown and vice versa.
Consumer Price Index Inflation	High inflation is one of the major roadblocks in the pathway of economic development. It is evident that low inflation creates a developmental investment environment.
Gross Fixed Capital Formation	GFCF quantifies the amount that the government spends on the capital formation of the country. Capital formations such as infrastructure building, land improvements, machinery, and equipment purchases, and so on.
Urban Population Growth Rate	To accommodate a higher influx of people, cities are expanding along their outskirts, leading to the growth of urban agglomerates.
Per Capita Electricity Consumption	Since electricity is one of the major drivers of growth thus per capita consumption of electricity is a good indicator of the demand for electricity in a region. Often, faster growth regions are associated with higher electricity demands.
Road Length Added	Roads are one of the basic regional development indicators. Better connectivity of a region helps in better transportation and thus provides impetus to growth.

Source: Yadav et al., 2019.

TABLE 1.6
Computed Markov Chain Transition Probabilities for 2001–2002, Learned HMM Transition Probabilities for 2014, Computed MC Transition Probabilities for 2014

Given LC class (2001)	Changed to LC class (2002)			Changed to LC class (2014) (HMM)			Changed to LC class (2014) (MC)		
	V	I	S	V	I	S	V	I	S
V	0.7920	0.1067	0.1013	**0.8710**	0.0030	0.1260	0.6787	0.1661	0.1542
I	0.0503	0.8996	0.0501	0.0001	**0.9610**	0.0389	0.1538	0.6484	0.1978
S	0.3058	0.1321	0.5621	0.0020	0.1710	**0.8270**	0.1372	0.0863	0.7765

Source: Yadav et al., 2019.

1.3.3.2 CO_2 Emissions

Carbon dioxide is one of the pollutants causing global warming and climate change. Urban and peri-urban centres account for 74% of the total carbon. Only urban centres excluding peri-urban centres account for 35% of the world's carbon. Most of the carbon from the urban areas is coming from transportation (46%). Some of the CO_2 emission statistics are astonishing-50% of the global emissions come from urban and semi-dense urban clusters occupying only 1% of the global surface (Crippa et al., 2021). Only one hundred cities in the world emit 18% of the world's carbon (Moran et al., 2018). Urban centres are observing growth over the years and so do CO_2 emissions. However, the annual increase in CO_2 emissions is more in developing countries than in developed countries. Figure 1.7 shows carbon emission trends by the economic status of the country. The classification of the countries is according to the "World economic situation prospects, 2014" (United Nations, 2014). The carbon emission data is gathered from the EDGAR report (Crippa et al., 2020). The countries were grouped according to their economic status as suggested by the "World economic situation prospects" and emissions were aggregated for the broad categories such as developing countries and so on. The global trend follows the trend of emissions by developing countries and countries with an upper-middle economic status.

Table 1.7 shows the 20 cities from the Carbon Disclosure Project (CDP, n.d.) (Aché, 2021) reporting the highest CO_2 in 2020. These cities were selected because most of them have reported scope 1 to scope 3 emissions from the city. The reported CO_2 emissions are prepared using the data inventory method, that is, all the activity data related to fossil field burning was collected and maintained for a year and then the right emission factors were applied. The data is not verified by any other means. As can be seen from the table, some of the large cities emit large amounts of carbon annually.

Carbon emissions are one of the important emissions from the city that are significant for local and global impact. The carbon emission models from a city at the macro or micro level are hence key ingredients of any urban metabolism framework. Now

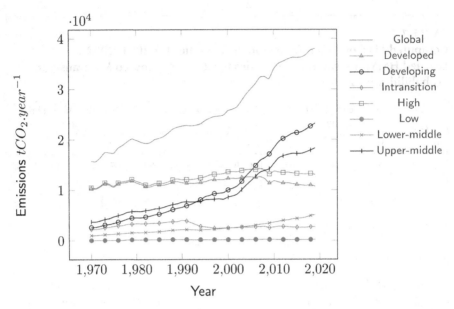

FIGURE 1.7 Annual carbon emission trends according to the economic status of the countries.

TABLE 1.7
Top Twenty Cities According to the Annual Scope 1 CO Emissions

Cities	CDP area (sq km)	Population density	Scope 1 emissions (1x10³) t	Scope 2 emissions (1x10³) t	Scope 3 emissions (1x10³) t
Hong Kong	1107.00	6781.75	40142.82	26250.67	NA
Steve Tshwete	3993.00	69.81	29803.31	2937.28	291.11
Tokyo	2193.96	6359.11	25161.98	36880.89	13939.36
Seoul	605.23	16540.79	22875.00	21918.00	1892.00
São Paulo	1521.11	7398.22	12989.95	2533.06	607.95
Honolulu	1545.00	630.79	12065.65	NA	4043.17
Calgary	848.00	1516.17	11928.50	6901.92	154.38
Chennai	426.00	18591.55	11922.07	8302.47	1.97
Montréal	500.00	4058.76	10793.70	43.93	296.49
Austin	845.70	1140.18	9651.48	5116.51	NA
Pretoria–Tshwane	6298.00	5249.60	9138.31	119473.51	340.70
City of Denizli	12153.00	84.57	5851.89	1650.78	NA
Amman	800.00	4750.00	5807.59	5073.79	1590.25
Hermosillo	1273.00	717.45	5351.76	1578.52	771.72
Tegucigalpa	1514.94	808.64	5034.29	NA	NA
São José dos Campos	1099.41	656.66	4180.38	152.27	39.60
Milano	182.00	7716.65	3299.64	2237.03	160.37
Curitiba	432036.00	4383.89	3045.25	368.88	90.91
Barcelona	102.15	16023.12	3002.19	723.64	69.18
Halifax	5600.00	75.35	2357.93	3120.82	305.74

Source: Aché, 2021.

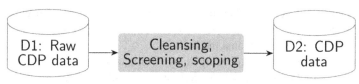

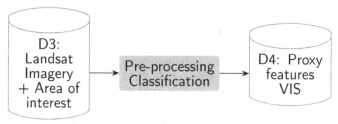

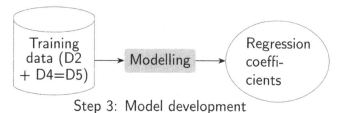

FIGURE 1.8 Carbon model development steps.

we will discuss a few methods to calculate CO_2 emissions from a city. We will discuss the method, as an example, which uses remote sensing data for calculating CO_2 so that it is useful in an urban metabolism framework. CO_2 is an example of an important emission from the city. The model development procedure is generic and can be used for any other target emission; for example, particulate matter, sulphur dioxide and so on. There are two main approaches:

1. the methods which calculate the emissions using CO_2 concentration (top-down approaches) in the atmosphere and pollutant transport model, also referred to as the inverse models, and
2. the methods which calculate CO_2 emissions based on the data inventory (bottom-up models), also referred to as the forward models (Methods for Remote Determination of CO_2 Emission, 2011).

The guidelines, which are used by world cities for calculating CO_2 emissions, recommend a bottom-up procedure; that is, the procedure using data inventory (Eggleston et al., 2006; Fong & Doust, 2014). Thus, the city administrative departments responsible for calculating and reporting the emissions must maintain the amount of fossil fuel burned by the city activities. Maintaining such data for a city is an effort-intensive exercise. The reliability of the data is another issue, especially for the cities in developing countries. Remote sensing methods using proxy features provide a viable alternative to conventional methods.

We present here a model developed by Deshpande et al. (2022) for carbon emission by a city. We provide the procedure for model development in detail. The steps are graphically illustrated in Figure 1.8. CDP data (CDP) was available from Ache (2021), for the development of the model. The data contain annual emissions reported by world cities for the years 2018, 2019, and 2020. A predefined format by the CDP and the International Council for Local Environmental Initiatives (ICLEI) is used by each city administrative unit for reporting emissions. The majority of the cities use Global Protocol for Communities (GPC) or IPCC guidelines for calculating carbon emissions (Deshpande et al., 2022). These methods are inventory-based methods. Other emissions such as CH_4, and particulate matter are also reported. The original data in the CSV files was converted into a useful table format representing each city record on a single row. Forty-six cities reporting all three scope emissions were selected. 2020 data was selected for the model development. Only scope 1 and 2 CO_2 emissions were considered. Other important attribute reported by the cities such as population was used for model development, in addition to the carbon data.

The proxy attributes considered for model development were broad-level VIS classes. The Landsat 8 data was used for extracting the VIS classes from the cities. Atmospherically corrected reflectance values were used for model development. The images of forty-six cities were downloaded and processed. Landsat was chosen for its high-quality 30 m data that provide a good combination of resolution and quick global coverage. Three high-level steps for model development were:

1. emission data preparation using raw CP data,
2. VIS feature extraction, and
3. model development for carbon emissions.

Complex classification methods were avoided for computational efficiency; for example, fractions were extracted using library pure pixels. This is because large data is required to be processed for model development. Library end members helped reduce the time taken for calculating fractions. Training pixels from each image were identified using visual interpretation. The Support Vector Machines (SVM) model was trained using the training data. The VIS fractions were calculated using a Fully Constrained Least-square Solution (FCLS) algorithm (Heinz & Chein-I-Chang, 2001).

The relation between the city attributes extracted from the Landsat images and the CO_2 emission from the CDP data was developed. The equations are as follows:

$$y1 = c_1 a_1 + c_2 a_2 + c_3 a_3 \ldots c_n a_n \qquad \text{Equation 1.5}$$

$$y2 = k_n . y1 \qquad \text{Equation 1.6}$$

where,

$y1$ and $y2$ = scope 1 and 2 CO_2 emissions in t.year^{-1},
a_1 to a_n = independent variables,
c_1 to c_n and are coefficients of regression.

Multiple experiments were performed with various combinations of independent variables and CO_2 to understand the useful relations between them. The total area of the VIS classes in km², the population, and the population density were the main independent variables considered for the model development. Different forms of regression equations, such as simple least-square analysis, regularized regression, and regression with non-negative constraints on coefficients were considered. The model was developed using thirty-six cities and the error analysis was performed using the ten remaining cities. The form of the model is given in Equation 1.7 and Equation 1.8. Table 1.8 summarizes the model coefficients. The average total carbon reported for the world by all models is remarkably close to the world total carbon as per the budget report (Friedlingstein & others, 2020). The total carbon reported by the carbon budget report average was 9.8 GtC.year^{-1} and 7.8 GtC.year^{-1} respectively. Regularized regression with impervious surfaces and population or population alone as the independent variables provided reliable results. The soil and vegetation components did not show any major impact on the emissions. For the details see (Deshpande et al., 2022).

$$\text{Scope 1 } CO_2 \text{ (Metric tonnes.year}^{-1}) = [\text{Impervious surface (I) factor}] \times I \text{ in sq. km.} + [\text{Population (Ppl) factor} \times Ppl] \quad \text{Equation 1.7}$$

$$\text{Scope } 2 = 0.63 \times \text{scope } 1 \quad \text{Equation 1.8}$$

TABLE 1.8
Coefficients of Different Models for Calculating Carbon Dioxide Emission from a City

Type	Model	I factor	Ppl factor	Scope 1	Scope 2	Total carbon
Bounding box[1] (BB)	I	14772.79		3.12	1.96	5.08
BB	I-Ppl	5917.85	1.72	4.81	3.03	7.84
BB	P		2.11	4.38	2.76	7.14
Map boundary (MB)	I	27575.25		5.82	3.66	9.48
MB	I-Ppl	13535.06	1.62	6.21	3.91	10.13
MB	P		2.11	4.38	2.76	7.14
			Avg.	4.78	3.02	7.80

Note:
[1] These models are developed considering the area of the cities according to the EOS bounding box (Land Viewer: EOS, 2023). The reference URL points to Pune city as an example. The user needs to type the city name in the text box at the north-east corner of the screen, and press enter key. The bounding box appears on the maps.

The framework for urban metabolism using remote sensing may sound farfetched in the beginning. However, there is a large body of work already present which fits this framework very naturally. Those isolated efforts integrated in this manner provide a powerful tool for understanding the interaction between urban activity and nature.

Figure 1.9 illustrates the conceptual links between sustainability, especially the SDG framework, and urban metabolism. The SDG framework provides an administrative outlook and identifies goals for humanity to achieve sustainable development. Whereas the framework is reasonably complete, it is a complex framework too. The indicators are intertwined and require a data inventory and a complex data model for organizing them. SEE Burden-Benefit-Vulnerabilities (Equation 1.3) outlook simplifies the assessment for complex entities. It provides concepts for organizing the indicators or assessing the sustainability of an enterprise in a generic sense by weighing its SEE burden and benefits. The urban metabolism model is a model for assessing the burden, benefits, and vulnerabilities of a large and complex enterprise such as a city.

1.4 REMOTE SENSING OPPORTUNITIES

Remote sensing has been one of the important tools to monitor land use and land cover in urban areas (Jensen, 1993; Jensen & Cowen, 1999; Gamba, 2013) and over the years, it has been used with multiple perspectives from mapping, planning, to ecological studies of urban and semi-urban places (Pickett et al., 2001). The remote sensing platforms and the urban area to be monitored, both have changed since their inception. Arial photographs have been replaced by sub-meter resolution satellite imagery and mapping requirements are replaced by requirements to understand and quantify biophysical parameters within a city area. Urban places are some of the most complex ecosystems and pose many challenges to using remote sensing. Whereas many researchers advocate the use of very high spatial resolution imagery for land use and land cover mapping (Jensen & Cowen, 1999; Thomas et al., 2003; Myint et al., 2011), high spatial resolution is not synonymous to improved results (Barnsley & Barr, 1996; Gamba, 2013). Considering the heterogeneity of land use and land cover in an urban area and equally heterogeneous goals for the analysis, integrated use of spectral and spatial multi-resolution imagery is desired (Plaza et al., 2009; Shafri et al., 2012; Gamba, 2013).

The advent of hyperspectral data provides new opportunities in urban remote sensing (Gamba, 2013; Rogan & Chen, 2004; Plaza et al., 2009). Hyperspectral data provides many advantages over conventional multispectral data for impervious surface detection in the urban environment. The minute details available in the spectral signature of a particular material are lost in broad-band data collection (Vane & Goetz, 1985; Goetz et al., 1985; Goetz, 2009). In contrast, the signatures provided by hyperspectral data are unique, most of the time, enabling discrimination of a large variety of natural and artificial materials.

Direct identification of a particular material by comparing the *target spectrum* with a *reference spectrum* in the library has thus become a viable option. Despite the advantages, hyperspectral data has not been used for urban studies as extensively as it is used for minerals, vegetation, and environmental monitoring studies (Xu & Gong,

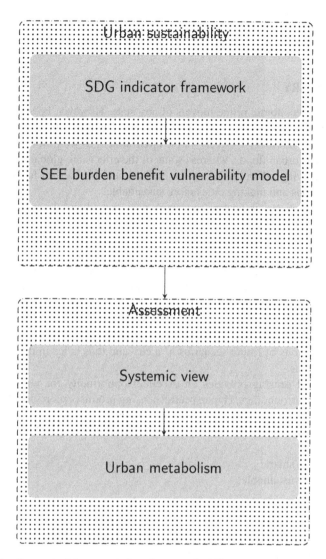

FIGURE 1.9 Conceptual framework for urban sustainability assessment.

2007; Weng et al., 2008; Shafri et al., 2012). Most of the hyperspectral studies for urban applications are performed using airborne Airborne Visible-Infrared Imaging Spectrometer (AVIRIS) sensor (Ridd et al., 1992; Ridd, 1995; Hepner et al., 1998; Chen & Hepner, 2001), and only a few using a space-borne sensor (Xu & Gong, 2007; Weng et al., 2008).

The situation is changing fast. Many hyperspectral satellites have been launched recently. Spaceborne hyperspectral data with various specifications would be available soon to researchers and practitioners. With the abilities of hyperspectral data, the hope is to identify richer and richer proxies for the urban metabolism models. With richer information about the urban systems derived using hyperspectral data, the hope is to restore the balance and make cities much more joyous places to stay – menace to meaningful again!

1.5 SUMMARY

Cities are socio-economic prime movers of any state. However, because of the high population density and scale of activities performed within urban centres, they are the prime reason for adverse effects on the environment as well. Some of the effects are local such as urban floods, whereas some of the effects are global such as climate change. Quantifying urban impact on the environment is essential for designing the right interventions and making cities more sustainable.

The United Nations' indicator framework provides a helpful perspective on sustainable development. However, sustainability assessment by that is a very data- and effort-intensive endeavour. A systemic view providing benefits, burdens, and vulnerabilities of the urban centres is helpful in reducing the complexity.

Many of the indicators require a synoptic view, which is provided by remote sensing alone. An urban metabolism model using remotely sensed data, based on the systemic view of the city, is useful in predicting the impact of urban activities on its surroundings. The urban metabolism framework is a collection of remote sensing models that provide estimates of urban consumption and emissions. It further provides predictions of future scenarios of a city and thus helps in the design of the right interventions.

Hyperspectral sensing provides an excellent opportunity for monitoring urban areas and their surroundings. Hyperspectral sensing promises more detailed information about the targets and hence is a key ingredient in unlocking the urban sustainability puzzle. The city is not "smart" unless and until it is sustainable.

Sustainable is smart,
    smart is sustainable!

NOTES

1. The Indian census 2021 is not yet published. The figures are based on the projections by WORLDBANK based on the past trend (THE WORLD BANK, 2023).
2. Scan Line Corrector was a mechanism to compensate the gaps between the scanning line. The mechanism failed on May 31, 2003. See (USGS, 2019) for more details.
3. Resource consumption and pollution (a and d) both have negative impact on SEE or they are burden to SEE, in general.
4. Compensation is proactive act of keeping ecological balance by replacing/replenishing/remunerating lost ecological entity in cash or kind, for example, lost tree cover is compensated by tree plantation in the same watershed.
5. If sustainability goals are achieved. SDGs can be seen as necessary conditions for sustainable enterprise.

WORKS CITED

Aché, M. (2021). *CDP–Unlocking climate solution.* Retrieved January 15, 2023, from Kaggle: www.kaggle.com/mathurinache/cdpunlockingclimatesolutions

Arnold, C. L., & Gibbons, C. J. (1996, Nov). Impervious surface coverage: The emergence of a key environmental indicator. *Journal of the American Planning Association, 62*(2), 243–258.

Barnsley, M. J., & Barr, S. L. (1996). Inferring urban land use from satellite sensor images using kernel-based spatial reclassification. *Photogrammetric Engineering & Remote Sensing, 62*(8), 949–958.

Basu, R., & Samet, J. (2002). Relation between elevated ambient temperature and mortality: A review of the epidemiologic evidence. *Epidemiologic Reviews, 24*(2).

BHUVAN. (2015). *Bhuvan: Gateway of Indian Earth Observation*. Retrieved September 1, 2015, from Bhuvan: http://bhuvan.nrsc.gov.in/data/download/index.php

Booth, D. B. (1991). Urbanization and the natural drainage system–impacts, solutions, and prognoses. *The Northwest Environmental Journal, 7*, 93–118.

CDP. (n.d.). *CDP*. Retrieved January 15, 2023, from www.cdp.net/en/

Chen, J., & Hepner, G. F. (2001). Investigation of imaging spectroscopy for discriminating urban land covers and surface materials. *AVIRIS Earth Science and Applications Workshop, Palo Alto, California. JPL Publication 02-1*. Jet Propulsion Laboratory, Pasadena, California. Retrieved from https://aviris.jpl.nasa.gov/proceedings/workshops/01_docs/2001Chen_web.pdf

Chen, X., Zhao, H., Li, P., & Yin, Z. (2006). Remote sensing image-based analysis of the relationship between urban heat island and land use/cover changes. *Remote Sensing of Environment, 104*(2), 133–146.

Clarke, K., Hoppen, S., & Gaydos, L. (1997). A self-modifying cellular automaton model of historical urbanization in the san francisco bay area. *Environment and Planning B: Planning and Design, 24*, 247–261. doi:10.1068/b240247

Crippa, M., Guizzardi, D., Muntean, M., Schaaf, E., Solazzo, E., Monforti-Ferrario, F., … Vignati, E. (2020). *Fossil {CO2} emissions of all world countries–2020 Report*. (EUR 30358 EN, Publications Office of the European Union, Luxembourg) Retrieved January 15, 2023, from https://edgar.jrc.ec.europa.eu/booklet/EDGARv5.0_FT2019_fossil_CO2_booklet2020.xls

Crippa, M., Guizzardi, D., Pisoni, E., Solazzo, E., Guion, A., Muntean, M., … Hutfilter, A. F. (2021). Global anthropogenic emissions in urban areas: patterns, trends, and challenges. *Environmental Research Letters, 16*(7), 074033.

Decker, E. H., Elliott, S., Smith, F. A., Blake, D. R., & Sherwood, R. F. (2000). Energy and material flow through the urban ecosystem. *Annual Review of Energy and the Environment, 25*, 685–740.

Deshpande, S. S., Banolia, C., & Balamuralidhar, P. (2022). Approximate and quick estimation of carbon emissions from a city using remotely sensed data. *IGARSS 2022–2022 IEEE International Geoscience and Remote Sensing Symposium* (pp. 4635–4638). doi:10.1109/IGARSS46834.2022.9883528

Duvigneaud, P., & Denaeyer-De Smet, S. (1977). L'Ecosystème urbs: l'écosystème urbain Bruxellois. In P. Duvigneaud, & P. Kestemont (Ed.), *Productivité Biologique en Belgique. Paris-Gembloux: Editions Duculot* (pp. 581–599).

Eastman, J. R. (2016). *TerrSet Geospatial monitoring and modelling system: Tutorial*. Retrieved January 15, 2023, from Clark Labs: https://clarklabs.org/wp-content/uploads/2016/10/TerrSet-Tutorial.pdf

Eggleston, S., Buendia, L., Miwa, K., Ngara, T., & Tanabe, K. (2006). *2006 IPCC Guidelines for National Greenhouse Gas Inventories*. Retrieved January 15, 2023, from IPCC: www.ipcc-nggip.iges.or.jp/public/2006gl/index.html

Elke, S. (2003). Hans Bernard Reichow and the concept of Stadtlandschaft in German planning. *Planning Perspective, 18*(2), 119–146.

Fong, W. K., & Doust, M. (2014). *Global Protocol for Community-Scale Greenhouse Gas Emission Inventories*. Retrieved January 15, 2023, from WRI, C40 CITIES, ICLEI: www.c40knowledgehub.org/s/article/The-Global-Protocol-for-Community-Scale-Greenhouse-Gas-Emission-Inventories-GPC?language=en_US

Friedlingstein, P., & others. (2020). Global Carbon Budget 2020. *Earth System Science Data, 12*(4), 3269–3340. doi:10.5194/essd-12-3269-2020

Gamba, P. (2013, March). Human settlements: a global challenge for EO data processing and interpretation. *Proceedings of the IEEE, 101*(3), 570–581.

GDAL_PROXIMITY. (2015, Sep 8). *GDAL_PROXIMITY*. Retrieved from GDAL: www.gdal.org/gdal_proximity.html

GDALDEM. (2015, Sep 8). *GDALDEM*. Retrieved from GDAL: www.gdal.org/gdaldem.html

Goetz, A. F. (2009). Three decades of hyperspectral remote sensing of the Earth: A personal view. *Remote Sensing of Environment , 113*, Supplement 1(0), S5–S16. Retrieved from www.sciencedirect.com/science/article/pii/S003442570900073X

Goetz, A. F., Vane, G., Solomon, J. E., & Rock, B. N. (1985, June). Imaging Spectrometry for Earth Remote Sensing. *Science, 228*(4704), 1147–1153.

Heinz, D., & Chein-I-Chang. (2001). Fully constrained least squares linear spectral mixture analysis method for material quantification in hyperspectral imagery. *IEEE Transactions on Geoscience and Remote Sensing, 39*(3), 529–545. doi:10.1109/36.911111

Hepner, G. F., Houshmand, B., Kulikov, l., & Bryant, N. (1998, August). Investigation of the integration of AVIRIS and IFSAR for urban analysis. *Photogrammetric Engineering & Remote Sensing, 64*(8), 813–820.

Hosseinzadeh, S. R. (2005). The effects of urbanization on the natural drainage patterns and the increase of urban floods: case study Metropolis of Mashhad-Iran. *WIT Transactions on Ecology and the Environment, 84*, 423–432.

Industrial Environmental Performance Metrics: Challenges and Opportunities. (1999). Washington, DC: The National Academies Press. doi:10.17226/9458

Institute of Chemical Engineers. (n.d.). *The sustainability metrics: Sustainable development progress metrics recommended for use in the process industries*. Retrieved January, 2023, from www.greenbiz.com/sites/default/files/document/O16F26202.pdf

Jensen, J. R. (1993). Urban/Suburban land use analysis. In J. R. Jensen, & R. N. Colwel (Eds.), *Manual of Remote Sensing, Second Edition* (pp. 1571–1666). Falls Church, Verginia: American Society of Photogrammetry.

Jensen, J. R., & Cowen, D. C. (1999). Remote sensing of urban/suburban infrastructure and socio-economic attributes. *Photogrammetric Engineering & Remote Sensing, 65*(5), 611–622.

Kennedy, C., Pincetl, S., & Bunje, P. (2010). The study of urban metabolism and its applications to urban planning and design. *Environmental Pollution, 189*, 1–9. doi:10.1016/j.envpol.2010.10.022

Khain, A. P., Benmoshe, N., & Pokrovsky, A. (2008). Factors determining the impact of aerosols on surface precipitation from clouds: an attempt at classification. *The Journal of Atmoshperic Scinces, 65*, 1721–1748.

Land Viewer: EOS. (2023). Retrieved from https://eos.com/landviewer/?lat=18.52465&lng=73.87670&z=11

Methods for Remote Determination of CO2 Emission. (2011, January). (JASON Program Office, 7515 Colshire Drive, McLean, Virginia 22102) Retrieved from JASON, The MITRE Corporation.

Mirzaei, P. (2015). Recent challenges in modeling of urban heat island. *Sustainable Cities and Society, 19*, 200–206.

Moran, D., Kanemoto, K., Jiborn, M., Wood, R., Többen, J., & Seto, K. C. (2018, June). Carbon footprints of 13000 cities. *Environmental Research Letters, 13*(6), 064041. Retrieved from https://doi.org/10.1088/1748-9326/aac72a

More, R., Kale, N., Kataria, G., Yadav, P., & Deshpande, S. (2015). Understanding thermal fluxes in and around Pune city using USING remotely sensed data. *Asian Conference on Remote Sensing*. Manila, Philippines.

Myint, S. W., Gober, P., Brazel, A., Grossman-Clarke, S., & Weng, Q. (2011). Per-pixel vs. object-based classification of urban land cover extraction using high spatial resolution imagery. *Remote Sensing of Environment, 115*, 1145–1161.

NASA. (2011). Landsat 7 Science Data Users Handbook. *National Aeronautics and Space Administration*. Retrieved September 1, 2015, from http://landsathandbook.gsfc.nasa.gov/pdfs/Landsat7_Handbook.pdf

National Institute of Urban Affairs. (2016, September). *India-Urban climate change fact sheet*. Retrieved August 12, 2017, from https://smartnet.niua.org/sites/default/files/resources/FS%203_Urban%20Flooding.pdf

Office of the Registrar General & Census Commissioner, India. (2011-1). *Provisional Population Totals Paper 2, Volume 2 of 2011: Maharashtra*. Retrieved June 2, 2014, from www.censusindia.gov.in/2011-prov-results/paper2-vol2/census2011_paper2.html

Office of the Registrar General & Census Commissioner, India. (2011-2). *Provisional Population Totals Paper 2 of 2011 India Series 1*. Retrieved July 6, 2017, from www.censusindia.gov.in/2011-prov-results/paper2/prov_results_paper2_india.html

Ogden, F. L., Raj Pradhan, N., Downer, C. W., & Zahner, J. A. (2011). Relative importance of impervious area, drainage density, width function, and subsurface storm drainage on flood runoff from an urbanized catchment. *Water Resources Research, 47*(12).

Oke, T. (1982). The energetic basis of the urban heat island. *Quarterly Journal of the Royal Meteorological Society, 108*(455).

OPENSTREETMAP. (2015, Sep 8). *OPENSTREETMAP*. Retrieved from www.openstreetmap.org

Oudin, L., Salavati, B., Furusho-Percot, C., Ribstein, P., & Saadi, M. (2018). Hydrological impacts of urbanization at the catchment scale. *Journal of Hydrology, 559*, 774–786.

Phelan, P., Kaloush, K., Miner, M., Golden, J., Phelan, B., Silva, H., & Taylor, R. (2015). Urban Heat Island: Mechanisms, Implications, and Possible Remedies. *Annual Review of Environment and Resources, 40*, 285–307.

Pickett, S. A., Cadenasso, M. L., Grove, J. M., Nilon, C. H., Pouyat, R. V., Zipperer, W. C., & Costanza, R. (2001). Urban ecological systems: linking terrestrial ecological, physical, and socioeconomic components of metropolitan areas. *Annual Review of Ecology and Systematics, 32*, 127–157.

Plaza, A., Benediktsson, J. A., Boardman, J. W., Brazile, J., Bruzzone, L., Camps-Valls, G., ... Trianni, G. (2009). Recent advances in techniques for hyperspectral image processing. *Remote Sensing of Environment, 113*, S110–S122.

Ray, K., Giri, R., Ray, S., Dimri, A., & Rajeevan, M. (2021). An assessment of long-term changes in mortalities due to extreme weather events in India: A study of 50 years' data, 1970–2019. *Weather and Climate Extremes, 32*.

Ridd, M. K. (1995). Exploring V-I-S (vegetation – impervious surface – soil) model for urban ecosystem analysis through remote sensing: comparative anatomy of the cities. *International Journal of Remote Sensing, 16*(12), 2165–2185.

Ridd, M. K., Ritter, N. D., Bryant, N. A., & Green, R. O. (April 1992). Neural network classification of AVIRIS data in the urban ecosystem. *Association of American Geographers Annual Meeting*. San Diego, California.

Rogan, J., & Chen, D. (2004). Remote sensing technology for mapping and monitoring land-cover and land-use change. *Progress in Planning, 61*, 301–325.

Schäffler, A., & Swilling, M. (2013). Valuing green infrastructure in an urban environment under pressure – The Johannesburg case. *Ecological Economics, 86*(C), 246–257. doi:10.1016/j.ecolecon.2012.0

Shafri, H. Z., Taherzadeh, E., Mansor, S., & Ashurov, R. (2012, June). Hyperspectral remote sensing of urban areas: an overview of techniques and applications. *Research Journal of Applied Sciences, Engineering and Technology, 4*(11), 1557–1565.

Streutker, D. (2003). Satellite-measured growth of the urban heat island of Houston, Texas. *Remote Sensing of Environment, 85*(3).

The World Bank Group. (2016). *Urban population (% of total)*. Retrieved Jul 4, 2016, from http://data.worldbank.org/indicator/SP.URB.TOTL.IN.ZS

THE WORLD BANK. (2023, February 11). *Urban population (% of total population)–India*. (United Nations Population Division. World Urbanization Prospects: 2018 Revision.) Retrieved from THE WORLD BANK: https://data.worldbank.org/indicator/SP.URB.TOTL.IN.ZS?locations=IN

Thomas, N., Hendrix, C., & Congalton, R. G. (2003, September). A comparison of urban mapping methods using high-resolution digital imagery. *Photogrammetric Engineering & Remote Sensing, 69*(9), 963–972.

United Nations. (1987). *Our common future (Brundtland report)*. Retrieved January 15, 2023, from www.are.admin.ch/dam/are/en/dokumente/nachhaltige_entwicklung/dokumente/bericht/our_common_futurebrundtlandreport1987.pdf.download.pdf/our_common_futurebrundtlandreport1987.pdf

United Nations. (2002). *World population prospects: The 2001 revision, data tables and highlights*. New York: (United Nations publication, Sales No. ESA/P/WP.173).

United Nations. (2014). *Statistical annex–Country classification, World economic situation prospects 2014*. (United Nations: Department of Economic and Social Affairs) Retrieved from www.un.org/development/desa/dpad/wp-content/uploads/sites/45/2013/12/WESP2014.pdf

United Nations. (2022). *SDG Indicators: Global indicator framework for the Sustainable Development Goals*. Retrieved January 15, 2023, from Sustainable Development Goals: https://unstats.un.org/sdgs/indicators/Global%20Indicator%20Framework%20after%202022%20refinement_Eng.pdf

United States Census Bureau. (2023, 1). *City and Town Population Totals: 2020–2021*. Retrieved from www.census.gov/data/tables/time-series/demo/popest/2020s-total-cities-and-towns.html#tables

USGS. (2015, Sep 8). *USGS*. Retrieved from Earth Explorer: http://earthexplorer.usgs.gov/

USGS. (2019, August 2). *Landsat 7 Scan Line Corrector Processing Algorithm Theoretical Basis*. Retrieved from www.usgs.gov/media/files/landsat-7-scan-line-corrector-processing-algorithm-theoretical-basis

Vane, G., & Goetz, A. F. (1985, April). Introduction to the Proceedings of the Airborne Imaging Spectrometer (AIS) Data Analysis Workshop. In G. Vane, & A. F. Goetz (Ed.), *Airborne Imaging Spectrometer Data Analysis. JPL Publication 85-41*. Jet Propulsion Laboratory, Pasadena, California USA.

Verma, P., Yadav, P., Deshpande, S., & Gubbi, J.(2016). Urban growth studies for Johannesburg city using remotely sensed data. *Asian Conference on Remote Sensing*. Colombo Sri Lanka.

Watershed, O.-G. (2015, Sep 8). *GRASS Watershed*. Retrieved from OSGEO: https://grass.osgeo.org/grass64/manuals/r.watershed.html

Weng, Q., Hu, X., & Lu, D. (2008, June). Extracting impervious surfaces from medium spatial resolution multispectral and hyperspectral imagery: a comparison. *International Journal of Remote Sensing, 29*(11), 3209–3232.

Wolman, A. (1965). The metabolism of cities. *Scientific American, 213*, 179–190.

Xu, B., & Gong, P. (2007, August). Land-use/land-cover classification with multispectral and hyperspectral EO-1 data. *Photogrammetric Engineering & Remote Sensing*, 73(8), 955–965.

Yadav, P., & Deshpande, S. S. (2015). Spatio-Temporal assessment of urban growth impact in Pune city using remotely sensed data. *Asian Conference on Remote Sensing*. Manila, Philippines.

Yadav, P., & Deshpande, S. S. (2016). Assessment of anticipated runoff because of impervious surface increase in Pune Urban Catchments, India: a remote sensing approach. *Proceedings of SPIE 10033, Eighth International Conference on Digital Image Processing (ICDIP 2016)*.

Yadav, P., Ladha, S., Deshpande, S., & Curry, E. (2019). Computational model for urban growth using socioeconomic latent parameters. *ECML PKDD 2018 Workshops: Nemesis 2018, UrbReas 2018, SoGood 2018, IWAISe 2018, and Green Data Mining 2018, Dublin, Ireland, September 10–14, 2018, Proceedings 18* (pp. 65–78). Dublin: Springer.

2 Introduction to Hyperspectral Imaging

This chapter will discuss:

- Hyperspectral data and its characteristics,
    - How it is different from multispectral data,
    - Spatial vs spectral
    - Advantages and disadvantages
- Terminology
- Quantum mechanical explanation of absorption
- What are the challenges (high dimensionality etc.)?
- Sensors and platforms
- General image analysis workflow
    - Goal
    - Visual interpretation and exploration
    - Image processing pipeline

2.1 HYPERSPECTRAL DATA

2.1.1 Definition and Characteristics

Hyperspectral data refers to the remotely sensed data collected over electromagnetic spectrum – usually invisible and infrared range from 400 nm to 2500 nm – in contiguous bands of ~10 nm width or less. Hyperspectral imaging is thus a particular imaging technique which samples the spectral power distribution of the object, that is, the graph of the reflected intensity of the light per wavelength contiguously. This implies that the image so formed will have a large number of bands within the spectral range of 400 nm to 2500 nm. The purpose of the hyperspectral imaging is to gather more detailed spectral information. A large number of contiguous bands describe the spectral behaviour of a target which otherwise is not available in a multispectral or three-band Red-Green-Blue (RGB) image.

This is illustrated in Figure 2.1. The figure shows the three-dimensional structure of the images. X and Y dimensions are columns and rows respectively and the third dimension Z indicates the number of bands in the image. A pixel having the same row and column index in each band is marked for all the bands of each image – dark square at the south-west corner of the image. If we read the value of each marked pixel from all the bands, it will form a D dimensional vector where D=number of bands in the image. These D-dimensional pixel vectors extracted from each type of

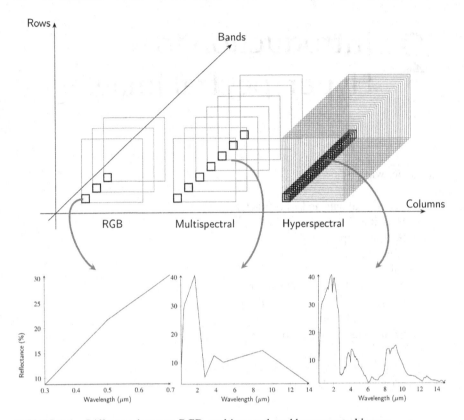

FIGURE 2.1 Difference between RGB, multispectral, and hyperspectral image.

image are displayed at the bottom. Note that the RGB image bands are limited to the visible range of the electromagnetic radiation. Thus, the plot shows the intensity of reflected light or grey values versus wavelength for a pixel of visible range image (RGB image), a multispectral image, and a hyperspectral image. These are the plots for a common concrete found in urban areas. As can be seen from the plots, the spectral resolution of the broadband multispectral data is too coarse to obtain the detailed interaction between light and matter.

The coarse spectral resolution of visible or multispectral imaging is incapable of recording highly structured reflectance and emittance determined by electromagnetic energy and vibrational energy states in the material (Vane & Goetz, 1985). As the interaction between the material and the electromagnetic energy is recorded at a much finer resolution, diagnostic features are available which can help detect the material. This is similar to spectroscopic methods performed in the laboratory for detecting material using its absorption or emission spectrum. A spectral width of 10 nm, 9.3 to be specific, in the spectral range of 400 nm to 2400 nm is sufficient to completely describe the diagnostic features of the materials (Goetz et al., 1985). Table 2.1 provides a comparison of broadband multispectral data and hyperspectral data.

TABLE 2.1
Comparison between Multispectral Data and Hyperspectral Data

	Hyperspectral	Multispectral
Data characteristics	Continues with ~10 nm width	Discrete, broad bands
Data processing and analysis	High dimensional data needs dimensionality reduction measures, size of the data also might need some special treatment	High dimensionality is not an issue
Unsupervised classification	True unsupervised classification is possible with spectral library	No such possibility
Material detection	Direct material detection is possible, spectral library can be built and used subsequently	Signature contains insufficient details for detecting materials
Quantification	Accurate quantification models can be developed, for example, pollutants in water	Models developed would not be as accurate as hyperspectral data because diagnostic absorption may not be available

2.1.2 Some Useful Terms

Some of the commonly used terms in the context of hyperspectral data acquisition and analysis are explained below. These are specific terms used in describing or analysing the hyperspectral data. All of them may not be necessary for understanding the remaining topics discussed in this book. However, a general idea about them would be helpful in the overall appreciation of the hyperspectral data and its analysis.

Reflectance – This is a unitless quantity indicating the ratio of reflected solar radiation by the target and incoming solar radiation. The reflectance properties of the land covers are of interest to remote sensing analysts as these basic properties form the spectral basis of land cover detection. The reflectance values are more significant in hyperspectral remote sensing as it forms the basis of spectral matching. There are minor differences in how the reflectance is calculated for remotely sensed data:

  a. approximate reflectance calculated using radiance recorded by satellite sensor and top of atmosphere irradiance ignoring atmospheric effects, referred to as Top of the Atmosphere (TOA) reflectance, and
  b. reflectance calculated using at surface irradiance and reflected energy by the surface by correcting the irradiance and at sensor radiance. This is called as "true reflectance" or "surface reflectance" or "ground reflectance". The terms "at sensor" and "at surface" are also used commonly for TOA and ground reflectance respectively.

Spectral signature – The spectral signature term is commonly used for a unique graph of the spectral response function of a substance. It is normally represented by

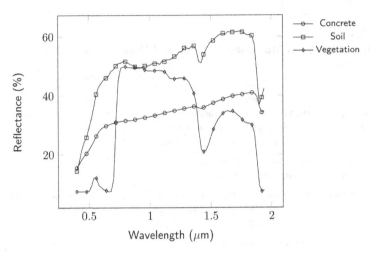

FIGURE 2.2 Signatures of the three basic urban materials.

a plot of reflectance values or simply reflectance versus wavelength. As the reflectance is measured with contiguous and high spectral resolution bands, the shape of the plot or the spectral curve is ideally unique for a given material, as if the shape is a signature of the material. This uniqueness of the spectral signature over a given wavelength range is the basis of remote sensing. If two materials have exactly similar signatures, they cannot be resolved spectrally. Additional spatial context to determine their identity is required in such cases. Figure 2.2 shows three basic urban materials namely vegetation, concrete, and soil. Concrete surface is impervious to water and hence these are dubbed as VIS – Vegetation, Impervious surface/s, and Soil (Ridd, 1995). Each one has unique spectral characteristics and can be easily discriminated from each other. The concrete signature is marked by the flat signature throughout the range, except blue bands. The signature shows absorption at ~1900 nm because of water. The water absorption at 1900 nm in concrete is highly correlated with its degradation (Arita et al., 2001).

Spectral line – The spectral line term is used to indicate the single wavelength corresponding to the peak of the plot of intensity and wavelength or frequency (Hollas, 2004). It is a way to describe the single wavelength responsible for the absorption of electromagnetic energy. Though the single wavelength is used to describe the spectral line, the plot of absorption intensity and wavelength is not infinitely narrow. It spreads out beyond the spectral line over a small section. The spreading of absorption beyond the central wavelength is called broadening. The shape of the absorption intensity and the wavelength plot is referred to as the line profile. Wavenumber is the preferred unit of radiation in spectroscopy as it is proportional to the energy of radiation (Pavia et al., 2001). The various parts of a spectral profile are displayed in Figure 2.3.

Full width half maximum (FWHM) – Most commonly called FWHM is a unit for expressing the spectral resolution of a sensor or a spectral line. It denotes the width

Introduction to Hyperspectral Imaging

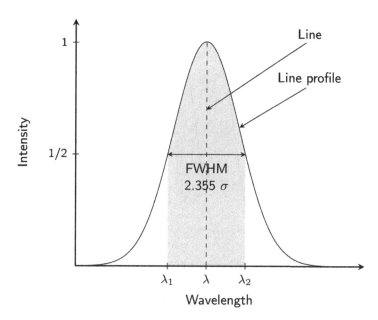

FIGURE 2.3 Spectral line profile and full-width half maximum (FWHM).

in nm at half of the maximum intensity of the spectral response of the sensor. FWHM and sampling interval determine the nature of a spectral signature provided by a particular sensor. FWHM determines the ability to resolve the absorption. Narrow FWHM can resolve fine absorptions, that is, the smaller the FWHM, the better it is. Figure 2.3 shows the FWHM of a spectral profile or a spectral response function.

Sampling interval – This is another quantity used for specifying the spectral signature collected by a spectrometer, imaging or otherwise. The sampling interval refers to the interval between the two consecutive band centres of the spectrometer (Figure 2.4). Like FWHM, it is expressed in nm. A finer sampling interval means the distance is narrower between two consecutive band centres and more would be the continuity in recording the signature. The ability of a sensor to record the diagnostic absorption over a wavelength range depends upon the sampling interval and FWHM of the sensor. Measurement of narrow absorption features requires finer sampling interval and FWHM.

Continuum – The envelope of a spectral signature without any absorptions. In other words, it is a spectral signature of background absorption (Clark & Roush, 1984).

Diagnostic absorption – The electromagnetic energy will be absorbed by the target material at specific wavelengths, which would indicate the presence of certain chemicals or groups of chemicals. These absorption/s confirm beyond any doubt the presence of a particular chemical and is hence called diagnostic absorption. The depth of absorption is measured by removing continuum absorptions or using continuum signature as

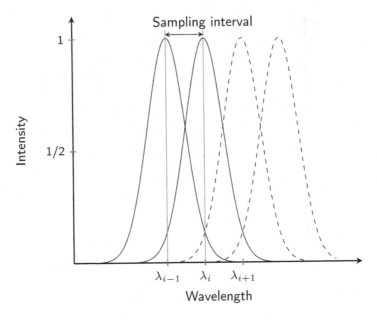

FIGURE 2.4 Sampling interval.

the reference. The diagnostic depth is further used for establishing quantitative relation between the concentration of the chemical and the diagnostic depth. Figure 2.5 shows the continuum and the diagnostic depth at ~1900 nm for the concrete sample.

Spectral range or coverage – This is the wavelength range of electromagnetic radiation covered by the sensor. For hyperspectral imaging, it is commonly denoted in nano metre (nm). For example, 400 nm to 1000 nm or 400 nm to 2500 nm. The useful spectral range changes as per the application. This is because the unique characteristics available in the signatures are present within the specific wavelength range. For example, for vegetation studies, the spectral range of 400 nm to 1000 nm may be sufficient. Thus, it is customary to have a dedicated sensor for a given application with limited spectral coverage or range. This is to optimize the sensor performance.

2.1.3 Quantum Mechanical Explanation of Radiation Absorption

There are three basic interactions possible between electromagnetic radiation and a module (Figure 2.6).

1. Induced absorption
2. Induced emission
3. Spontaneous emission (Hollas, 2004)

Induced absorption is an absorption process in which a molecule or atom absorbs a quantum of energy of the radiation and goes from a quantum mechanical state m to

Introduction to Hyperspectral Imaging

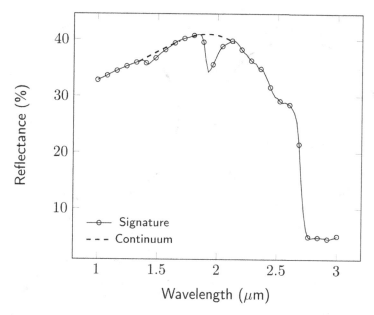

FIGURE 2.5 Diagnostic absorption and depth.

n, where n indicates the higher state of quantum mechanical energy and m indicates lower. This is referred to as induced as it requires the presence of radiation.

Induced emission is one in which a molecule in state n emits a quantum of radiation and goes to the state m. A quantum of radiation is required to induce this emission.

Spontaneous emission is the emission in which a quantum of radiation is emitted spontaneously.

As discussed in earlier sections, hyperspectral imaging attempts to measure molecular or atomic absorptions using high spectral resolution. The interaction of electromagnetic radiation such as a light and the matter is completely determined by quantum electrodynamics. The electromagnetic energy is absorbed by the molecules if the energy of the radiation at specific wavelengths matches the difference of energy between to quantum mechanical states of the molecule.

2.2 MULTISPECTRAL, SUPERSPECTRAL, HYPERSPECTRAL, ULTRASPECTRAL

"Multi", "Super", "Hyper", "Ultra" are some of the common adjectives used in the context of acquired remote sensing data. Multispectral imaging is very well understood and used extensively since beginning of the satellite imagery beyond visible range. "Super" and "Ultra" are recently introduced terms. Of these, "superspectral" is ambiguous at times as there are imaging sensors capable of recording images in 35 to

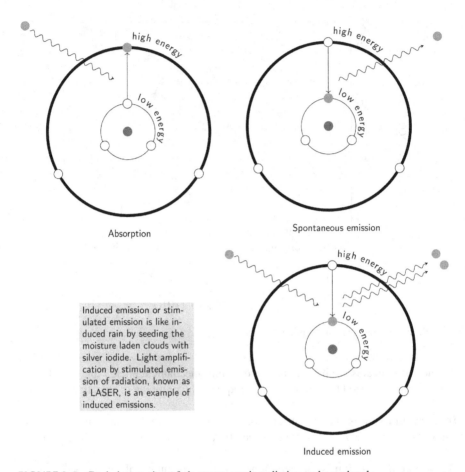

FIGURE 2.6 Basic interaction of electromagnetic radiation and a molecule.

40 bands in a contiguous manner and in a discrete manner as multispectral imaging (Polder & Gowen, 2020). "Super", "Hyper", "Ultra" adjectives indicate an increasing number of bands. However, excluding the latter two, a higher number of bands may not be correlated with the fine resolution absorption features which require spectroscopic methods for analysis. A higher number of narrow bands in superspectral imaging does not necessarily mean the bands can measure diagnostic absorptions. It would entirely depend upon the specifications of the sensor.

Thus, it is better to see an increasing number of bands and the nature of sampling (continuous or discrete) as the conditions demanding specific analysis techniques for processing the concerned image. The motivation for hyperspectral imaging is to acquire the finer details of the energy matter interaction and use these spectral features for target detection or quantitative analysis. In principle, this is like spectroscopic analysis done in the laboratory. Hence, hyperspectral imagery is also referred to as "imaging spectroscopy". As such, the terms "hyperspectral data" and "imaging spectrometric data" are used interchangeably in the literature. Jet Propulsion

Introduction to Hyperspectral Imaging

Laboratory (JPL) prefers "imaging spectroscopy" and "imaging spectrometric data" to "hyperspectral imaging" or "hyperspectral data" to make it amenable to other science disciplines such as physics, chemistry, biology (Jet Propulsion Laboratory, 2017-a). We use "hyperspectral data" and "hyperspectral imaging" to indicate the data and imaging process that have sufficient spectroscopic features in the signal.

The motivations for processing multispectral data and hyperspectral data are different. Multispectral data is useful in providing broad-level land use land cover category labels for the pixels. Whereas the hyperspectral data, because of the availability of minute spectral details in the spectral signature, is used for providing fine-level land cover category labels for the pixels. Specifically, it attempts to detect the target material using the analysis of its spectral characteristics. For example, spectral matching techniques can be used to label the pixel in the image. Further to this, it may be used for developing quantitative relations with the subject of interest. For example, for soil quality studies, multispectral data can detect the broader soil quality zones, whereas hyperspectral data is more helpful in developing a quantitative relation between the diagnostic depth and concentration of, say, nitrates in the soil. Certainly, this is subjected to the right spectral resolution of the sensor. However, generally speaking, the direction of investigation is to assess the potential of the spectral signature by a given sensor in detecting a target material directly. Thus, while multispectral data is helpful in differentiating between two land covers at a broader level, hyperspectral data goes a step further in helping to identify their material or chemical composition.

Additionally, multispectral data provides higher spatial resolution given all other sensor characteristics remain the same. Thus, on many occasions, the focus of the multispectral data analysis is to leverage the spatial information available in the image and perform more object-based analysis. That is, the focus is on detecting objects in the image. For example, counting the number of trees in an urban scene using submeter multispectral data. If detecting the object alone is the only objective of the image analysis, then high-resolution multispectral data is useful. However, it is not useful if we want to seek information about its chemical nature. The information about the chemical composition of the object can only be resolved by high-resolution spectral information providing the unique spectral signature. Table 2.2 summarizes the spectral characteristics available at different imaging techniques. Table 2.3 shows some of the examples of hyperspectral remote sensing applications. These are indicative applications. Hyperspectral remote sensing is a basic tool for detecting target material and/or its chemical composition. Thus, hyperspectral imaging is useful in all the cases where land cover or land use can be resolved spectrally and/or further chemometric analysis is desired.

2.3 CHALLENGES

2.3.1 Coarse Spatial Resolution

The spectral and spatial resolution are conflicting goals. The amount of light reaching a sensor is limited to narrow bands in hyperspectral imaging and hence the energy reaching the sensor is limited too. Thus, for having sufficient signal strength, the

TABLE 2.2
Type of Sensor and Their Spectral Characteristics

Type	~Bandwidth in nm	No. of bands	Spectroscopic features
Panchromatic	300	1	No spectroscopic feature is available, broad level reflectance characteristics are available
RGB	100	3	Colour
Multispectral	100	12–16	Colour
Multispectral narrow bands[1]	10–30	25–30	Some broad band absorptions may be available, the spectroscopic techniques such as spectral matching may not be useful, narrow band absorptions and diagnostic depth analysis may be available depending upon the band resolution
Hyperspectral	10–20	100–200	Spectral matching and diagnostic depth analysis can be deployed
Hyperspectral with very high spectral resolution[2]	<10	1000–2000	Spectral matching is useful, the fine diagnostic absorptions would be available

Notes:
[1] is called "superspectral" sometimes, [2] is called "ultraspectral". These terms are not as common as multispectral and hyperspectral. Moreover, the current adjectives such as "multi" and "hyper" can define all the characteristics for the range of sensors. In our opinion, terms like "superspectral" and "ultraspectral" should be avoided till they are systematically defined or conventionally accepted like multispectral or hyperspectral.

TABLE 2.3
Hyperspectral Remote Sensing Applications

Domain	Application	Examples
Urban studies	Impervious surface detection, urban growth studies	Detection of roof materials, road materials etc.
Environment	Pollution monitoring	Quantification of turbidity, total suspended solids, a particular pollutant in the water, gas leaks
Agricultural	Yield prediction, precision agriculture	Soil nutrient analysis, crop detection, health assessment
Machine vision	Object recognition	Detection of hazardous material
Resource exploration	Prospecting, ore quality assessment	Identification of minerals or group of minerals
Asset management	Pipeline monitoring	Leakage detection, encroachment detection
Defence	Camouflage detection, threat assessment	Detection of surface materials, chemicals

Introduction to Hyperspectral Imaging

reflected energy from a large ground sample is needed. Hence, given the same conditions, hyperspectral data would have medium to coarse ground resolution – say 20 m to 30 m for spaceborne platforms. Because of the coarse spatial resolution, spatial information such as the contour of smaller objects (shape of the object) is not available in the image. For example, the exact shape of a tree crown in an urban scene would be difficult to extract from 30 m resolution imagery. Thus, the hyperspectral data analysis depends upon the degree of spectral information available. In case spectral information is not sufficient for the task at hand, it needs to be augmented with the neighbourhood information or information from external sources. For example, statistics of spatial association of the land covers may be helpful in identifying the land cover of a given pixel without any ambiguity.

2.3.2 Increased Data Size

With an increasing number of bands, the size of the image increases too. This may not be a problem for processing a single image on modern personal computers, but processing many images for change detection or trend analysis of any kind would be a challenging task. In addition to the processing time, the size of the data increases the burden of data communication and data storage. For example, the Landsat 8 image with 10 bands, covering 185 km x 180 km area has a size of 1.20 GB whereas EO-1 Hyperion hyperspectral image of 242 bands covering 180 km x 7.7 km has a size of 403 MB (USGS, 2014, Landsat 8 Scene: LC81680612014354LGN00, for Target Path 168, and Target Row 61, 2014; USGS, 2002, EO1 Hyperion Scene: EO1H1680602002153110PZ_1R, for Target Path 168, and Target Row 60, 2002). The other details of both images are provided in Table 2.4. If the hyperspectral image was to cover the same area, the size of the image would have been ~24 times the size of its corresponding multispectral image with the same radiometric resolution.

2.3.3 Small Number of Samples and High Dimensions

A large number of spectral bands means a large number of dimensions for a D dimensional hyperspectral pixel vector, where D = *number bands*. This increases the computational complexity of the machine learning algorithms deployed for analysis of the image. The specific techniques to reduce the dimensionality of the data, ideally without any loss of spectral information, are required. This is particularly essential

TABLE 2.4
Landsat 8 and EO-1 Hyperion Image Size Comparison

	Landsat 8	EO-1 Hyperion
Image file name	LC81680612014354LGN00	EO1H1680602002153110PZ_1R
Bands	10	242
Scene size	180 km X 185 km	180 km X 7.7 km
Dimensions	7751 rows X 7591 columns	6703 rows X 256 columns
Radiometric resolution	16 bit	16 bit

in cases where the number of samples for a given class is less than the number of dimensions. One of the most common approaches for reducing the dimensions is to remove the bands that are redundant or bands that can be recreated using correlation with the other bands. Selecting the only bands that are helpful in increasing the accuracy of further processing is another common strategy too.

2.3.4 LACK OF UNIQUE SPECTRAL SIGNATURE

This is one of the most challenging problems to solve. The hyperspectral data advantages are with the premise that the spectral signature acquired is unique (most of the time) for a given material. In case this is not true, spectral matching fails to resolve the identity of the target material or spectral matching may provide confusing results. The problem of a similar spectral signature is difficult to overcome without any additional information.

2.4 SENSORS AND PLATFORMS

The use of hyperspectral sensors is increasing day by day. Earlier, the use of hyperspectral sensors was limited to remote sensing applications such as earth observations. However, the reduction in size and the easy availability of the sensors have resulted in many industrial applications as well. New airborne platforms such as unmanned aerial vehicles (UAV) and drones provide an added advantage of sensing at much higher spatial resolution and actuation, which is not possible easily through satellite platforms. For example, the effective application of pesticides over a large agricultural area is better managed by drones or UAVs than satellites. The excessive cost of the hyperspectral sensor is still discouraging. Mass production of the sensors fuelled by increasing industrial indoor and outdoor applications is expected to change the situation. The snapshot hyperspectral cameras for industrial applications at moderate cost have made their entry into the commercial market (SPECIM, 2018).

As can be seen from the discussion so far, the hyperspectral sensor applications are diverse and hence there is large variation in the sensor specifications as well. We review some of the important airborne and spaceborne sensors and platforms that are commonly used or will start sharing the data very soon. They include sensors on spaceborne platforms, airborne platforms, and commercial sensors that are designed for UAVs and drone platforms.

2.4.1 SPACEBORNE PLATFORMS

2.4.1.1 Earth Observing-1 (EO-1)-Hyperion

National Aeronautics and Space Administration's (NASA) EO-1 satellite was launched on November 21, 2000, as a part of a one-year technology demonstration mission. Though the mission was completed, NASA and the United States Geological Survey (USGS) reached an agreement to allow the continuation of the EO-1 program as an Extended Mission. EO-1 was a tasking satellite with an

Introduction to Hyperspectral Imaging 53

onboard hyperspectral sensor Hyperion. It could service twelve requests per day. It also collected information in broad bands with Advanced Land Imager (ALI) sensor simultaneously (USGS, 2011-2). Since January 6, 2017, EO-1 has stopped accepting Data Acquisition Requests (DAR) because of the end of normal operations. The historical record of EO-1 data will be available through Earth Explorer https://earthexplorer.usgs.gov/. The EO-1 satellite was decommissioned effective February 22, 2017 (EO-1, 2017).

2.4.1.2 Spaceborne Hyperspectral Applicative Land and Ocean Mission (SHALOM)

The mission, founded and managed by the Israeli Space Agency (ISA) and the Italian Space Agency (ASI), aims at commercializing hyperspectral sensing by providing higher spectral-spatial and temporal resolution. The designed less than 10 m spatial resolution is expected to enhance the accuracy of urban applications. Different data products such as L1 – radiometrically corrected, L2 – atmospherically corrected, and L3 – geometrically corrected would be available to the intended users.

In addition to providing thematic maps of natural resources, the main objective of the mission is to provide high spectral data on the lithosphere, atmosphere, and oceanophyte. According to Feingersh and Dor (Feingersh & Dor, 2015) SHALOM was supposed to be launched around 2020. SHALOM is supposed to cover a 200,000 sq. km area on a daily basis acquiring 6696 Gbit (~7 GB) of hyperspectral data (Feingersh & Dor, 2015).

2.4.1.3 PROBA-1-CHRIS (Project for On-Board Autonomy-1 (PROBA-1, Compact High Resolution Imaging Spectrometer)

PROBA, renamed PROBA-1, is a unique platform with an onboard sensor Compact High Resolution Imaging Spectrometer (CHRIS) (ESA, About PROBA-1). Its operations are fully automated and require little effort from the ground station. CHRIS is a completely programable hyperspectral sensor for collecting hyperspectral data with a 17 m spatial resolution. It has the capability of taking five pictures of a target area from five distinctive look angles in a single orbital overlap (along-track pointing angle). It can look at the same target from fifteen different angles in a few days from multiple orbital overpasses (using across-track pointing). The capability of five views along the track and fifteen views across the track for the same target helps CHRIS to measure the Bidirectional Reflectance Distribution Function (BRDF) of the target (ESA, CHRIS Overview).

CHRIS weighs just 14 kg. The onboard CHRIS was developed by SIRA technologies Ltd (ESA, Mission background). The data is acquired in five modes. Each mode determines the number of bands, band centres, spatial resolution, and spectral resolution of the data (Cutter, 2008). The mission is still operational – it recently completed twenty operational years (ESA, 2022).

2.4.1.4 PRISMA (PRecursore IperSpettrale della Missione Applicativa)

The satellite is launched and operated by ASI. The platform acquires hyperspectral images using a pushbroom sensor, with a 30 m spatial resolution and ~12 nm spectral resolution. The revisit period for the sensor is seven days for an off-nadir viewing angle. The satellite was launched in low orbit by Vega on March 22, 2019. The designed service life of the sensor is five years and the data is available to users on registration (Guarini et al., 2018).

2.4.1.5 Hyperspectral Imager Suite (HISUI)

HISUI was launched in 2019 and was deployed on Japan's Experimental Module (also known as kibo) on the International Space Station (ISS) (METI, 2019). The satellite is developed by Japan's Ministry of Economy Trade and Industry (METI) (Japan Space Systems, 2021). HISUI will provide ~20 m resolution data with ~10 nm spectral resolution. The data of 10 GB corresponding to a 30000 km^2 area would be transmitted on a daily basis. The rest of the data, that is, 300 GB/day corresponding to the area of 90000 km^2, will be recorded by the mission data recorder-pressurized module on the ISS kibo. It will be sent back to earth by cargo ships three to four times a year (Matsunaga et al., 2017). HISUI captured its first image on September 4, 2020. The data policy is under consideration. The satellite was about to start its operations by Nov. 2020 and begin its application demonstration by Mar. 2021 (Japan Space Systems, 2020)

2.4.1.6 DLR Earth Sensing Imaging Spectrometer (DESIS)

DESIS is developed and operated by Teledyne Brown Engineering (TBE) located in Huntsville, Alabama, USA, and the German Aerospace Centre (DLR), Germany. DESIS is a very high spectral resolution sensor providing 30 m spatial resolution imagery of the earth. Its design life is 3.5 years. DESIS is the first instrument to utilize the Multi-User-System for Earth Sensing (MUSES) platform's external payload accommodation of the ISS. The commissioning and validation phase was finished in March 2019. DESIS is operational now and would continue its operations at least until the end of 2023 (DLR, n.d.).

DESIS can point, using its Pointing Unit (POI), in a forward and backward viewing direction within the ±15° range with respect to nadir. This capability is achieved by two fixed and one rotating mirror in the front of the entrance slit. The POI has the capability of pointing with static and dynamic modes. The instrument can be operated with 3° angle steps for the viewing direction. It can be operated in dynamic mode with up to 1.5° change in viewing direction per second. This allows acquiring stereo or BRDF products (Image Stereo Mode), in addition to normal mode (Image Strip Mode) (DLR, n.d.).

New DESIS images can be obtained, for previously submitted and evaluated science proposals, using Teledyne Brown Engineering company's web platform, whereas EOWEB GeoPortal can be used for obtaining archived images without any proposal (DLR, 2019).

2.4.1.7 The Environmental Mapping and Analysis Program (EnMAP)

EnMAP is a German hyperspectral sensor with the designed objective of monitoring the earth's environment. EnMAP will provide data related to human activity and climate change once it is operational. It can acquire an image with 6.5 nm spectral resolution, and 30 m spatial resolution within the spectral range of 420 to 1500 nm.

Introduction to Hyperspectral Imaging

30° tilt pointing angle enables it to take an image of a target area every fourth day (DLR, 2022).

The EnMAP mission is divided into five phases of mission progress. The EnMAP mission is in the development and production phase (Phase D) as of June 2020. The ground segment for in-orbit operations is ready. The utilization of ground segment for the in-orbit operation has been authorized. Further verification and validation of all the mission components is to be performed. The designed life of the instrument is ~5 years (DLR, 2022).

2.4.1.8 Copernicus Hyperspectral Imaging Mission for the Environment (CHIME)

The Copernicus infrastructure did not support the monitoring themes like CO_2. Six High priority Candidate Missions (HPCM) were proposed by ESA to meet the priority needs of the users monitoring themes such as CO_2. CHIME is one of them. The main objective of the mission is to provide support to European Union (EU) policies for natural resources management. The hyperspectral capabilities of CHIME are envisaged to enhance services for food security, agriculture, and raw materials (Buschkamp et al., 2021).

The feasibility and preliminary definition for CHIME is completed and the mission has entered in B2 phase. The preliminary definition suggests a pushbroom sensor acquiring images with less than 10 nm spectral resolution with 30 m spatial resolution over a 400 nm to 2500 nm range. The revisit period for a given target area would be 12.5 with two satellites in orbit (Buschkamp et al., 2021).

2.4.1.9 OrbitalSidekick

With its Global Hyperspectral Observation Satellite constellation (GHOSt) Orbital Sidekick (OSK) provides the best spatial resolution of 8 m for spaceborne hyperspectral imagery. Six more satellites are about to join the present constellation, between 2022–23 (OSK, n.d.). 100 kg satellites are designed by Astro Digital, and Maverick Space Systems provides mission integration and management services. OSK builds upon its prior experience in collecting and analysing hyperspectral data with HEIST, a mission mounted on the ISS in 2019 (GHOSt Constellation Deployment Plans Finalized By Orbital Sidekick, 2021; "Orbital sidekick announces the upcoming launch of its newest global hyperspectral earth observation constellation": GHOSt, 2021).

2.4.1.10 Pixxel

The Pixxel TD-2, a commercial hyperspectral satellite, was launched into orbit by SpaceX on April 1, 2022. The 15 kg satellite will provide 10 m imagery on a daily basis. The images would be captured in 150 spectral bands. The main objective of the mission is to monitor the health of the earth's ecosystem. Some of the specific objectives are to monitor pollution, forest health, crop health and so on. This launch is a precursor to the first commercial launch planned in early 2023 and subsequent commercial data sale. Pixxel is about to launch six more satellites in sun-synchronized orbits to complete the constellation. The goal of daily coverage would be achieved by 2024 as per the sources (Pixxel, 2022).

2.4.1.11 Other Ventures (Wyvern, HySpeqIQ, Constellr)

Wyvern is a Canadian company founded in 2018. They are planning to launch a satellite capable of providing 1 m imagery in the Visible and Near-infrared (VNIR) region, and 5 m in the Short-wave Infrared (SWIR) region. The first-generation satellite was launched in 2022. The innovative telescopic system unfolds in space and thus allows for a compact instrument at launch time. The focus of the system would be to provide high-quality data to various industrial sectors such as forestry, agricultural, energy, defence and so on. A satellite constellation with a 1 m spatial resolution capability is expected to be launched in 2023 (Wyvern, 2022).

HySpecIQ had planned to deploy hyperspectral imaging instruments on two Boeing satellites but had to cancel the plans in 2015. The company recently received an order for a study contract from the National Reconnaissance Office (NRO), Sep. 2019 (Erwin, 2019). HySpecIQ believes that they would be able to launch two to three satellites in orbit within the next 24 to 36 months. HySpecIQ was founded in 2013 (HySpecIQ, 2021; Erwin, 2019).

Constellr is one of the leading space systems companies providing thermal data for monitoring the food and agricultural industry. They have recently acquired Belgian hyperspectral firm ScanWorld. The fusion of hyperspectral scanning with the Constellr's existing thermal scanning capabilities would lead to precision monitoring of various crop management activities (Wolf, 2022). The summary of space-borne platforms is provided in Table 2.5.

2.4.2 Airborne Sensors

2.4.2.1 Airborne Visible InfraRed Imaging Spectrometer (AVIRIS-classic)

One of the early sensors collecting hyperspectral data is Airborne Visible InfraRed Imaging Spectrometer (AVIRIS). It has been flying since 1986 (JPL, AVIRIS next generation, https://avirisng.jpl.nasa.gov/aviris-ng.html) AVIRIS uses four aircraft – National Aeronautics and Space Administration's (NASA) ER-2 with a flight height of ~20 km above sea level, at 730 km.hr^{-1} and Twin Otter International's turboprop aircraft with a flight height of ~4 km, at 130 km.hr^{-1}. The other two aircraft are Scaled Composites' Proteus, and NASA's WB-57 (Jet Propulsion Laboratory, 2017-b). AVIRIS is a whiskbroom type of scanner with an Instantaneous Field of View (IFOV) of 1.0 mrad (Jet Propulsion Laboratory, 2017-c).

2.4.2.2 Airborne Visible InfraRed Imaging Spectrometer – Next Generation (AVIRIS-NG)

AVIRIS-NG is designed for continued access to high spatial and spectral resolution data for science applications. It is expected to replace the AVIRIS-Classic instrument that has been flying since 1986 (JPL, 2021). The current AVRIS NG flights are conducted by King Air B200 operated by Dynamic Aviation Group Inc. from Bridgewater, Virginia. The maximum altitude for the King Air B200 platform is 10.6 km. The other available platform options are the ER-2 operated by NASA's Armstrong Flight Research Centre, and the DHC-6 operated by Twin Otter International Ltd (TOIL).

TABLE 2.5
Spaceborne Hyperspectral Platforms

Satellite	PROBA-1-CHRIS	DESIS	PRISMA	HISUI	SHALOM	OrbitalSidekick	EnMAP	CHIME
Agency	ESA	DLR	ASI	JSS, METI	ISI, ASI	OSK	DLR	ESA
Sensor type	Pushbroom	Pushbroom	Pushbroom	Pushbroom	Pushbroom	Pushbroom	Pushbroom	Pushbroom
Orbit (km)	615	ISS	615	ISS	640	NA	653	625
Coverage (nm)	400–1050	402–1000	400–2500	400–2500	400–2500	400–2500	420–2450	400–2500
Bands	18/62[1]	235[2]	237	185	275	>500	222	250
FWHM (nm)	1.5–11.25[3]	3.5–7	<=12	10–12.5[4]	<10	3.5	6.5–10[5]	<10
Spatial resolution (m)	17/34	30 m and 1024	30	20 x 30[6]	<10, 2.5–5[7]	8, 3[8]	30	30
Temporal resolution (days)	<7	3–5[9]	7/29[10]	ISS	<4days	NA	4/21[11]	10–12.5/20–25[12]
Radiometric resolution	NA	13	12 bit	12 bit	12 bit	NA	>=14	14
Swath	14/18	30	30	20	>10	2[13]	30 km SWATH	130 km swath
Launch year	2001	2018	2019	2019	2020	2021–22	2022	NA
Data availability	Available	Available on submitting proposal	Available by registering the project[14]	Data policy under consideration	NA	Commercial	NA	NA

Notes:

[1] Mode dependent., [2] Without binning, [3] Mode dependent, [4] VNIR and SWIR respectively, [5] VNIR and SWIR respectively, [6] Cross track x along the track, [7] Spectral image and panchromatic image respectively, [8] Spectral image and panchromatic image respectively, [9] International Space Station tracks, [10] Off nadir viewing and nadir viewing respectively, [11] 30° tilt and 5° tilt, [12] With one satellite and with two respectively, [13] Swath for panchromatic mode, [14] https://prismauserregistration.asi.it/.

The ER2 is a high-altitude platform (21.3 km) (NASA, 2022), whereas DHC-6 is a mid-altitude platform (JPL, 2021-3).

2.4.2.3 The Hyperspectral Thermal Emission Spectrometer (HyTES)

The Hyperspectral Thermal Emission Spectrometer (HyTES) is an airborne imaging spectrometer, which covers a wavelength range from 7.5 μm to 12 μm in 256 spectral bands and 512 pixels across the track. The main objective of the HyTES instrument is to support the Hyperspectral Infrared Imager (HyspIRI) mission. HyTES is designed to provide data at much higher spectral and spatial resolution than HyspIRI. Thus, the HyTES data would be helpful in determining the optimum band positions for the HyspIRI-TIR instrument. The first flight was completed by HyTES in July 2012 (JPL, n.d.). HyTES completed its latest flight (first flight in Sweden) on Aug 17, 2021, in a joint campaign with NASA, ESA, King College London, Natural Environmental Research Council (NERC), National Centre for Earth Observation (NECO) on a British Atlantic Survey (BAS) aircraft (NASA/JPL/ESA/KCL, 2021).

2.4.2.4 HyMap

HyMap is a commercial hyperspectral sensor manufactured by Integrated Spectronics Pty. Ltd. and operated by HyVista Corporation, Australia. It uses a whiskbroom type of scanner and utilizes 32-element detector arrays for recording data. HyVista started providing hyperspectral data to government, academic, and commercial customers all over the word since 1999 (Kruse et al., 2011). The summary of the airborne sensors is provided in Table 2.6. The summary of commercial off-the-shelf sensors for UAV and drone platforms is provided in Table 2.7.

TABLE 2.6
Airborne Hyperspectral Sensors

Airborne	AVIRIS-NG (2012)	AVIRIS-classic (1986)	Hyperspectral Thermal Emission Spectrometer (HyTES)	HyMap
Agency	NASA	NASA	NASA	Hymap
Coverage (nm)	380–2510	400–2500	7500–12000	450–2500
Bands	224	224	256	128
FWHM (nm)	5	10	17	15–20
Spatial resolution (m)	2, 4, 6[1]	4, 20[2]	3.41, 34.1[3]	03–10 m
Temporal resolution (days)	ground speed 70, 100, 128kts (130, 185, 237 kmph)	NA	NA	NA
SWATH	NA	11 km, 1.9 km	512 pixels across track; 1868.33, 18683.31	~1.5–~5.0 km (512 pixels)

Notes:
[1] for ~2 km, 4 km, 6 km flight altitude respectively, [2] For 4 km and 20 km flight altitude respectively, [3] at 2 km and 20 km respectively

2.4.3 STANDARD DATA SETS FOR RESEARCH

Just like spaceborne and airborne hyperspectral platforms are required for ongoing earth observations, readily available hyperspectral images with ground truth are an invaluable resource for research. Hyperspectral image processing research is well supported by the standard data sets in the public domain. The standard data sets are helpful in bench marking the results and comparing the performance of various methods. There are many data sets according to the applications. For example, two of the most popular data sets, Pavia University and Indian Pines, are focused on urban land cover detection and crop monitoring respectively. Some of the data sets are dedicated to unmixing analysis, for example, Urban data set. There is a lot of variation in the available resolution as well. The users need to keep in mind these characteristics and interpret the results in that context. For example, the Pavia data set is having high resolution and hence the classifiers dominated by spatial features (more in Chapter 5) perform well on the Pavia data. However, the same technique may not provide the same performance level on moderate-resolution imagery, say 30 m EO1-Hyperion or Indian Pines. These data sets are available with ground truth. Table 2.8 provides available details of the data sets available in the public domain.

There are many EO1-Hyperion images available in the public domain that do not have ground truth available for them. Or even if it is collected it may not be in the public domain. However, these data sets can be used for qualitative hyper spectral image processing studies. For example, classification and clustering can be performed, and the images can be interpreted with the help of same-date high-resolution imagery. For some of the EO1-Hyperion or AVIRIS images in the public domain, the data from other sensor networks such as AERONET (AERONET, 2015) would be available too. In that case, such data sets would be extremely useful even for quantitative studies.

2.5 VISUAL INTERPRETATION OF HYPERSPECTRAL IMAGES

The process of visually interpreting hyperspectral images remains the same as multispectral images. However, there is added multichannel perspective which is not so useful in multispectral images. The common objectives of the visual interpretation of remotely sensed data are to:

a. Identify spatial signatures of some common land use and land covers – using colour composite or side-by-side multichannel view. This is required for identifying training areas or pixels for reference signatures for land use land cover classification if classification is the goal,
b. Identify spectral signatures of target land covers and land use. The side-by-side channel view is particularly useful for this purpose. The other alternative for this is plotting the D dimensional pixel vector as illustrated in Figure 2.1 and Figure 2.5, and
c. Identify potential dark and flat objects within the scene. This is required for image-based atmospheric correction methods (discussed in a separate chapter later).

TABLE 2.7
Commercial Off-the-Shelf Sensors for UAV and Drone Platforms

Sensor	HySpex VNIR-1800	HySpex SWIR-640	Specim Aisa FENIX	Specim Eagle	Specim FX10	Cubert Firefly	Rikola FS4	Rikola X series VNIR	Headwall MV.X	Headwall SWIR	Headwall X series
Spectral range (nm)	400–1000	960–2500	380–2500	400–970	400–1000	450–950 / 500–900	350–2500	550–1650	400–1000	900–1700	900–1700
Number of bands	186	360	348/174/87/274	60–488	224	125 / 380	2151	96	301	134/67	320
FWHM (nm)	4.5	7	3.5–10	3.3	5.5	8 / 10	3.0–8.0	6.0–12.0	6	10	6.0–12.0
Sampling (nm)	3.26	4.38	1.7/3.4/6.8/5.7	1.15–4.60	2.7	4 / 1	1.4–1.1	-	2	4.8	-
Radiometric resolution (bit)	16	16	12 and 16	12	12	12–14	NA	NA	14	12	14	14
IFOV/FOV	17°	16°	32.3°	0.029	38°	30, 20, 13, 7	36.5	25	-	-	-
FPS	260	140	100	160–30	327	25	0.728	-	100	120/346	100
Weight (kg)	5	4.1	15	6.5	-	0.47	5.44	0.75	3	1.9/0.9	0.75
Dimensions (cm)	39–9.9–15	36–11–15	38.7–22.3–5	14.6–14.6–34.7	15–8.5–7.1	20.0–6.7–6.0	8.0–9.7–15.9	-	-	25.0–13.0–13.6	25.0–13.0–13.6	-

Note: These are the primary providers of reliable hyperspectral sensors. because of the advent of UAV platforms and industrial applications, the number of vendors providing hyperspectral sensors is increasing. An exhaustive survey of all such sensors is beyond the scope of this chapter.

Sources:
www.hyspex.com/hyspex-products/hyspex-classic/hyspex-vnir-1800/
www.hyspex.com/hyspex-products/hyspex-classic/hyspex-swir-640/
www.specim.com/wp-content/uploads/2020/03/AisaFENIX-ver1-2020.pdf
www.specim.com/wp-content/uploads/2020/02/Specim-FX10-Technical-Datasheet-04.pdf
www.adept.net.au/cameras/specim/systems/Aisa_eagle.shtml
www.cubert-hyperspectral.com/products/fireflyeye-185
www.researchgate.net/publication/343721676_Ice_Detection_on_Aircraft_Surface_Using_Machine_Learning_Approaches_Based_on_Hyperspectral_and_Multispectral_Images
www.malvernpanalytical.com/en/products/product-range/asd-range/fieldspec-range/fieldspec4-hi-res-high-resolution-spectroradiometer#specs
www.headwallphotonics.com/hubfs/Headwall_MVX_Datasheet_04May2022.pdf
www.headwallphotonics.com/hubfs/MicroHyperspec%20Sensor%2004May2022.pdf
www.polytec.com/fileadmin/website/optical-systems/spezialkameras/pdf/PH_HWP_NIR.pdf

TABLE 2.8
Hyperspectral Data Sets Available in the Public Domain

Name	Sensor	Spectral Range	Spectral Resolution in nm	Spatial Resolution in m	Number of Bands	Classes	Atmosphere corrected	Radiance/Reflectance	Format	Application
Pavia University	ROSIS	430–860	4	1.3	103	9	Yes	Reflectance	MATLAB data files	Urban classification
Pavia Centre	ROSIS	430–860	4	1.3	102	9	Yes	Reflectance	MATLAB data files	Urban classification
Indian Pines	AVIRIS	400–2500	10	20	200	16	Yes	Reflectance	MATLAB data files	Vegetation classification
Salinas	AVIRIS	400–2500	10	3.7	204	16	Yes	Radiance	MATLAB data files	Classification
Cuprite	AVIRIS	370–2480	10	15.5	188	12	Yes	Reflectance	Reflectance file, MATLAB data file, ENVI file	Unmixing
Kennedy Space Centre	AVIRIS	400–2500	10	18	176	13	Yes	Reflectance	MATLAB data files	Classification
Washington DC Mall	HYDICE	401.2881 ~ 2473.16	10	2.8	191	7	Yes	Reflectance	Tif	Classification
Houston		380–1050		2.5	144	15				Classification
Urban	HYDICE	400–2500	10	2	162	4, 5, 6	Yes	Reflectance	MATLAB data files	Unmixing
Botswana	Hyperion	400–2500	10	30	145	14	Yes	Reflectance	MATLAB data files	Classification
Jasper Ridge		380–2500	9.46		198	4	Yes	Reflectance	MATLAB data files	Unmixing
Samson		401–889			156	3	Yes		MATLAB data files	Unmixing
Chikusei	Headwall Hyperspec-VNIR-C	363–1018	6	2.5	128	19	Yes	Reflectance	MATLAB data file, ENVI file	Vegetation and Impervious surfaces classification

Sources:
www.ehu.eus/ccwintco/index.php/Hyperspectral_Remote_Sensing_Scenes
www.ehu.eus/ccwintco/index.php/Hyperspectral_Remote_Sensing_Scenes
www.ehu.eus/ccwintco/index.php/Hyperspectral_Remote_Sensing_Scenes
www.ehu.eus/ccwintco/index.php/Hyperspectral_Remote_Sensing_Scenes
https://rslab.ut.ac.ir/data
www.ehu.eus/ccwintco/index.php/Hyperspectral_Remote_Sensing_Scenes
www.researchgate.net/publication/342660656_Dimensionality_reduction_method_for_hyperspectral_image_analysis_based_on_rough_set_theory,
https://hyperspectral.ee.uh.edu/?page_id=459; https://engineering.purdue.edu/~biehl/MultiSpec/hyperspectral.html
http://lesun.weebly.com/hyperspectral-data-set.html; https://rslab.ut.ac.ir/data
www.ehu.eus/ccwintco/index.php/Hyperspectral_Remote_Sensing_Scenes
http://lesun.weebly.com/hyperspectral-data-set.html
http://lesun.weebly.com/hyperspectral-data-set.html
https://paperswithcode.com/dataset/chikusei-dataset

The spatial and spectral signatures of some of the urban features are described in this section. The imagery used for explanation is of Apr. 22 and Apr. 27 EO-1-Hyperion images (USGS, 2013-1; USGS, 2013-2). The example images are decent quality cloud-free images with good brightness and contrast. The description is based on a false colour composite (FCC) and side-by-side display of all the channels using a multichannel view provided by Multispec (Biehl & Landgrebe, 2002) for visual interpretation of imagery. The default FCC assignments of Multispec were used for displaying the images. These assignments are, the red colour is assigned to the band no. 50, green colour is assigned to the band no 27, and blue colour is assigned to the band no. 17.

Common features such as vegetation, built-up area – comprised of road networks and building blocks, open grounds with different fractions of dry grass, water bodies and so on, are easily separable. Further analysis of signatures with the help of spatial association cues yields subclasses as well. For example, a red, saturated patch in the middle of a water channel is water hyacinth and large building complexes in the outskirts of a city are industrial units or warehouses. Below is a detailed description of observed visual cues, spectral and spatial, of a few common important land use and land covers in urban set up:

2.5.1 Vegetation

Vegetation appears red with a different intensity and different saturation for different classes of vegetation. The degree of saturation depends upon the fraction of vegetation and the thickness of the canopy. For example, the isolated patches of two to three trees are unsaturated because of mixing with other urban classes (Figure 2.7).

Farms appear unsaturated red with a distinct texture resulting from rectangular farm plots and are easily separable from irregular-shaped tree areas. Further, their association with river basins and canals reaffirm the agricultural land use. Water hyacinth appears saturated red with medium to low brightness (similar to the red of dense tree cover) and is easily separable as it is associated with river channels flowing through urban settlements (Figure 2.7).

Lawns such as golf course lawns and other grassy patches near water bodies appear saturated red with more brightness than trees. There is a minor shift towards a yellowish hue. Their spatial association with either a structure or water body helps in identifying the type of grass. Dry grass has a yellow ochre signature with a smooth texture. Yellowness indicates the degree of dry grass fraction on the ground (Figure 2.7).

Thick patches of deep forest or dense trees exhibit more saturated signatures than the vegetation classes discussed earlier. The saturation is sometimes comparable with saturation provided by lush green grass such as a golf course. However, a thick tree canopy can be easily identified by its more reddish hue. Deciduous forests on the hill slopes around the city appear highly unsaturated red, as they are not in full blossom. The shrub cover appears as red unsaturated tints with a dominant yellow hue (Figure 2.7).

Introduction to Hyperspectral Imaging

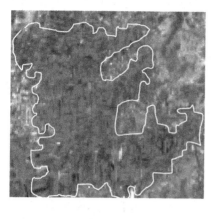

a Agricultural land

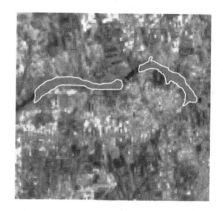

b Water hyacinth

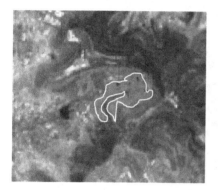

c Golf course lawn

d Thick forest-canopy

FIGURE 2.7 Visual signatures of vegetation.

2.5.2 Soil

Open areas appear as unsaturated (towards yellow ochre) yellow with high brightness. The degree of yellowness increases with an increase in grass fraction. Other open areas in and around the city also have a dry grass fraction with varying degrees. Though the dry grass density is reduced in some places, the signatures remain the same as in areas with dense dry grass. This is because the intertwined soil area is covered with shredded shoots of dry grass creating a thin blanket (Figure 2.8).

Open areas cleared for construction or open playgrounds have a very bright, saturated sky-blue signature. Open grounds such as playgrounds are the brightest spots in the image because of their bright signature and bigger size than building blocks (Figure 2.8).

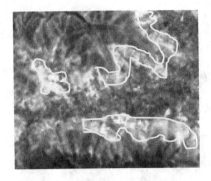
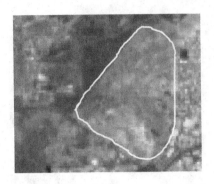

a Dry grass   b Bare soil

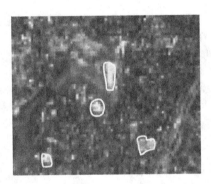

c Play grounds

FIGURE 2.8 Visual signatures of soil.

2.5.3 Impervious Surfaces (Built-up)

Concrete buildings appear bluish in colour. The saturation depends upon the age of the building, vegetation in the vicinity and the height of the buildings and so on. Fully developed settlements because of their tree fractions in the area show unsaturated blue purple, or red purple in colour. Medium-to-low-economy household areas show saturated blue colour and distinct medium-grained rectangular texture because of metal/concrete rooftops. The distinct texture is because of a lack of tree cover and a poor road network. Very new settlements on the fringes of the city also show similar spectral and spatial signatures. However, they can be differentiated from low economic zones using texture and size information (Figure 2.9).

A road network is easily identified because of its linear shape. The road network is separable from other linear features such as river channels based on their size and regular shape. The challenge is to differentiate between concrete and asphalt surfaces (Figure 2.9).

Introduction to Hyperspectral Imaging

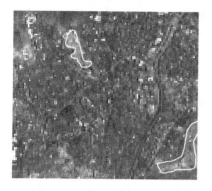

a Slums and developed township

b New settlement

c Industrial roofs

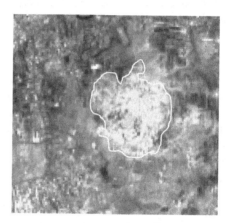

d Earth excavation-quary

FIGURE 2.9 Visual signatures of impervious surfaces.

Industrial roofs provide very bright blue signatures: they appear white at times because of white coating or specular reflection. They have a larger block size than residential or other commercial buildings (Figure 2.9). The summary of visual cues for interpreting hyperspectral images is provided in Table 2.9.

2.6 COMMON WORKFLOW FOR HYPERSPECTRAL IMAGE PROCESSING

2.6.1 Convert the Digital Numbers (DN) to Radiances

Often, the first step is the conversion of a digital number to radiance values or simply radiance. For example, for 8-bit data the digital numbers would vary from 0 to 255. They correspond to radiant energy, which was converted to digital numbers during analogue to digital conversion. They are converted back, using a calibration constant given by a sensor, to radiant energy. The radiance is a physically correlated quantity that can be related to the bio-physical parameter of our interest. Radiance is commonly measured as energy in watts or microWatts per area (for example, µw. cm^{-2}).

2.6.2 Solar Elevation Angle Correction

Solar elevation angle correction may not be required for analysing a single image. However, normalizing illumination differences while analysing multi-date data becomes necessary. The illumination differences are normalized by assuming that if the sun was at the zenith for each image. This is achieved by dividing the radiance of each pixel by sine of solar elevation angle or cosine of zenith angle. Normalization of illumination differences because of sun-earth distance is essential as well. It is performed by dividing the irradiance by squared sun-earth distance. This step can be integrated with the elaborate atmospheric correction procedure followed in the next step.

2.6.3 Convert Radiance to Reflectance

This is a particularly key step. The spectral signature is revealed when the radiance values are converted to a unitless quantity reflectance. Reflectance is a ratio of incident solar radiation per wavelength to reflected solar radiation per wavelength. The conversion of radiance to ground or at surface reflectance or true reflectance requires atmospheric correction. This is because the radiant energy received by the satellite sensor is a combination of the energy radiated by the atmosphere and the energy coming from the target. The energy coming from the atmosphere and the absorption effect of the atmosphere on the direct energy at the sensor is removed to get at surface radiance. The energy reflected by the atmosphere in the direction of the satellite sensor has an overall additive effect on the energy recorded at the sensor, that is, the actual energy is lower than the quantity recorded by the sensor. The remaining part of the energy is some fraction of the light reflected from the surface as the light energy is scattered or absorbed by the atmosphere during its upward path towards the sensor. This is a multiplicative factor. Similarly, the effect of scattering and absorption is removed from the light in the direction of the surface from the top of the atmosphere to recover at surface solar radiation. The ratio of these quantities, that is, at surface incoming radiation to at surface reflected radiation is calculated.

2.6.4 Extract the Features

The spectral signature (reflectance), that is the D dimensional pixel vector where D = number of bands, can be directly used for further processing if it is used for spectral matching. In the case of spectral matching, the reflectance values of each band (also called grey values in the literature) are treated as the features. If the chemometric analysis is to be performed then selecting a few bands and using the corresponding grey values is essential. If the diagnostic feature is known, then the diagnostic depth

TABLE 2.9
Visual Cues for Interpretation of Hyperspectral Image (30 m Spatial Resolution, EO-1 Hyperion Image) of an Urban Area

Feature (object)	Spectral signature	Spatial signatures
Trees	Red hue different degree of saturation – indicating density	Irregular shape, follow contours, with fine point texture
Crops	Red hue	Predominantly associated with river basin and canal, coarse rectangular texture,
Lawns/Grasses	Red with high intensity	Specific shape depending upon use, for example, cricket ground has circular shape, so does the traffic island lawns
Dry grass	Unsaturated yellow hue (yellow ochre) – indicating fraction of dry grass with high intensity	Lower slopes, fine to smooth textured
Bare soil	Saturated bright blue	Rectangular shape and noticeable size at 30 m resolution
Soil with different degree of dry grass	Unsaturated yellow with medium to high brightness	irregular shape, dry grass on slopes follows contours and is associated with lower level
Built-up	Saturated blue with varying degree of brightness depending upon age	Medium block texture

is calculated for further use. Removing continuum absorption before calculating the diagnostic depth is an essential step.

2.6.5 USE SUPERVISED OR UNSUPERVISED LEARNING APPROACH

Depending upon the goal of the study the suitable features are used for drawing final inferences. For example, target material identity can use band reflectance as a feature, whereas quantification relation between spectral features and the biophysical parameter of interest would require the diagnostic depth. Spectral matching methods such as Spectral Angle Mapper (SAM) or machine learning methods such as Support Vector Machines (SVM) random forest are a few examples of the methods used for drawing inferences. Regression is one of the popular methods for developing quantification relations between grey values and biophysical parameters.

2.6.6 DISSEMINATION

Depending upon the end use, the results can be directly consumed by any system, or a little post-processing may be required. In the case of single-date images, post-processing can involve simple packing and sending the results to the storage system. It can involve removing unwanted parts of the images other than the target pixels, drawing a bounding box, and so on for better visualization of the results. In some cases, the post-processing may involve integrating the analysis results from multi-date images and creating structured data for further processing such as time series analysis. It can involve the comparison of analysis results from multi-date images to assess the changes. Both the single-date and multi-date image analysis results can be packaged and stored in Geographic Information System (GIS) compatible data in the form of a thematic map. Depending upon the GIS use, the raster data resulting from the image analysis is converted to vector data for GIS consumption.

This is a simplified version of common tasks. The summary is illustrated in Figure 2.10. It is not complete and does not discuss all the options. It is for a general understanding of the common steps and the processing involved in each case. Note that this workflow focuses on spectral features. The spatial feature steps remain the same at

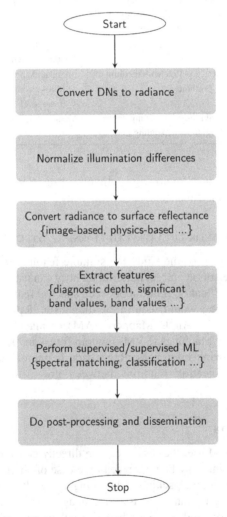

FIGURE 2.10 Workflow of common image processing steps.

Introduction to Hyperspectral Imaging

the broad level. However, the feature extraction process would be different from spectral. It may involve extracting texture, object contours, or features like SIFT and HOG. Modern deep learning may learn the features during the process and may not require any handcrafted features like mentioned previously. It would take an image as a three-dimensional array input (of reflectance or DNs depending upon the goal) and process it further.

The steps remain almost similar for analysing the measurements by a field spectrometer (and some of them are already covered). The conversion of radiance to reflectance is internal to the measurement process. The standard reference plate or calibration readings are taken, and the signature is saved to the connected personal computer or device memory. The objective of the field measurements is also slightly different than image processing: they are used for building a library of reference spectra, and they are used for detailed spectral analysis to observe the statistically significant diagnostic absorptions or reflection of the target materials. This is because the current field spectrometers have a better FWHM and sampling interval. In addition to the qualitative analysis, the laboratory signatures are analysed for diagnostic absorption and its relationship with a target chemical constituent. Furthermore, the high-resolution laboratory spectra are used for creating spectra with different resolutions by resampling them with the required spectral response functions (SRF) and interval required. In that case, the spectra are convolved with a set of SRFs and sampling intervals are kept ready for further comparison.

2.7 SUMMARY

Hyperspectral imaging captures the interaction between electromagnetic radiation and matter by narrow and contiguous bands. This results in the signature with more detailed information about the target material as compared to the broad band multispectral signature. This helps in the direct identification of the target material by matching the pixel spectrum from the image with the spectrum in the reference library of field or laboratory spectra. The quantification models using hyperspectral data are more accurate than multispectral data models too.

Because of the technical advances in recent years and the utility of the hyperspectral data in monitoring earth resources more accurately, many spaceborne platforms are being launched or are due for launching. Some of the important missions are: SHALOM, PRISMA, HISUI and so on. Commercial hyperspectral missions are on the horizon as well. OrbitalSidekick and Pixxel are good examples of them.

Some of the processing steps of hyperspectral data are the same as the processing of multispectral data. However, hyperspectral data processing focuses more on extracting spectral features, especially diagnostic absorption, and depths. Conversion of radiances to reflectance is a critical step in the processing of the hyperspectral data. The high dimensionality of the hyperspectral data is another issue which needs a separate treatment.

WORKS CITED

AERONET. (2015). *AERONET – Aerosol robotic network*. Retrieved December 16, 2015, from http://aeronet.gsfc.nasa.gov/

Arita, J., Sasaki, K.-i., Endo, T., & Yasuoka, Y. (2001). Assessment of concrete degradation with hyperspectral remote sensing. *22nd Asian Conference on Remote Sensing, November 5–9.*

Biehl, L., & Landgrebe, D. (2002, December). MultiSpec-a tool for multispectral-hyperspectral image data analysis. *Computers and Geosciences, 28*(10), 1153–1159. Retrieved from https://engineering.purdue.edu/~biehl/MultiSpec/

Buschkamp, P., Sang, B., P., P., Pieraccini, S., Geiss, M. J., Roth, C., ... Nieke, J. (2021). CHIME's hyperspectral imaging spectrometer design result from phase A/B1. In B. Cugny, Z. Sodnik, & N. Karafolas (Ed.), *International Conference on Space Optics – ICSO 2020* (pp. 1091–1105). SPIE. doi:10.1117/12.2599428

Clark, R. N., & Roush, T. L. (1984). Reflectance spectroscopy: Quantitative analysis techniques for remote sensing applications. *Journal of Geophysical Research: Solid Earth, 89*(B7), 6329–6340.

Cutter, M. (2008, July 7). CHRIS data format. Guildford, Surrey, UK: Surrey Satellite Technology Limited. Retrieved July 11, 2022, from https://earth.esa.int/eogateway/documents/20142/37627/CHRIS-Data-Format.pdf

DLR. (2019, October 18). *DESIS data from the ISS now available for science and research.* Retrieved July 11, 2022, from Earth Observation Center: www.dlr.de/eoc/en/desktop default.aspx/tabid-13247/23165_read-59014/

DLR. (2022). *Mission.* Retrieved from EnMap: www.enmap.org/mission/

DLR. (n.d.). *Instrument.* Retrieved July 12, 2022, from Earth Observation Center: www.dlr.de/eoc/en/desktopdefault.aspx/tabid-13622/23667_read-54280/

DLR. (n.d.). *Mission.* Retrieved July 12, 2022, from Earth Observation Center: www.dlr.de/eoc/en/desktopdefault.aspx/tabid-13618/23664_read-54267/

EO-1. (2017, Jun 22). (U.S. Geological Survey) Retrieved August 23, 2018, from Earth Observing 1 (EO-1): http://eo1.usgs.gov/

Erwin, S. (2019, September 23). *NRO awards first commercial contract for hyperspectral imaging from space.* Retrieved July 12, 2022, from SpaceNews: https://spacenews.com/nro-awards-first-commercial-contract-f+R6or-hyperspectral-imaging-from-space/

ESA. (2022, October 22). *Proba-1 celebrates 20th birthday in orbit.* Retrieved from The European Space Agency: www.esa.int/Enabling_Support/Space_Engineering_Technology/Shaping_the_Future/Proba-1_Celebrates_20th_Birthday_In_Orbit

ESA. (About PROBA-1). *About PROBA-1.* Retrieved from ESA earth online: https://earth.esa.int/eogateway/missions/proba-1

ESA. (CHRIS Overview). *CHRIS Overview.* Retrieved July 11, 2022, from ESA earth online: https://earth.esa.int/eogateway/instruments/chris/description

ESA. (Mission background). *Mission background.* Retrieved Jul 11, 2022, from ESA earth online: https://earth.esa.int/eogateway/missions/proba-1/description

Feingersh, T., & Dor, E. B. (2015). SHALOM–A Commercial Hyperspectral Space Mission. In *Optical Payloads for Space Missions* (pp. 247–263). John Wiley & Sons, Ltd. doi:https://doi.org/10.1002/9781118945179.ch11

GHOSt Constellation Deployment Plans Finalized By Orbital Sidekick. (2021, February 4). Retrieved July 12, 2022, from Satnews: https://news.satnews.com/2021/02/04/ghost-constellation-deployment-plans-finalized-by-orbital-sidekick/

Goetz, A. F., Vane, G., Solomon, J. E., & Rock, B. N. (1985, June). Imaging Spectrometry for Earth Remote Sensing. *Science, 228*(4704), 1147–1153.

Guarini, R., Loizzo, R., Facchinetti, C., Longo, F., Ponticelli, B., Faraci, M., ... others. (2018). PRISMA hyperspectral mission products. *IGARSS 2018-2018 IEEE International Geoscience and Remote Sensing Symposium* (pp. 179–182).

Hollas, J. M. (2004). Electromagnetic radiation and its interaction with atoms and molecules. In J. M. Hollas, *Modern Spectroscopy* (Fourth ed.), pp. 28–34. Wiley.

HySpecIQ. (2021). Retrieved July 12, 2022, from HySpecIQ: https://hyspeciq.com/

Japan Space Systems. (2020, September 28). *The first-light image of the Hyperspectral Imager SUIte (HISUI)*. Retrieved July 11, 2022, from Japan Space Systems: www.jspacesystems.or.jp/jss/files/2021/06/HISUI_first-light-image_1030E.pdf

Japan Space Systems. (2021). *Solutions–HISUI Hyper-spectral Imager SUIte*. Retrieved July 11, 2022, from Japan Space Systems: www.jspacesystems.or.jp/en/project/observation/hisui/

Jet Propulsion Laboratory (n.d.). *Welcome to Hyperspectral Thermal Emission Spectrometer website*. Retrieved July 12, 2022, from Hytes Hyperspectral Thermal Emission Spectrometer: NASA, Jet Propulsion Laboratory, California Institute of Technology: https://hytes.jpl.nasa.gov/

Jet Propulsion Laboratory. (2017-a, January). *AVIRIS data*. Retrieved August 2, 2017, from NASA, Jet Propulsion Laboratory, California Institute of Technology: https://aviris.jpl.nasa.gov/data/index.html

Jet Propulsion Laboratory. (2017-b, January). *AVIRIS overview*. Retrieved September 2, 2017, from NASA, Jet Propulsion Laboratory, California Institute of Technology: https://aviris.jpl.nasa.gov/aviris/index.html

Jet Propulsion Laboratory. (2017-c, January 31). *AVIRIS data–New data acquisitions*. Retrieved September 2, 2017, from NASA, Jet Propulsion Laboratory, California Institute of Technology: https://aviris.jpl.nasa.gov/data/newdata.html

Jet Propulsion Laboratory. (2021, May 6). *AVIRIS-NG OVERVIEW*. Retrieved July 12, 2022, from Airborne Visible InfraRed Imaging Spectrometer–AVIRIS Next generation: NASA, Jet Propulsion Laboratory, California Institute of Technology: https://avirisng.jpl.nasa.gov/aviris-ng.html

Jet Propulsion Laboratory. (2021-2, May 6). *SPECIFICATIONS*. Retrieved July 12, 2022, from Airborne Visible InfraRed Imaging Spectrometer–AVIRIS Next generation: NASA, Jet Propulsion Laboratory, California Institute of Technology: https://avirisng.jpl.nasa.gov/specifications.html

Jet Propulsion Laboratory. (2021-3, May 6). *PLATFORM*. Retrieved July 12, 2022, from AVIRIS Next generation: NASA, Jet Propulsion Laboratory, California Institute of Technology: https://avirisng.jpl.nasa.gov/platform.html

Kruse, F. A., Boardman, J. W., Lefkoff, A. B., Young, J. M., Kierein-Young, K. S., Cocks, T. D., ... Cocks, P. A. (2011). *HyMap: an Australian hyperspectral sensor solving global problems–Results from USA HyMap data acquisitions*. Retrieved September 6, 2017, from www.hyvista.com/wp_11/wp-content/uploads/2011/02/10ARSPC_hymap.pdf

Matsunaga, T., Iwasaki, A., Tsuchida, S., Iwao, K., Tanii, J., Kashimura, O., ... Tachikawa, T. (2017). Current status of Hyperspectral Imager Suite (HISUI) onboard International Space Station (ISS). *2017 IEEE International Geoscience and Remote Sensing Symposium (IGARSS)* (pp. 443–446). doi:10.1109/IGARSS.2017.8126989

METI. (2019, December 6). *Hyperspectral Imager Suite (HISUI) Successfully Launched for Space Demonstration Tests*. Retrieved from METI: www.meti.go.jp/english/press/2019/1206_001.html

NASA. (2022, January 10). *Aircraft List*. Retrieved July 12, 2022, from NASA airborne science program: https://airbornescience.nasa.gov/aircraft/ER-2_-_AFRC

NASA/JPL/ESA/KCL. (2021, August 17). *2021 HyTES Joint Campaign Status*. Retrieved from Hytes Hyperspectral Thermal Emission Spectrometer: https://hytes.jpl.nasa.gov/2021-campaign-status

Orbital sidekick announces upcoming launch of its newest global hyperspectral earth observation constellation: GHOSt. (2021, February 4). Retrieved July 12, 2022, from CISION PR Newswire: www.prnewswire.com/news-releases/orbital-sidekick-announces-upcoming-launch-of-its-newest-global-hyperspectral-earth-observation-constellation-ghost-301222258.html

OSK. (n.d.). *Technology.* Retrieved July 12, 2022, from OSK Orbital Sidekick: www.orbitalsidekick.com/technology

Pavia, D. L., Lampman, G. M., & Kriz, G. S. (2001). Infrared spectroscopy. In D. L. Pavia, G. M. Lampman, & G. S. Kriz, *Introduction to spectroscopy* (Third ed.), p. 14.

Pixxel. (2022, April 1). *Launching Towards Pixxel's Hyperspectral Vision: Earth's First Health Monitoring Constellation.* Retrieved July 12, 2022, from Pixxel: www.pixxel.space/blogs/launching-towards-pixxels-hyperspectral-vision

Polder, G., & Gowen, A. (2020). The hype in spectral imaging. *Journal of Spectral Imaging, 9*(1), a4. doi:10.1255/jsi.2020.a4

Ridd, M. K. (1995). Exploring V-I-S (vegetation–impervious surface–soil) model for urban ecosystem analysis through remote sensing: comparative anatomy of the cities. *International Journal of Remote Sensing, 16*(12), 2165–2185.

SPECIM. (2018). *specim IQ–Hyperspectral goes mobile.* Retrieved July 12, 2022, from SPECIM: www.specim.fi/iq/

USGS. (2002, June 2). EO1 Hyperion Scene: EO1H1680602002153110PZ_1R, for Target Path 168, and Target Row 60. Sioux Falls, SD, USA: U. S. Geological Survey. Retrieved from http://earthexplorer.usgs.gov

USGS. (2013-1, April 22). EO1 Hyperion Scene: EO1H1470472013112110KZ_PF1_01, for Target Path 147, and Target Row 47. *EO1 Hyperion Scene: EO1H1470472013112110KZ_PF1_01, for Target Path 147, and Target Row 47.* Sioux Falls, SD USA: U. S. Geological Survey. Retrieved from http://earthexplorer.usgs.gov

USGS. (2013-2, April 27). EO1 Hyperion Scene: EO1H1470472013117110PZ_PF2_01, for Target Path 147, and Target Row 47. Sioux Falls, SD USA, USA: U.S. Geological Survey. Retrieved from http://earthexplorer.usgs.gov

USGS. (2014, December 20). Landsat 8 Scene: LC81680612014354LGN00, for Target Path 168, and Target Row 61. Sioux Falls, SD, USA: U. S. Geological Survey. Retrieved from http://earthexplorer.usgs.gov

Vane, G., & Goetz, A. F. (1985, April). Introduction to the Proceedings of the Airborne Imaging Spectrometer (AIS) Data Analysis Workshop. In G. Vane, & A. F. Goetz (Ed.), *Airborne Imaging Spectrometer Data Analysis. JPL Publication 85–41.* Jet Propulsion Laboratory, California Institute of Technology, Pasadena, California USA.

Wolf, S. P. (2022, April 11). *Constellr Acquires Hyperspectral Firm ScanWorld in a Move Set to Disrupt the Use of Beyond-Visual Data in Smart Farming Worldwide.* Retrieved July 12, 2022, from https://constellr.space/blog.html#Scanworld_ad

Wyvern. (2022). *The future of hyperspectral imaging.* Retrieved July 12, 2022, from Wyvern: https://wyvern.space/media/

3 Reflectance Calibration

This chapter will discuss:
- Importance of atmospheric correction
- Atmospheric effects
- Image-based methods
  - Dark object,
  - Improved dark object,
  - Flat field method,
  - Internal area relative reflectance,
- Physics-based method – RTC code, 6SV
- Working example for image based on physics-based methods

3.1 REFLECTANCE CALIBRATION

Conversion of atsensor radiance values to reflectance values (reflectance calibration) is a fundamental step in the analysis of hyperspectral data (Kruse, 1994; Schowengerdt, 1997). The reflectance signature is used for studying the unique spectral response of a given material, and for identifying the target material by searching a match in a spectral library of field spectra or laboratory spectra. The conversion of atsensor radiance values to reflectance values is more important if the in-depth spectroscopic analysis is desired to achieve the objective. For example, the objective is to develop a quantification model between the diagnostic depth and the dependent variable such as nutrient in a soil. As the reflectance is a ratio of the insolation and the reflected light per wavelength, removing the effects of the atmosphere from the incoming solar radiation and the radiation reflected by the surface reaching the satellite sensor is required.

There are two main methods for removing atmospheric effects: a) empirical and/or image-based methods, and b) physics-based methods. A combination of these two methods is possible. It is possible to use one of the methods as a primary method of correction and use the other method to improve the results further. For example, Clark et al., 2002 use a physics-based method as the primary method for correction and then use an image-based method for improving the results further.

The image-based methods use the data available within the image for modelling the atmospheric effects. For example, the additive component of the atmospheric effects such as path radiance is derived from the image pixels having the specific properties. The physics-based methods model the scattering and absorption of light by atmospheric gases and aerosols, using first principles. Naturally, the physics-based methods require accurate atmospheric data such as the concentration of gases in the atmosphere, aerosols and so on. Thus, both methods have advantages and disadvantages. The physics-based methods are accurate but require additional atmospheric data. The additional data might not be easily available at every site and hence they are difficult to deploy. Computational complexity of the algorithms may add to the difficulties further. On the other hand, the image-based methods are approximate but all the required data for atmospheric correction is available within the image. The algorithms are easy to understand, and computational complexity is extremely low. There is ample evidence in the literature on successful implementation of either of these methods. We favour image-based methods for their simplicity and ease of use for correcting atmospheric effects. The image-based correction methods provide a practical advantage over physics-based methods as most of the required information is available within the image (Gao et al., 2009; Chavez, 1996; Griffin & Burke, 2003).

There are many studies conducted in past using hyperspectral data for urban area that do not correct images for atmospheric effects (Ridd et al., 1992; Ridd, 1995; Hepner et al., 1998; Chen & Hepner, 2001; Weng, Hu, & Lu, 2008; Myint et al., 2011). Interestingly, the atmospheric correction is not performed during many of the AVIRIS studies for the other applications (Harsanyi & Chang, 1994; Hoffbeck & Landgrebe, 1996; Melgani & Bruzzone, 2004; Ham et al., 2005).

Though atmospheric effects do not influence the classification accuracies by large margins and are not mandatory in certain conditions such as single-date multispectral data, the narrow spectral resolution of the hyperspectral data requires sound atmospheric correction for accurate results. The results by Song, Woodcock, Seto, Lenney, & Macomber (2001) have led some of the later studies into using hyperspectral data without any atmospheric corrections. For example, Xu and Gong (2007) use the conclusions by Song et al. (2001) and do not correct the spaceborne hyperspectral images for atmospheric effects. Another important reason for ignoring atmospheric effects is the lack of in situ measurements required for the physics-based methods (Yang et al., 2003). It should be noted though that the atmospheric corrections are critical for hyperspectral data analysis because of its high spectral resolution and should be performed before any further analysis.

Further sections provide details of the image-based and the physics-based methods. Before we discuss them, we will take a detailed look at the atmosphere and the light interaction. In the second chapter we have seen the three basic interactions between light and matter. In the next section, we discuss how the interaction takes place for the solar radiation from the source to the surface and then the surface to the sensor. Understanding this interaction is important as we may make some relaxing assumptions under certain conditions and may ignore a particular component of the interaction for simplifying reflectance calibration.

3.1.1 Atmospheric Effects

The goal of the reflectance calibration process is to convert the radiance recorded by the satellite sensor to, ideally, true reflectance. This requires estimating quantities of the incoming light, and the light leaving the surface based on the light measured by the satellite sensor. The downward solar radiation, that is, the solar radiation from the top of the atmosphere to the surface first interacts with the atmosphere before illuminating the surface. Then it interacts with the surface. Finally, the light interacts with

Reflectance Calibration

the atmosphere in its upward path, that is, light reflected from the surface towards the satellite sensor before reaching the satellite sensor. Thus, to estimate the atsurface reflectance, modelling of the light and the matter interaction in its upward and downward path is required. The nature of the main interaction remains the absorption and scattering of the energy by the atmosphere and the surface.

The model for true reflectance is built by estimating the quantities of light illuminating the surface and the sensor. The surface is illuminated by three different pathways:

1. The direct solar radiation incident on the surface.
2. The light scattered by the atmosphere in the direction of the surface.
3. The light reflected by the neighbourhood areas (pixels) scattered back in the direction of the surface.

Thus, the first source is the solar illumination directly by the light which passes through the atmosphere in a downward direction, the second source is the light scattered by the atmosphere in the direction of the target (pathways 2, 3, see Figure 3.1). The atmosphere itself can be illuminated by direct sunlight or the light reflected from the neighbourhood area.

The sensor is illuminated by three major pathways:

1. The direct solar radiation reflected by the surface in the direction of the sensor (that is, incident on the sensor).
2. The light scattered by the atmosphere in the direction of the sensor.
3. The light scattered by the atmosphere in the direction of the sensor where the illumination source is the light reflected from the neighbourhood areas (pixels) in the direction of the atmosphere.

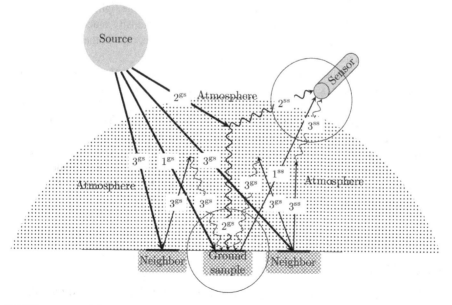

FIGURE 3.1 Atmospheric effects.

Thus, the first source is the direct illumination by reflected solar light from the surface in the direction of sensor. The second source is the light scattered by atmosphere in the direction of the sensor (causing 2, 3, see Figure 3.1). The atmosphere itself can be illuminated by direct sunlight or the light reflected from the neighbourhood area.

To summarize, the surface of the sensor is illuminated directly by the source and the atmosphere. The atmosphere is illuminated in two ways; directly by the light coming from the source (that is sun) or the surface, and the light coming from the neighbouring surface (pixel). Thus, the atmosphere not only attenuates the light energy from the source to the surface and the surface to the sensor but also illuminates the sensor or the surface by scattering the light from the various sources.

Figure 3.1 shows the various pathways illuminating the surface of the ground sample and the sensor. The diagram is a schematic representation and is not to scale. The thickness of the arrow approximately indicates the amount of energy. For example, the light reflected by the surface is less than the light incident on the surface as the surface absorbs some intensity. The zig-zag arrow indicates the light scattered by the atmosphere. The atmosphere is not homogeneous in the way the diagram may indicate, and the atmospheric scattering does not happen only at one layer as represented in the diagram. The superscript indicates the light pathway that illuminates the ground sample (gs) or the satellite sensor (ss). The ground sample is illuminated by pathways 1^{gs}, 2^{gs}, and 3^{gs}. Similarly, the satellite sensor is illuminated by 1^{ss}, 2^{ss}, and 3^{ss}. 2^{ss} and 3^{ss} are unwanted signals and do not carry any information related to the target. The goal of the atmospheric correction is to remove 2^{ss} and 3^{ss} from the total signal recorded by the satellite sensor and obtain 1^{ss}. Then correct 1^{ss} for the multiplicative effect of the atmosphere on the light in the path of the ground sample to the sensor in an upward direction.

Out of the light-reaching sensor, the light pathway 1^{ss} carries the signal. Other quantities reaching the sensor do not contain any information about the surface material. The effect of the atmosphere on the light from the surface to the sensor (or the source to the surface) is of multiplicative nature, that is, the light is attenuated by the atmosphere and some fraction of the original light intensity reaches the destination (the surface of the sensor). Whereas the 2^{ss} is additive, that is, the total radiance at the sensor is an addition of 1^{ss} and 2^{ss} and 3^{ss}.

Depending upon the availability of the data, the goal, and the desired accuracy of the reflectance values, some of the effects are ignored under certain assumptions. For example, it is customary to assume a clear atmosphere, ignore the multiplicative effect, and consider the path radiance alone for the correction. Thus, the correction step involves subtraction of the path radiance from the radiance at the sensor for the atmospheric corrections. In this case, TOA irradiance and atsensor radiance after removing path radiance are sufficient for calculating the atsurface reflectance. Thus, the reflectance calibration may involve detailed atmospheric correction or a simple procedure which uses all the information from the image itself.

3.2 PHYSICS-BASED METHODS

3.2.1 EO-1 Hyperion Equation for Clear Atmosphere

The Unites States Geological Survey (USGS, 2011-1) provides a simplified formula for converting radiance values to reflectance values in relatively clear atmospheric conditions:

Reflectance Calibration

$$\rho_p = \frac{\pi . L_\lambda d^2}{ESUN_\lambda . \cos\theta_s} \qquad \text{Equation 3.1}$$

where,
ρ_p is reflectance,
L_λ is spectral radiance at the sensor,
d^2 is Earth-Sun distance in astronomical units,
$ESUN_\lambda$ is Hyperion solar irradiance,
θ_s is solar zenith angle in degrees (USGS, 2011-1).

It can be seen from Equation 3.1 that it is a simplified version for correcting the atmospheric effects in which only the irradiance is normalized for the difference in sun-earth distance and solar zenith angle. The ignored components are path radiance, absorption by the atmospheric gases, and the scattering of the light on the way upward and downward direction. However, this is an easy method for correcting the image and hence the useful physics-based method in the absence of any other data.

3.2.2 Second Simulation of a Satellite Signal in the Solar Spectrum-Vector (6SV)

6SV is a widely used radiative transfer code that provides a simulation of satellite signals accounting for elevated targets (Vermote et al., 2006). One of the practical advantages of 6S is that it provides standard atmosphere and aerosol models. These models are used for atmospheric correction in absence of the accurate field measurements. 6S provides comparable or better results than the similar radiative transfer codes available (Kruse F. A., 2004; Zhao, Tamura, & Takahashi, 2000; Lee, Yeom, Lee, Kim, & Han, 2015). The Python interface is available for (Wilson, 2012) calculating the reflectance calibration constants namely xa, xb, and xc (Vermote, et al., 2006; Zhao, Tamura, & Takahashi, 2000). Corrected reflectance is calculated using the following formula:

$$y = xa \times L_i - xb \qquad \text{Equation 3.2}$$

$$acr = \frac{y}{1 + xc \times y} \qquad \text{Equation 3.3}$$

where L_i (W m^{-2} SR^{-1} μm^{-1}) is measured radiance for a given satellite band i. 6SV is one of the important physics-based methods and is used for comparative assessment with image-based methods. Hence, the method is explained in detail.

3.2.2.1 Atmospheric Correction Procedure by 6SV

First, the user must download the 6SV program from (Vermote et al., 2006) and install it on a machine or Google Colab according to the preferred runtime. The steps for calculating the atmospheric correction factors using 6SV are given below.

1. Install the 6SV and compile to create the executable.

2. Create an input file for the 6SV program. This file is a plain text file containing all the parameters required for atmospheric correction. They are discussed in the next section after the step-by-step process.
3. Run the 6SV with the input file. It will create three files xa, xb, and xc for the wavelength range mentioned in the input file.
4. Using the output files created by 6SV calculate the multiplicative correction factor and the additive correction factor.
5. Apply the correction factors for correcting the radiance values measured at satellite sensor. Use Equation 3.2 and Equation 3.3.

The user can create the input file using two methods. If the user knows well all the parameters required for the input file and can specify the variable names according to the requirement, the user can use any text editor to create this file. The other option is using a Py6S interface for 6SV (Wilson, 2012). This would need a separate installation of the Py6S components. Py6S provides an easy-to-use application programming interface for the 6SV code. The user can create a 6SV object with all the required parameters. The various model options will appear in a dropdown list during writing the script in the Python editor. This takes off the cognitive load of remembering all the options for creating the file. Once the object is created, Py6S will create an input file for the user after calling an appropriate function in the script. The sample files are made available to the reader of this book and the details are provided in the appendix.

The atmospheric correction by physics-based methods is an involved activity. It requires many atmospheric parameters to calculate the scattering and absorption coefficients of the atmospheric constituents. We provide here just enough information so that the user can make the right choices during the atmospheric correction process. The complete treatment of the topic can be found in some of the important references cited in this chapter; for example Vermote et al., 2006. A brief description of the input parameters required is given below:

The first parameters of the 6SV input files are the geometric properties of the illumination source and the sensor. They determine the amount of light incident on a particular surface. They are required for normalizing the amount of light received by the sensor in the case of image-based methods as well. 6SV takes predefined values based on the known satellite; for example, Landsat, or user-defined values as well. The values required (in the same order) are solar zenith angle, solar azimuth angle, view zenith angle, view azimuth angle, day, and month of the image capture. For the hyperspectral sensor, the user-defined option is required as there are no hyperspectral platforms defined in 6SV at present. For a Hyperion image, these values can be found in a metadata file, which is a part of the bundle of user downloads from the USGS earth explorer.

Next, is the selection of the atmospheric model. This model refers to the gases' composition of the target area. The gases considered are ozone (O_3), water vapour (H_2O), oxygen (O_2), carbon dioxide (CO_2), methane (CH_4), and nitrous oxide (N_2O). The last four gases in this list, that is, oxygen (O_2), carbon dioxide (CO_2), methane (CH_4), and nitrous oxide (N_2O) are assumed to have a uniform distribution over the globe and hence are not required as the inputs. However, the water vapour and ozone concentration need to be provided as an input to the 6SV program. In this case as well there are options for using a predefined composition or user-defined composition. The predefined models follow MODTRAN®/FLAASH model nomenclature (available at

TABLE 3.1
Aerosol Models and Their Constituents

Model name	Water-soluble	Dust	Oceanic	Soot
Average continental	0.93877	2.27×10^{-6}	0	0.06123
Urban	0.5945	1.67×10^{-7}	0	0.4055
Maritime	0.99958	0	0.00042	0
Continental	0.93876	2.27×10^{-6}	0	0.06123
Urban or industrial	0.5945	1.66×1^{-7}	0	0.4055

Source: Levoni, Cervino, Guzzi, & Torricella, 1997.

www.l3harrisgeospatial.com/docs/FLAASH.html); for example, the Tropical model uses 5119 std atm-cm, and 4.11 g.cm^{-2} water vapour values, and a temperature of 27° Celsius. Other models are Sub-Arctic Winter (SAW), Mid-Latitude Winter (MLW), U.S. Standard (US), Sub-Arctic Summer (SAS), and Mid-Latitude Summer (MLS). Remember that the unit of ozone concentration is cm.atm and water vapour is gm.cm^{-2}.

Next, one of the most important parameters is the aerosol model. Aerosol properties, especially for the boundary layer, show a large variation from place to place and needs to be carefully chosen. In the case of the aerosol model using a user-defined model, the user expresses the aerosol model as a mixture of four basic components (more later) such as dust, soluble aerosols, oceanic, and soot. A few specific predefined models are available and the user needs to specify the name of the model. The mixture components for these predefined models are listed in Table 3.1. For urban areas, a natural choice for the model is "urban". However, depending upon the location and the amount of fossil fuel burned by the city, the soot properties can be increased further. The same is true for dust. Many cities in developing countries may show more than a standard dust component because of vigorous construction activities.

Other parameters required are the altitude of the sensor platform, the target altitude above mean sea level in km, and the nature of the surface and reflection assumption such as Lambertian. These values are easy to get from the satellite metadata and information about the target is assumed to be available to the user.

This information (and the help in the appendix) should be sufficient to create an input file manually or by using Py6S. The procedure given in the earlier section can be followed easily to get the xa, xb and xc. The sample input file is available in the appendix. The aerosol models are discussed in the next section.

3.2.2.2 Aerosol Models

Aerosols are significant atmospheric constituents scattering light. A vertical layer of atmosphere exhibiting vertically uniform optical properties is referred to as an aerosol model. For aerosol model consideration, the atmosphere is divided into four major layers (Shettle & Fenn, 1976):

1. Boundary layer models 0 km to 2 km
2. Upper troposphere 2 km to 10 km

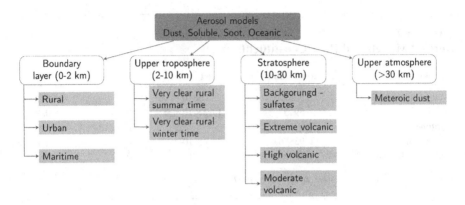

FIGURE 3.2 Atmospheric layers and aerosol models.

3. Stratospheric models 10 km to 30 km
4. Upper atmosphere > 30 km

Aerosols in each of these layers are a combination of a few basic aerosol generation processes namely dust, soluble aerosols, soot, and oceanic. The aerosol model for each layer is expressed as any combination of these four basic components (for the boundary layer). Soot aerosols are the result of the combustion process in an urban area because of residential and industrial fossil fuel burning. Oceanic aerosols are the salt particles created by the evaporation of sea spray (Figure 3.2).

1. **Boundary layer**

   The aerosols in the boundary layer, the layer within 0–2 km of the earth's surface, are affected by natural and anthropogenic activities. Because of that, the aerosols' microphysical properties vary according to the locations. Thus, it is imperative to choose the right atmospheric properties for calculating the effect of the atmosphere on the signal reaching satellite sensor.

   Rural model: This is a gas and dust particle mixture. The water-soluble component is 70% and the dust component is 30%. Water-soluble components constitute ammonium and calcium sulphate, and organic compounds.

   Urban model: The urban model is a rural model added with combustion products. The fractions of soot and rural aerosol are 35% and 65% respectively. Thus, the urban model is a rural model with soot particles.

   Maritime: The main aerosols in the boundary layer in this model are salt particles resulting from the evaporation of sea spray droplets. Overall, the model is assumed to have two components – sea spray and continental component, which is assumed to be like rural (Shettle & Fenn, 1976).

2. **Upper troposphere (2–10 km)**

   This layer shows less variation in atmospheric aerosol properties as compared to the boundary layer. The aerosol properties are more uniform in this layer and hence it can be represented by a single model. The rural model without

Reflectance Calibration

large dust particles is well suited for this layer. However, based on the empirical evidence there can be two models: one for summertime and one for winter (Shettle & Fenn, 1976). Summer-time aerosol concentration is found to be more than wintertime aerosol concentration.

3. **Stratosphere (10–30 km)**
   This region also shows uniform concentration globally. Aerosols in this region are mainly because of sulphates formed by a photochemical reaction. This concentration in increased in the case of a large volcanic eruption.

4. **Stratosphere (>30 km)**
   The concentration of aerosols in this layer is negligible as compared to the other layers. The aerosols in this region are because of meteoric dust (Shettle & Fenn, 1976).

3.2.3 MODerate Resolution Atmospheric TRANsmission (MODTRAN®)

MODerate resolution atmospheric TRANsmission (MODTRAN®) is a widely used commercial radiative transfer code. The MODTRAN code is jointly developed by Spectral Sciences, Incorporated (SSI) and the Air Force Research Laboratory (AFRL) (Spectral Sciences Incorporated, 2016-1). A few other atmospheric correction software such as Atmosphere CORrection Now (ACORN) and Fast Line-of-sight Atmospheric Analysis of Hypercubes (FLAASH) are based on MODTRAN code (Gao et al., 2009).

MODTRAN computes line-of-site (LOS) atmospheric transmittances and radiances over a spectral range from 200 nm to 100000 nm. The code provides a physics-based model for a stratified horizontally homogenous atmosphere. The vertical profile of the atmosphere can be specified by the user using atmospheric data or built-in atmospheric condition models. The absorption, scattering, and emittance by the atmospheric elements are calculated using a band model with a spectral resolution of 0.1 cm^{-1}. MODTRAN6 is a major upgrade from its previous version and incorporates modern software architecture. MODTRAN6 is made more user-friendly by providing a graphic interface and application programming interface for easy integration with the other applications (Berk et al., 2014; Spectral Sciences Incorporated, 2016-2).

3.2.4 Discrete Ordinated Radiative Transfer Code (DISORT)

Discrete ordinated radiative transfer code (DISORT) is another general-purpose algorithm for calculating the transfer calculation in a vertically inhomogeneous plane parallel medium. The spectral range from ultraviolet to radar is supported. The FORTRAN code is made available to the interested users.

We have summarized a few important free and commercial radiative transfer codes. There are numerous other radiative transfer codes (Stamnes, 1988). A Wiki page provides a convenient summary of most of the codes (Atmospheric radiative transfer codes, 2018) as well.

3.3 IMAGE-BASED METHODS

As discussed in the earlier sections, the image-based methods use all the data available within the image and use it for the atmospheric corrections. The image-based methods correct both the additive and multiplicative effects. Path radiance removal is sufficient in some situations, whereas removal of the atmospheric absorptions and scattering is required in addition to the path radiance removal. The following sections would describe these procedures.

3.3.1 Dark Object Subtraction

The Dark Object (DO) subtraction technique, as the name suggests, uses the radiance (or DN) recorded at a dark pixel within the image for correcting the path radiance or "haze removal" as it is called in the literature. Given a sufficiently large image (for example, EO1 scene would have ~ 1 million pixels), the chances of a few pixels in the image not reflecting any of the solar radiation back to the sensor are very high. Such a pixel or pixels should have zero radiance (henceforth referred to as the "dark pixel" or the "dark object"). The dark pixel may exist because of the deep shadows or because of the physical property of the surface. For example, water bodies do not reflect any energy in the near infrared region. The radiance value recorded for such a pixel or pixels is assumed to be because of the path radiance. Thus, the correction of the path radiance is performed by subtracting the constant radiance value at the dark object from all the pixel radiance values within the image. The implicit assumption in this method is that the atmosphere is homogenous over the image scene.

The operation is performed for each band of the hyperspectral image, that is, the path radiance is selected for each band separately. The path radiance is commonly selected using the histogram. The histogram of the band will start at some nonzero value. In addition, there would be a sharp increase in the frequency of the pixels at a particular radiance value (or DN). The radiance value at this inflection point is selected as the path radiance value for that band. The same procedure is repeated for all the spectral bands (Chavez, 1988). The other method is using the radiance values recorded at the known dark objects, like water bodies, in the infrared region (Lillesand & Kiefer, 1999).

Figure 3.3 shows the path radiance calculated for the EO-1 Hyperion image acquired on April 22, 2013. The path radiance value is selected from each spectral band of Hyperion and plotted against the bandcenters (wavelengths). The simple algorithm was used to select the minimum nonzero value from each band. The path radiance is shown along with the atsensor radiance, and after subtracting the path radiance from each band atsensor radiance values.

3.3.2 Improved Dark Object Subtraction

One of the problems, in addition to the assumptions of spatially homogeneous atmosphere in the scene, and at least a few dark pixels, is that the path radiance selected for each band may not represent the true path radiance as per the physics of scattering. For example, if we select a value from the band at say 450 nm, and

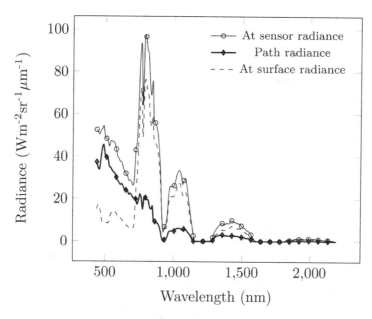

FIGURE 3.3 Path radiance for EO-1 Hyperion image calculated using histogram method.

the atmospheric scattering is inversely proportional to the fourth power of the wavelength, then the value selected at say 500 nm and 450 nm should be consistent with the atmospheric scattering model. This may not happen because of the selection of the path radiance from each band by the histogram method. The selection of the path radiance from the band is subjected to similar problems, in addition to the subjectivity in the path radiance selection process. The pixel in dark shadows selected for the path radiance calculation may not be completely dark and may have a radiance of its own. Thus, the path radiance selected by the histogram method or from the known dark object method may result in over correction of the atsensor radiance values for some or all the bands. To avoid this Chavez (1988) developed an improved dark object technique.

The improved dark object technique (Chavez, 1988), unlike the simple dark object technique, assumes that the relation between the path radiance and the wavelength is different for different atmospheric conditions. The relation between path radiance and wavelength is referred to as a relative scattering model. The procedure suggests that instead of selecting the path radiance for all the bands from the histogram, it can be selected from a single band. It can be selected using the predefined dark object as well. In the dark object case, it is easier to select the correct path radiance values as the dark object is determined by the physical properties of the object. Then, these path radiance values should be scaled up or down as per the relation between wavelength and the path radiance according to the atmospheric conditions at the time of data collection. Once the path radiance values are calculated for all the bands, they can be subtracted from all the radiance values of all the pixels in the image. Because of this, the path radiance calculated should remain close to the physics of scattering

TABLE 3.2
Atmospheric Models Source

Atmospheric condition	Scattering model
Very clear	λ^{-4}
Clear	λ^{-2}
Moderate	λ^{-1}
Hazy	$\lambda^{-0.7}$
Very hazy	$\lambda^{-0.5}$

Source: Chavez, 1988.

and overcome the drawback of the simple dark object technique. Table 3.2. shows the relative scattering models that are used in the improved dark object technique.

The generic improved DO technique procedure is as follows: To begin, select the relative scattering model according to the atmospheric condition at the time of data collection and then select the base path radiance from a selected band using a histogram (or another predefined object with known spectral properties).

Then choose the model based on the weather conditions on the day and time of data collection. The atmospheric conditions considered are Very clear, Clear, Moderate, Hazy, and Very hazy (Table 3.2.). Due attention should be given to spatial and temporal variability in the atmospheric conditions while choosing a model. For example, atmospheric conditions in Indian cities in winter would be very hazy, while in summer they would provide a very clear atmosphere.

In the next step, path radiance for the other bands is calculated by scaling the base path radiance. The scaling factor (propagation factor) is determined by the wavelength of the new band centre, the wavelength of the base band centre, and the selected scattering model. For example, for a simple model where the path radiance is inversely proportional to the wavelength (that is, the atmospheric conditions are assumed to be "Moderate" as per Table 3.2.), the path radiance of the band 0.300 μm is 2.67 ({1/.300}/{1/.800}) times path radiance of 0.800 μm band. So, if the base path radiance of value one is selected from the band of 0.800 μm wavelength, then the path radiance of the band with wavelength 0.300 μm is 2.67.

After the calculation of the scaling factors for all the bands, the base path radiance is multiplied by the scaling factor to calculate the final path radiance to be removed from each pixel. The flowchart (Figure 3.4) summarizes these steps.

The improved dark object technique has been successfully used for Landsat data (Chavez, 1996). Though the improved dark object technique has been extensively used for multispectral data, it is easily extended to any hyperspectral data. Often, a lookup table for the scaling factors for a given platform is prepared and then used as all the inputs other than base path radiance are determined by the relative scattering model and band centres of the sensor. Modern computation on personal computers or high-end machines makes the explicit step of creating a lookup table redundant. This is because the scaling factor depending upon the relative scattering model and the band centres of the sensor is easily calculated in the software program and applied to

Reflectance Calibration

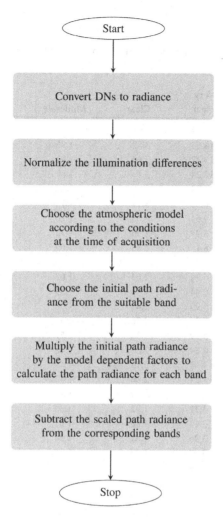

FIGURE 3.4 Improved dark object technique.

the base path radiance. However, to get a general idea about how the scaling factors vary and what is the impact of choosing a base path radiance from a particular band, a lookup table for EO-1 Hyperion is given in Table 3.3. Note that the scaling factors are provided for the VNIR range for calibrated bands only (Beck, 2003). The base path radiance is selected from band no 52, that is, at band centre 874.53 nm. As can be seen from the table, the base path radiance value should be carefully selected from the near infrared band. The wrong values of base path radiance from the near infrared band overcorrects the visible spectrum. The turbid water bodies show reflectance above zero and all the recorded radiance for such a pixel is not because of haze only. Thus, the value selected from the near infrared band should be adjusted subjectively according to the turbidity of water.

TABLE 3.3
Scaling Factors for the Hyperion Band Centres for Different Scattering Models

EO-1 Hyperion Bandcenters (nm)	Very clear	Clear	Moderate	Hazy	Very hazy
436.99	16.04	4.01	2.00	1.63	1.41
447.17	14.63	3.82	1.96	1.6	1.4
457.34	13.37	3.66	1.91	1.57	1.38
467.52	12.24	3.5	1.87	1.55	1.37
477.69	11.23	3.35	1.83	1.53	1.35
487.87	10.32	3.21	1.79	1.5	1.34
498.04	9.51	3.08	1.76	1.48	1.33
508.22	8.77	2.96	1.72	1.46	1.31
518.39	8.1	2.85	1.69	1.44	1.3
528.57	7.49	2.74	1.65	1.42	1.29
538.74	6.94	2.64	1.62	1.4	1.27
548.92	6.44	2.54	1.59	1.39	1.26
559.09	5.99	2.45	1.56	1.37	1.25
569.27	5.57	2.36	1.54	1.35	1.24
579.45	5.19	2.28	1.51	1.33	1.23
589.62	4.84	2.2	1.48	1.32	1.22
599.8	4.52	2.13	1.46	1.3	1.21
609.97	4.23	2.06	1.43	1.29	1.2
620.15	3.95	1.99	1.41	1.27	1.19
630.32	3.71	1.92	1.39	1.26	1.18
640.5	3.48	1.86	1.37	1.24	1.17
650.67	3.26	1.81	1.34	1.23	1.16
660.85	3.07	1.75	1.32	1.22	1.15
671.02	2.89	1.7	1.3	1.2	1.14
681.2	2.72	1.65	1.28	1.19	1.13
691.37	2.56	1.6	1.26	1.18	1.12
701.55	2.41	1.55	1.25	1.17	1.12
711.72	2.28	1.51	1.23	1.16	1.11
721.9	2.15	1.47	1.21	1.14	1.1
732.07	2.04	1.43	1.19	1.13	1.09
742.25	1.93	1.39	1.18	1.12	1.09
752.43	1.82	1.35	1.16	1.11	1.08
762.6	1.73	1.32	1.15	1.1	1.07
772.78	1.64	1.28	1.13	1.09	1.06
782.95	1.56	1.25	1.12	1.08	1.06
793.13	1.48	1.22	1.1	1.07	1.05
803.3	1.4	1.19	1.09	1.06	1.04
813.48	1.34	1.16	1.08	1.05	1.04
823.65	1.27	1.13	1.06	1.04	1.03
833.83	1.21	1.1	1.05	1.03	1.02
844	1.15	1.07	1.04	1.03	1.02
854.18	1.1	1.05	1.02	1.02	1.01

TABLE 3.3 (Continued)
Scaling Factors for the Hyperion Band Centres for Different Scattering Models

EO-1 Hyperion Bandcenters (nm)	Very clear	Clear	Moderate	Hazy	Very hazy
864.35	1.05	1.02	1.01	1.01	1.01
874.53	1	1	1	1	1
884.7	0.95	0.98	0.99	0.99	0.99
894.88	0.91	0.96	0.98	0.98	0.99
905.05	0.87	0.93	0.97	0.98	0.98
915.23	0.83	0.91	0.96	0.97	0.98
925.41	0.8	0.89	0.95	0.96	0.97

3.3.3 IMPROVED DARK OBJECT WITH OVERLAP CORRECTION

The improved dark object technique was initially developed for multispectral data like Thematic Mapper of Landsat. The number of bands in the multispectral data is limited to 8–16. However, for the hyperspectral data, the number of bands is commonly larger than two hundred. In addition, the hyperspectral data is collected using contagious overlapping bands. The effect of the data collection and its implication for the improved dark object procedure needs to be understood properly for more accurate path radiance calculations.

The improved dark object technique selects the base path radiance from a particular single band and then propagates the base path radiance to all the other bands using a band-dependent scattering model (Figure 3.4). The path radiance thus calculated for all the hyperspectral bands can be seen as a spectrum of haze, just like any other target material. This spectrum is then subtracted from the spectrum of each pixel (bandwise subtraction) respectively to remove the additive component of the atmospheric scattering. Thus, it is important to choose the correct base path radiance value to begin the improved dark object subtraction procedure. As the propagation factors for the correction procedure are ratios of power functions, the small errors introduced in the base path radiance are amplified for the other bands. As would be the case if we choose the path radiance value from a particular single band of a hyperspectral spectrum. Hyperspectral data is acquired by sampling the reflected energy with overlapping narrow bands (Goetz, 1985; Vane & Goetz, 1985). The spectral response function (SRF) of a given band is overlapped by the preceding and the following band's SRFs. Because of the overlapping SRFs, path radiance recorded by two adjacent bands is also measured by the central band as well. The contribution from adjacent bands would depend upon the hyperspectral sensor design. Hence, a more accurate procedure for the improved dark object technique is needed to correct the base path radiance for overlapping SRFs so that the propagated spectrum is consistent with the improved dark object procedure.

The objective of the work (Deshpande et al., 2018) was to improve the improved dark object subtraction technique further and extend it to the hyperspectral data of

EO-1-Hyperion sensor and use it as an atmospheric correction step for further classification experiments. They examined the effect of SRF overlap on the propagated path radiance for hyperspectral data using an improved dark object technique. Comparative assessment of the path radiance calculated by the improved dark object technique with and without overlap correction, and path radiance calculated by an 6SV algorithm (Vermote et al., 2006) was an important goal of this research. The study addressed a few important questions: What is the effect of choosing a particular initial band for path radiance on path radiance error because of SRF overlap? What are the important atmospheric conditions that require overlap correction? Can overlap correction be ignored in certain conditions without affecting path radiance calculation by a large margin? Which wavelength range is more suitable for choosing initial path radiance?

The intuition and the fundamentals of the method are given in the section below. The first few sections provide the background information about the spectral characteristics and then the detailed procedure is explained along with the working example.

3.3.3.1 Spectral Sampling of Haze Function

At sensor signal is a function of solar irradiance, atmospheric interference, object properties, and sensor properties. Mathematically the process is expressed as:

$$\rho_{g\lambda} = \frac{\left[\pi\left(L_{s\lambda} - L_{d\lambda\uparrow}\right)/\tau_{v\lambda}\right]}{\left[E_{o\lambda}\cos\theta_z \tau_{z\lambda} + E_{d\lambda\downarrow}\right]} \qquad \text{Equation 3.4}$$

where,

$\rho_{g\lambda}$ = spectral reflectance of surface,

$L_{s\lambda}$ = at sensor radiance,

$\tau_{v\lambda}$, $\tau_{z\lambda}$ = atmospheric transmittance along the upwelling and downwelling paths,

$L_{d\lambda\uparrow}$ = at sensor upwelling (path radiance) spectral radiance in the direction of sensor,

$E_{d\lambda\downarrow}$ = at surface downwelling spectral irradiance due to atmospheric scattering in the direction of target,

$E_{o\lambda}$ = solar irradiance at the top of atmosphere,

θ_z = the angle of the solar direct flux to the surface normal (Moran et al., 1992).

Atsensor radiance constitutes radiance from the target filtered by the atmosphere ($\tau_{v\lambda}$), and the haze ($L_{d\lambda\uparrow}$). The haze imparts an additive effect to the radiance and hence the signal reaching sensor is an addition of the radiance from the target surface and the haze. This path radiance is sampled by the sensors' Spectral Response Function (SRF). Hence, the objective of the path radiance removal procedure is to approximate the path radiance spectrum (from a histogram or simple dark object) and subtract the path radiance spectrum from the signal to recover the true atsensor radiance (Figure 3.3).

3.3.3.2 Spectral Characteristics of Hyperion Sensor

Earth Observing-1 (EO-1) is a tasking satellite with an onboard hyperspectral sensor Hyperion and broadband sensor Advanced Land Imager (USGS, 2013-1). Both

Reflectance Calibration

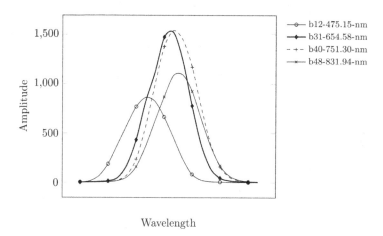

FIGURE 3.5 Sample Hyperion spectral response functions (ftp.eoc.csiro.au).

sensors can record the images simultaneously. The Hyperion sensor records the data in 242 bands with a 30 m spatial resolution and ~10 nm (Full Width Half Maximum–FWHM) spectral resolution over 355–2577 nm. The central wavelengths of Hyperion bands are separated by ~10 nm and the SRF of each band overlaps by ~12% with the SRF of each single preceding and following bands.

Only 198 bands out of 242 are calibrated (7–57 and 77–224) (Beck, 2003). Further removal of water-vapour-affected bands results in 141 useful bands. Gaussian spectral response function with central wavelength and FWHM as per the Hyperion documentation provide a close approximation to the observed spectral responses of Hyperion bands (Figure 3.5).

3.3.3.3 Spectral Deconvolution for Correcting Band Overlap

The plot (Figure 3.6) shows the overlapping response functions of adjacent bands i-1, i, i+1, where i is the central band, i-1 is the preceding band, and i+1 is the following band respectively. i-1, i, i+1, indicate increasing wavelengths of the bandcentres. As a result, reflected energy from a target within the overlapping wavelength regions is recorded by the two adjacent bands. The incident energy in the shaded region (Figure 3.6) is recorded by band i-1 and band i and hence contributes to the DNs of the band i-1 and i respectively. The same process repeats at the overlap of i and i+1 band.

As a result, the base path radiance selected from a particular band is affected by the overlap too. Using the original method, the propagated path radiance would be higher than the true values. The effect would vary according to the equation, for example, for λ^{-4} the effect will be more pronounced in shorter wavelengths. To correct this, the energy recorded by band i in the overlapping regions with i-1 and i+1 should be removed from band i.

Sampling of the reflected energy by the overlapping bands is slightly different from convolution where a function continuously samples the response. Hence, the standard deconvolution techniques are not found to be suitable to correct the effects

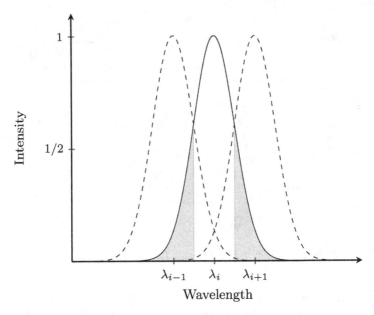

FIGURE 3.6 Overlapping spectral response functions of adjacent bands and their effect on measurements.

of the overlap (Schläpfer, Boerner, & Schaepman, 1999). Schläpfer et al. (1999) used a weighted SRF approach to correct the effect of simultaneous recording before resampling the low-resolution signal to higher resolution signal for APEX simulation studies. A similar approach has been used to increase the effective resolution of IMS (Senthil Kumar et al., 2013). Deshpande et al. (2018) used the initial step of the complete procedure for increasing spectral resolution (that is weighted deconvolution) for removing the overlap effect of a given band. The overlap corrected base path radiance was then propagated to the other bands for correction of the full spectrum. Deconvolved spectral radiance is given by the formula below (Schläpfer et al., 1999):

$$L_{k,dec} = \left(L_k - L_{k+1}W_{k+1} - L_{k-1}W_{k-1}\right)c \qquad \text{Equation 3.5}$$

where,
$L_{k,dec}$ = deconvolved spectral radiance of kth band,
L_i = spectral radiance of ith band and $i = k, k+1, k-1$,
$c = \dfrac{R_k}{R_k - W_{k+1} - W_{k-1}}$ = adjustment factor for new SRF,
$R_k = \int r_k(\lambda)d\lambda$ = integrated spectral response of kth band,

and $W_i = \dfrac{\int \min(r_i, r_{i+1})}{\int r_i}$ = weight factor between two adjacent bands.

Reflectance Calibration

As can be seen from Equation 3.5, if the responses from adjacent bands are equal to the response of the central band, that is, there is no absorption or reflection feature at a given position, the central band values are not corrected. However, it would accentuate the absorption or reflection feature at a given position present in the original signal as the absorption feature or reflection feature means lower and higher response values than adjacent bands respectively. Weighted deconvolution is not meant for constructing new information (Schläpfer et al., 1999). The calculation steps remain the same as the original improved dark object technique (Figure 3.4). However, the propagation factors or scaling factors are corrected for overlap. To do that the deconvolution weights of the adjacent bands are calculated using Equation 3.5.

To perform the improved dark object subtraction with overlap correction, perform the first steps until the base path radiance is selected. Then, calculate the scaling factors for all the bands to produce a spectrum of the scaling factors. Then, perform the weighted deconvolution (Equation 3.5) and correct the scaling factors for all the bands. Follow the remaining steps like the original improved dark object technique – that is multiply base path radiance to calculate path radiance for the other bands. Another alternative is to choose the base path radiance from three adjacent bands and then correct the base path radiance for the central band first and then use original scaling factors to propagate the path radiance for other bands. The second alternative involves additional efforts of choosing the base path radiance from the adjacent bands as well. Use the relation between FWHM and A for a normal distribution to calculate the weight factor:

$$FWHM = 2.355\sigma \qquad \text{Equation 3.6}$$

where, σ is the standard deviation (Weisstein, 2017). FWHM is ~10 nm for EO-1 and the distance between two adjacent band centres is 2.355 σ (is an approximation as the distance between two bands is not exactly 10 nm).

The worked example is given below. The example calculations are performed considering EO-1 Hyperion sensor:

The area under the curve (Figure 3.6, shaded area) is 12.1% and that is the minimum overlap of the SRFs in the overlapping region. The area under the SRF after removal of the overlap is 75.8. Hence the weighting factor is

$$W_{i+1} = 12.1/75.8 = 0.15963. \qquad \text{Equation 3.7}$$

The weighting factor is the same for all the EO-1 bands as they have similar SRF. Thus, the adjusted scaling factor by removing the band overlap for band 52 for model λ^{-4} is:

$$\text{Corrected factor } c = 1 - (1.05 \times 0.16) - (0.95 \times 0.16) = 0.68. \qquad \text{Equation 3.8}$$

For more accurate results, perform the calculation on an average distance between band centres (10.12 nm), and average FWHM over 242 bands (that is 10.94 nm = 2.17σ), the factor c is $1 - 1.05 * 0.19 - 0.95 * 0.19 = 0.62$. In practice, it is convenient to represent the scaling factors for a particular atmospheric model, with and without overlap correction, as a continuous function of wavelength. This makes it easy to prepare the lookup table for scaling factors depending on the wavelength, the atmospheric model, and the band for base path radiance. During the correction procedure, one can select bands to be corrected according to the application and corresponding scaling factors.

3.3.3.4 Comparison of Scaling Factors with and without Overlap Correction

Figure 3.7 shows the scaling factors for the various atmospheric scattering models with and without overlap corrections. The factors are calculated according to the procedure described in described in 3.3.3.3. It is the plot of the propagation factors on y axis and the bandcenters on x axis for the different atmospheric scattering models. The series names indicate the band for the base path radiance selection. For example, "Very clear 52" is a series of the propagation factors for "very clear" atmospheric scattering model (Table 3.2) and the base path radiance is selected from EO-1 Hyperion band 52.

The overlap corrected scaling factors follow trends similar to the atmospheric models (Figure 3.7). The overlap correction reduces the magnitude of the scaling factors. The margin depends upon the model and the initial band for the path radiance selection. In general, the application of the improved dark object technique without the overlap correction yields the overcorrected path radiance. The decrease in the corrected path radiance propagation factor would be determined by the SRF overlap of the adjacent bands. As can be seen from Figure 3.7, the overlap affects the clear

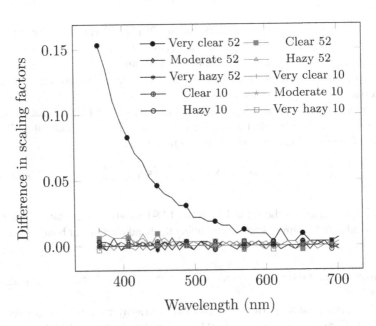

FIGURE 3.7 Propagation factors for various atmospheric scattering models with overlap correction and without overlap correction.

atmosphere models the most. As the scaling factors are multiplicative, they amplify the base path radiance selected from a particular band.

In addition to the atmospheric model, the values of the scaling factors are largely determined by the band of the base path radiance. If we choose the base path radiance from the infrared band (for example, Hyperion band no. 52) the overlap correction is more pronounced as compared with the path radiance selected from the visible range (Hyperion band no. 10). For other models that are "clear" to "very hazy", the differences between the overlap corrected factors and the uncorrected factors are marginal to negligible and can be ignored.

3.3.3.5 Comparison of Different Path Radiance Values with 6SV

Figure 3.8 shows the path radiance calculated by 6SV with the path radiance resulting from the different atmospheric models and the path radiance as recorded by the histogram of each band. Note that the path radiance is referred to as intrinsic atmospheric radiance in the 6SV code (Deshpande et al., 2018). Each series name indicates the overlap correction flag, the atmospheric model, and the band no. of the base path radiance. The series name format is <overlap correction flag, c=corrected for overlap><atmospheric model><band for base path radiance>. For example, "C-Clear 52 indicates" overlap corrected path radiance using "clear" model, and the base path radiance is selected from "band no. 52"). As can be seen in Figure 3.8, the 6SV path radiance for each Hyperion band varies a lot as compared to the histogram-recorded path radiance. The recorded path radiance show similarity with the path radiance resulting from the clear atmospheric condition, and the base path radiance selected from band no. 52. Though the clear model and the base path

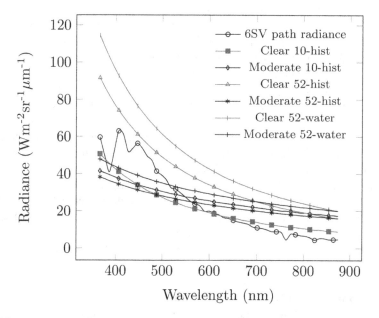

FIGURE 3.8 The path radiance comparison for Apr. 22, 2013, image using different correction procedures.

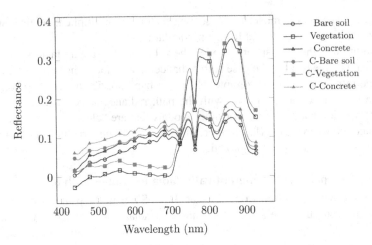

FIGURE 3.9 Comparison of common urban material signatures extracted using a dark object with overlap correction and without correction.

radiance from band no. 52 overestimate the path radiance values than the recorded values, the overall trend of the path radiance remains similar (similar slope values). In contrast, the 6SV path radiance values decrease more rapidly after 500 nm than the recorded path radiance values.

The path radiance by all the methods, including that recorded by the sensor which is represented by a histogram are the estimates of the path radiance and do not provide actual path radiance values. Considering the complex phenomenon of scattering the exact value of the path radiance may not be available for the correction. The best strategy for estimating the path radiance is hence to take a weighted summation of all the path radiance values. The probability or the confidence level associated with each model can be considered as the weights. The confidence level provides an advantage as the expert inputs are incorporated into the calculation. This proposed method of calculating path radiance using a weighted average over all the models requires further investigation.

Figure 3.9 displays the reflectance signatures of common vegetation, bare soil, and concrete materials in urban settings. The reflectance signatures are created using the equation provided by the United States Geological Survey (USGS, 2011-1) in combination with a dark object subtraction technique. The radiance values were first corrected for the path radiance using the dark object subtraction technique with and without overlap correction, and then converted to the reflectance values. The base path radiance of 16 W m^{-2} SR^{-1} μm^{-1} was chosen from band no. 52. As can be seen, the standard improved dark object technique overcorrects signatures in the visible spectrum.

3.3.3.6 Summary of Dark Object Subtraction

Many applications of the remotely sensed data require compensation for both the additive and the multiplicative components of atmospheric effects. Dark object subtraction is one of the important steps in the correction of remotely sensed data using image-based methods. The image-based techniques of the atmospheric correction for hyperspectral data (Gao et al., 2009) normalize the radiance value to the relative reflectance values using a reference signature within the scene, either average spectrum over all the pixels, or a predefined area with known spectral properties. In such methods, a separate procedure for removing the additive component is required initially. On the other hand, removing the additive component of atmospheric effects is sufficient for some of the analysis techniques (Vincent et al., 2004; Penn, 2002). The improved dark object technique by Chavez (1988) provides an improvement over the simple dark object technique (Vincent, 1972). However, the application of the improved dark object method to hyperspectral data requires further consideration for the SRF overlap. The overlap compensation is vital, as the scaling factors in the improved dark object technique are ratios of power functions; small errors in factors result in large compensation for path radiance. Some of the important observations are (Deshpande et al., 2018):

1. Overlap correction reduces the magnitude of scaling factors depending upon the model and initial band for path radiance selection. Hence, the application of the improved dark object technique without overlap correction yields overcorrected path radiance.
2. In addition to the atmospheric model, the values of scaling factors are largely determined by the band chosen for the selection of the base path radiance.
3. If we choose the base path radiance from an infrared band (in this case Hyperion band no. 52) the overlap correction is more pronounced as compared with a path radiance selected from a visible range (Hyperion band no. 10).

The overlap correction procedure should be considered in the following conditions:

1. Very clear to clear atmosphere model, this is a particularly important observation as users of remotely sensed data typically choose images acquired during relatively clear atmospheric conditions.
2. If the base path radiance is selected from infrared bands because of a predefined dark object such as a water body.
3. Further, visible bands are more suitable for the application of improved dark object subtraction techniques to hyperspectral images. In this case, the effect of the overlap of adjacent SRFs is negligible and can be safely ignored. However, the user needs to be aware of the strong dark object assumptions.

3.3.4 IAR AND FAR

Internal Area (average) Relative Reflectance (IAR) and Flat Field Relative Reflectance (FAR) are two prime image-based reflectance calibration methods (Green & Craig, 1985; Roberts & Yamaguchi, 1986). IAR and FAR remove only the multiplicative component of the atmospheric effects and hence a separate correction is required to remove the additive component, that is, the path radiance. Once the path radiance is removed, the multiplicative component can be removed by dividing each pixel by a

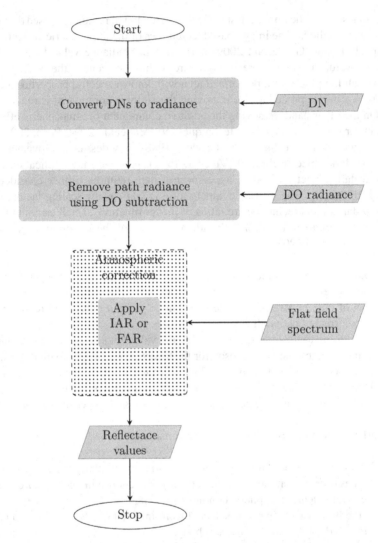

FIGURE 3.10 Procedure for calculating the relative reflectance of a pixel in the image using empirical methods such as IAR and FAR.

suitable reference spectrum. The averages over all the pixels within the image or over the pixels in a predefined area are considered as a reference spectrum for IAR and FAR, respectively (Figure 3.10).

Note that IAR and FAR do not convert the radiances to the absolute reflectance values, the reflectance values are relative to the reference spectrum. However, the spectral signature of the relative reflectance values resembles laboratory spectra. Knowing the absolute reflectance spectra of reference spectra, the relative reflectance values can be converted to absolute reflectance values. As IAR and FAR create the spectrum of relative reflectance they are considered normalization methods rather than absolute reflectance calibration

Reflectance Calibration

methods. Log residual is another such normalization method. However, the spectrum created by the normalization method does not resemble laboratory spectra and hence is not useful as IAR and FAR (Green & Craig, 1985).

3.3.4.1 Method

IAR and FAR calculate the relative reflectance by dividing each pixel in the image by a calibration reference. IAR uses the average spectrum over the entire image area for the calibration whereas FAR uses the average spectrum over a predetermined area with certain desirable properties like spectral blandness and high reflectance; for example, a playa. The average spectrum of total image area (IAR) in ideal situations where there is no vegetation or any other dominant material in the scene or a given predefined area in the case of FAR, indicates absorptions due to the atmosphere and hence is used as a reference spectrum to calculate the reflectance values. Formally, for the image of m rows and n columns with k bands:

$$IAR = \left(P_{ij}\right) / \frac{1}{mxn} \sum_{i=1}^{i=m} \sum_{j=1}^{j=n} P_{ij} \qquad \text{Equation 3.9}$$

where, P_{ij} is a k dimensional pixel vector corresponding to i^{th} row and j^{th} column.

Now, consider r rows and c columns of known Flat Field within the $m \times n$ image such that $0 < r < m$ and $0 < c < n$ then Flat Field Relative Reflectance (FAR),

$$FAR = \left(P_{ij}\right) / \frac{1}{rxc} \sum_{i=1}^{i=r} \sum_{j=1}^{j=c} P_{ij} \qquad \text{Equation 3.10}$$

It is to be noted that Flat Field need not be regular in shape as indicated by Equation 3.10.

3.3.4.2 Challenges

Though the image-based methods are easy to implement, both IAR and FAR face certain challenges in urban conditions. If the area under the study is covered with a dominant feature having a particular spectral characteristic, the IAR method is not suitable (Green & Craig, 1985). The modern urban spaces are dominated by concrete and hence the study needs to pay attention to the image's spatial coverage. A sufficiently large image area outside the urban boundaries may reduce the dominance of urban materials such as concrete. In the case of doubtful IAR results, identifying the Flat Fields within the scene becomes critical to the image-based methods and the hybrid methods (Clark et al., 2002). The hybrid methods use the Flat Fields within the scene to remove the residual atmospheric absorption and the scattering after the application of the radiative transfer calibration procedure (Clark et al., 2002). Finding the Flat Fields within the scene is critical for the other calibration methods too. The success of empirical line method for reflectance calibration depends upon the identification of the suitable Flat Field (Conel et al., 1987). Similarly, the reflectance-based vicarious calibration method for the post-launch sensor calibration requires the identification of the Flat Field within the scene (Slater et al., 1987; Slater et al., 1996; Secker et al., 2001; Biggar, Thome, & Wisniewski, 2003; Thome, 2004; NCAVEO, 2007).

There are two difficulties in finding the ideal Flat Field in an urban scene:

a) Finding the Flat Field requires exploratory work because of a lack of an effective algorithm to do so. Zhang et al. (2003) attempted to identify the Flat Field in a scene automatically. As the Flat Field is assumed to be proportional to the average spectrum of a given scene – the approach suffers from all the drawbacks of IAR. Another approach could be to compare the spectral blandness of the extracted spectral signatures and then select the most spectrally neutral reference as a Flat Field.

b) The ideal Flat Field might not be available within the study scene and in such a case researchers need to accept the best available Flat Field, as in the case of urban studies, finding the ideal Flat Field covering a sufficient number of pixels is difficult. The Ideal Flat Fields such as playa, salt flats, dam faces, sand beaches (Clark et al., 2002) are not quite common elements of any urban scene. The salt flats and the sandy beaches might be a part of some coastal cities but that is not the case for inland cities.

Because of these difficulties, urban covers such as open areas, playgrounds, building construction sites, concrete pavements, and residential and industrial roofs are a few important candidates for the Flat Fields within the urban scene (Deshpande et al., 2013). (Deshpande et al., 2017). There are only limited studies for the Flat Field assessments (Clark et al., 2002) and a few in urban settings.

3.4 EMPIRICAL LINE METHOD

The empirical line method is one of the simplest methods for reflectance calibration. Though it is simple and the data within the image (DNs) is used for calculating the reflectance values, it requires synchronized field measurements for signatures of at least a dark and a bright object (Flat Field) in the scene. Thus, it is effort intensive though it does not require any additional atmospheric data.

The steps are as follows:

1. Identify the reference sites that are easily identified in the image. The sites should be such that the surface is homogeneous and occupy more than nine pixels (Aspinall, Marcus, & Boardman, 2002).
2. Take the field measurements of the reference sites with a suitable spectrometer, synchronized with the satellite pass. Take multiple readings of the dark and the bright sites.
3. Extract the DNs from the pixels corresponding to the field measurement sites.
4. Develop a regression equation using linear fit between DNs and reflectance field measurements. The regression equations are developed for each band.
5. Apply the gain and offset values (coefficients of the regression) to convert the DNs to reflectance values. The gain represents a multiplicative correction factor and offsets indicate the additive component (path radiance) (Conel et al., 1987).

Reflectance Calibration

Of all the empirical methods, this method generates the reflectance signature which is remarkably close to the laboratory spectra or the field spectra (Aspinall et al., 2002). However, efforts are required to identify good reference sites. The field measurements must be taken within a reasonable time frame around the satellite pass to avoid any large deviation from the atmospheric conditions at the time of the satellite pass (Aspinall et al., 2002).

This method is suitable in case the site is going to be studied very often. In case the surface properties of the reference sites change over time, the relation needs to be developed again by taking new field measurements. The results of the empirical line method are site-specific and the same regression coefficients cannot be used to convert the DNs to reflectance values for other study areas, unless and until the atmospheric conditions outside the area are similar to the study area (Gao et al., 2009).

3.5 SIMPLIFYING ASSUMPTIONS AND THE METHODS

All the atmospheric correction procedures are with some implicit assumptions. These assumptions are essential for simplifying the procedure. We discuss the implicit assumptions in each of the methods discussed so far and their implications. Understanding them well is helpful in designing effective correction procedures.

Path radiance calculated from a dark object represents a path radiance at the location of the dark object. We may not get well-distributed dark objects in the image for calculating the path radiance at distinct parts of the image. Thus, the path radiance selected from a few pixels is subtracted from the band of each pixel assuming a spatially homogeneous atmosphere over the scene. An additional assumption in the simple dark object subtraction technique is that the dark object path radiance selected from each band of the dark object behaves according to the atmospheric model. Though the improved dark object technique is an improvement over the simple dark object technique, for compensating deviations from the ideal atmospheric interactions, it still is affected by spatially homogeneous atmosphere assumption. Improved or otherwise, the dark object path radiance is assumed to be entirely because of atmospheric back scatter. This is not true, especially for some of the difficult targets like water. In case the water is turbid, it may result in some reflectance in infrared bands and that can be mistaken as a path radiance. The same thing can happen for, say, a tree in the shadow; though it is in the shadow it still may have some residual signature which may result in wrong path radiance for different bands.

The empirical line method assumes a linear relation between the field radiance of the reference sites and the radiance measured by a satellite. It also assumes the atmosphere to be homogeneous as the gains and offsets worked from reference sites are applied for every pixel in the image. In addition to spatial homogeneity, it assumes temporal homogeneity too, that is, the atmosphere is assumed to be stable within the time window of satellite pass and the field measurements. This is not true. In the case of a turbulent unsteady atmosphere, this method may produce incorrect results. It is also assumed that the field reference surface remains the same as it was at the time of the satellite pass (at least care should be taken that it remains the same). This again results in inconstancy between the field readings and satellite readings. All the differences in the satellite-recorded radiance and field radiance in such dynamic conditions cannot be explained by atmospheric effects alone. Moreover, there is no way to retrieve the original field signature unless and until it is contentiously monitored.

In some of the simplified physics-based methods the multiplicative factors and additive factors are ignored completely – the atmosphere is assumed to have no effect at all. In such cases, the reflectance is calculated as a ratio of bandwise insolation at the top of the atmosphere and the radiance received at the sensor. This never is the case. In detailed physics methods as well, many sources illuminating the ground sample are ignored as the quantities are small as compared to the other sources. For example, the target ground sample is illuminated by the light from the atmosphere in the direction of the ground sample and the source of atmospheric light is direct sunlight reflected by the adjacent ground pixel (3^{gs}). The quantity is small as compared with the direct sunlight incident on the surface and the light scattered by the atmosphere in the direction of the ground sample (1^{gs}, 2^{gs} respectively). Similarly, 3^{gs} can be ignored too. Path three can be completely ignored even in the case of physics-based methods.

Even though the physics-based method in principle can derive all the quantities in detail effectively they are also affected by the homogeneous atmosphere assumption, in a separate way. The physics-based methods require the concentration of the many atmospheric gases such as NO_2 O_3 and so on. These are point readings only and represent the concentration over a small area. In such cases, the concatenation over a particular ground sample is used for correcting an entire image assuming the homogeneous atmosphere. The multiplication factor for the downward path and upward path are different. However, those factors are also considered to be the same many times.

In image-based methods, each pixel is divided by the scene average pixel or a signature from a flat field. In the average pixel case it is assumed that the area is not dominated by a single land cover, especially vegetation. The absorption features of the dominant land cover affect the correction procedure.

3.6 WORKING EXAMPLE

The image-based methods were extensively studied by (Deshpande et al., 2017) for urban land use land cover classification. We provide the details in the following sections as a working example.

3.6.1 DATA

3.6.1.1 EO-1 Hyperspectral Data

Earth Observing-1 (EO-1) is a tasking satellite with an onboard hyperspectral sensor Hyperion and broadband sensor Advanced Land Imager (USGS, 2011-2). Both sensors can record the images simultaneously. The Hyperion sensor records the data in 242 bands with a 30 m spatial resolution and ~10 nm (Full Width Half Maximum – FWHM) spectral resolution over 355–2577 nm. The central wavelengths of Hyperion bands are separated by ~10 nm.

Only 198 bands out of 242 are calibrated (7–57 and 77–224) (Beck, 2003). Further removal of water vapour-affected bands results in 141 useful bands. A Gaussian spectral response function with central wavelength and FWHM as per the Hyperion

Reflectance Calibration

documentation provides a close approximation to the observed spectral responses of Hyperion bands (Figure 3.5).

The method described in section 3.6.2 uses a Hyperion image on path 147 and row 47 with the scene centres 18.5020 N, 73.8151 E and 18.5020 N, 73.7457 E, dated April 22, 2013 (USGS, 2013-1). The image covers Pune city in the state of Maharashtra, India. The image is 7.7 wide and 102 km in length. The first image is 102 km and the second image is 107 km long. Pune city has seen tremendous urbanization in the recent past because of emerging information technology hubs (Deshpande et al., 2017; Deshpande et al., 2013). The study image can be downloaded from the EarthExplorer https://earthexplorer.usgs.gov/. Use path 147 and row 47 as one of the data search criteria, in addition to the date.

3.6.1.2 Reference Data

All the reference signatures are taken from pure pixels within the Hyperion image (USGS, 2013-1) and are verified by a suitable combination of field visit and high-resolution imagery. A few mixture signatures for the thematic classes also are extracted intentionally. Table 3.4 and Table 3.5 provide details of each sample. We describe the classes and their general characteristics below:

TABLE 3.4
Landcover References from the Hyperion Image Used in the Study

Code	Class	Description
BGCH	Burnt grass (BG)	Burnt grass near Chakan area
BGKH	Burnt grass (BG)	Burnt grass near Khadakwasla
BGKO	Burnt grass (BG)	Burnt grass near Kothrud
COCP	Concrete (CO)	Concrete roof of a commercial building (City Pride)
CONP	Concrete (CO)	A large parking lot at a government institute (NDA parking)
DGKA	Dry grass (DG)	Dry grass on hill slopes (Kalyan)
DGKA'	Dry grass (DG)	Dry grass on hill slopes
DGKO	Dry grass (DG)	Kondhana
GGHB	Green grass (GG)	Hyacinth in Bhama River
GGHI	Green grass (GG)	Hyacinth in Indrayani River
GGPA	Green grass (GG)	Marsh shrubs near the lake (Pashaan)
GGRF	Green grass (GG)	Agricultural area near Rajguru
IRCR	Indutrial roof (IR)	Industrial area at Chakan road
IRCU	Indutrial roof (IR)	Industrial roof of Cummins
IRKI	Indutrial roof (IR)	Industrial roof of Kirloskar factory, near Mula–Mutha Sangam
IRMR	Indutrial roof (IR)	Industrial area at MIDC Road
IRNR	Indutrial roof (IR)	Industrial area at Nashik Road
IRNR2	Indutrial roof (IR)	Industrial area at Nashik Road 2
IRTA	Indutrial roof (IR)	Industrial roof (Tata Motors)+C20
OGNP	Open ground (OG)	NDA parade ground (bitumen pavement)
PGAN	Playground (PG)	Playground at Agricultural Institute, New ground
PGAO	Playground (PG)	Agricultural Institute Old ground

(*Continued*)

TABLE 3.4 (Continued)
Landcover References from the Hyperion Image Used in the Study

Code	Class	Description
PGFC	Playground (PG)	Ferguson College playground
PGMO	Playground (PG)	Construction site of Mercedes-Benz open area
PGND	Playground (PG)	Playground at NDA (covered with a little grass)
PGPA	Playground (PG)	Playground at Pashan ARDE
PLCI	Plain (PL)	Chakan Indrayani open area/plain
PLMU	Plain (PL)	Mutha plains
PLS12	Plain (PL)	Sector 12 Chakan industrial area
PLS12'	Plain (PL)	Sector 12 right half, Chakan industrial area
PLS5	Plain (PL)	Sector 5, Chakan industrial area
QAMO	Quarry (QA)	Moshi quarry
REBN	Residential (RE)	Bhim Nagar
REDN	Residential (RE)	Datta Nagar
REGN	Residential (RE)	Ganesh Nagar
REHN	Residential (RE)	Hanumaan Nagar
RENS	Residential (RE)	Residential mid-low economy, New Sangavi
RENS2	Residential (RE)	New Sangavi 2
REOS	Residential (RE)	Old Snagavi
REPG	Residential (RE)	Sinhagad medical institute, Narhe road
RERN	Residential (RE)	Raam Nagar
RUFC	Residential (RE)	Ferguson College road
RULC	Residential (RE)	Law College road
RUMO	Residential (RE)	Model colony
RUMO2	Residential (RE)	Model colony 2
RUNS	Residential (RE)	Nava Sahyadri
RUNS2	Residential (RE)	Nava Sahyadri
TRBS	Tree (TR)	Trees near Baner STP
TRCH	Tree (TR)	Chattushrungi
TRLP	Tree (TR)	Leftbank Pashaan
TRLP2	Tree (TR)	Leftbank Pashaan 2–near bypass road
TRRP	Tree (TR)	Rightbank Pashan
TRSN	Tree (TR)	Shastri Nagar military farm
TRSN2	Tree (TR)	Shastrinagar miltary farm along the river
TRUN	Tree (TR)	University trees
TRUN2	Tree (TR)	University trees–North side
TRVG	Tree (TR)	Chittaranjan Vatika Garden
WBCM	Water Body (WB)	Chakan MIDC lake
WBKH	Water Body (WB)	Khadakvasala lake
WBPA	Water Body (WB)	Pashaan lake
WBPA2	Water Body (WB)	Pashaan lake 2

Notes:
* First two letters of the code indicate broader classes such REsidential, Green Grass, Dry Grass, Industrial Roof and so on, as given in column "Class". Next two letters of the code are the abbreviation of a specific location. For example, KH stands for Khdakwasla, a particular locality of the study area. In case the location is indicated by two words then first letter of each word is taken as a part of code. For example, NS stands for New Sangavi.

TABLE 3.5
Row and Column Coordinates of the Reference Sites Used in the Study

Code	R1	R2	C1	C2	Number of pixels
BGCH	1249	1257	206	219	126
BGKH	2370	2384	142	157	240
BGKO	2095	2127	16	20	165
COCP	2056	2057	182	183	4
CONP	2148	2150	55	55	3
DGKA	2628	2632	119	125	35
DGKA'	2616	2626	114	116	33
DGKO	2576	2580	180	184	25
GGHB	790	801	134	136	36
GGHI	1378	1380	67	88	66
GGPA	1976	1979	28	31	16
GGRF	873	876	153	158	24
IRCR	1182	1250	67	133	4623
IRCU	2073	2077	136	141	30
IRKI	1787	1792	195	197	18
IRMR	1494	1548	105	163	3245
IRNR	1496	1517	40	65	572
IRNR2	1313	1342	94	127	1020
IRTA	1498	1505	48	50	24
OGNP	2244	2247	9	11	12
PGAN	1885	1888	214	218	20
PGAO	1910	1913	232	235	16
PGFC	1955	1962	218	219	16
PGMO	1217	1225	62	69	72
PGND	2255	2257	4	9	18
PGPA	1931	1934	87	91	20
PLCI	1362	1371	71	76	60
PLMU	2205	2208	113	117	20
PLS12	1471	1479	138	146	81
PLS12'	1476	1490	114	121	120
PLS5	1508	1523	87	95	144
QAMO	1448	1456	208	218	99
REBN	1767	1781	155	165	165
REDN	1990	2003	123	133	154
REGN	1697	1721	241	252	300
REHN	2002	2010	173	179	63
RENS	1748	1764	94	109	272
RENS2	1744	1764	87	106	420
REOS	1757	1769	89	99	143
REPG	2224	2233	229	238	100
RERN	1577	1606	188	218	930
RUFC	1970	2008	190	236	1833

(Continued)

TABLE 3.5 (Continued)
Row and Column Coordinates of the Reference Sites Used in the Study

Code	R1	R2	C1	C2	Number of pixels
RULC	1977	2007	198	225	868
RUMO	1927	1940	208	222	210
RUMO2	1913	1941	178	243	1914
RUNS	2047	2064	200	218	342
RUNS2	1981	1998	203	223	378
TRBS	1847	1850	48	50	12
TRCH	1934	1937	196	199	16
TRLP	1974	1978	5	11	35
TRLP2	1976	1983	18	21	32
TRRP	1975	1981	53	62	70
TRSN	1690	1700	54	60	77
TRSN2	1700	1702	49	60	36
TRUN	1857	1862	166	169	24
TRUN2	1846	1851	165	166	12
TRVG	1926	1930	212	216	25
WBCM	1559	1563	39	43	25
WBKH	2318	2332	48	58	165
WBPA	1958	1961	41	44	16
WBPA2	1957	1964	34	46	104

1. Trees – Urban tree cover, the trees comprise raintree, mango tree, subabul and banyan tree, commonly found in the cities of Western Maharashtra.
2. Farms, green grass etc. – Farms, green grass, and other miscellaneous low-lying vegetation.
3. Open ground – An open ground with compact bare soil without grass cover, like playgrounds.
4. Dry grass – Open areas covered with dry grass on slopes, sometimes cut to the ground level.
5. Forest fire marks – Burned grass pastures and/or meadows at various locations.
6. Concrete – Bright road-grade cement concrete.
7. Industrial roof – Commonly used industrial-grade metal roofs.
8. Water bodies.
9. Quarry – A large stone quarry area with an exposed basaltic rock surface.
10. Plains – Common open areas with brown soil, typically with varying degrees of dry grass cover. This class is a mixture of dry grass and soil cover.
11. Residential – A mixture signature of areas covered predominantly with concrete slab roofs without little vegetation.
12. Residential 2 – A mixture signature of upmarket housing with good tree cover.
13. Industrial – A mixture of open areas and industrial roof covers.

Classes one to eight are material classes or land cover classes and nine to thirteen are land use classes. Some of the classes are interchangeable, for example, plains or common open areas represent both the land cover and the land use class. Not all the land use classes can be identified with a single land cover; for example, residential 2 is an upmarket residential area and is a mixture of tree and concrete land covers. Whereas some of the land uses are synonymous with land cover; for example, the mid-economy residential is repented by the concrete land cover. Some of these sites are used for Flat Field experiments as well. Each candidate Flat Field is selected based on the additional criteria such as it a) exhibited the bright pure signature, b) covered a sufficient number of pixels (ideally, >16 to reduce the error) (Clark et al., 2002), c) is well separated from the neighbouring pixels visually, d) is accessible – as far as possible – for synchronized field measurements in the future if any.

3.6.2 Method

The procedure for calibration and classification begins with a conversion of the DNs to the radiance values for the selected Hyperion bands. The channels removed for lack of calibration and water absorption were: 1–8, 58–81, 98–101, 119–134, 164–187, and 218–242 (Deshpande et al., 2017). This pre-processing step is subjected to change according to the sensor. Water absorption channels should be removed. However, the band number of the waste absorption would vary. Selected channels within the range 1–70 were divided by 40 and 71–242 were divided by 80 to convert the DNs to the radiance (W.m^{-2} SR^{-1} µm^{-1}) (Beck, 2003). After converting the DNs to the radiance values, an additive component of atmospheric effects is removed using an extension of the improved dark object technique (Chavez, 1988). For this example, a "very clear" atmospheric model is selected, that is the path radiance is inversely proportional to the fourth power of the wavelength, and channel 52 (874.53 nm) is selected for the initial path radiance value. The acquired scene contains large water bodies that should indicate zero radiance from channel 52. Hence, in this case, the dark bodies represented water bodies. 0.5 (W.m^{-2} SR^{-1} µm^{-1}) is selected as a starting path radiance value to avoid overcorrection of the channels beyond a visible range. Correction factors for all the wavelengths are calculated for "very clear" atmospheric conditions. The initial path radiance value is then multiplied by a correction factor for each channel to arrive at the radiance value that needs to be subtracted from each channel, respectively (Deshpande et al. 2017).

Next, FAR and IAR are used for calculating the reflectance values of each pixel in the image. Log Residuals (Green & Craig, 1985) is one more image-based method, which corrects the atmospheric effects in an equivalent manner. However, the method does not create relative reflectance signatures that resemble laboratory spectrum and hence was not considered for comparison.

Next, the Spectral Angle Mapper (SAM) (Kruse et al., 1993) is used to detect VIS classes in the acquired images. The algorithm calculates the spectral angle between each reference spectrum and a pixel in the image. The label of the reference with a minimum spectral angle with the pixel is assigned to the pixel.

Finally, the accuracy of the results is measured for all the experiments using a standard confusion matrix and further calculated Producer's and User's accuracy. Producer's accuracy indicates the fraction of reference pixels that are correctly identified by a classifier (omission errors), and User's accuracy indicates the fraction of pixels assigned to a particular class that are true class pixels (commission errors) (Congalton, 1991). For example, Experiment 1 reference signatures for VIS classes are extracted from, GGPA, REPG, and PLMU (Table 3.4 and Table 3.5), and accuracy is calculated using TRUN, RENS, and PLCI regions (Table 3.4 and Table 3.5), respectively. For example, the overall accuracy of classification for experiment 1 is: 94% (IAR) with Producer's and User's accuracy of 100% for the Vegetation class for both IAR and FAR, User's accuracy and Producer's accuracy of 100% for Impervious Surface class for both IAR and FAR and so on (Table 3.9, Table 3.10, Table 3.11). The steps are summarized in the flow chart in Figure 3.11.

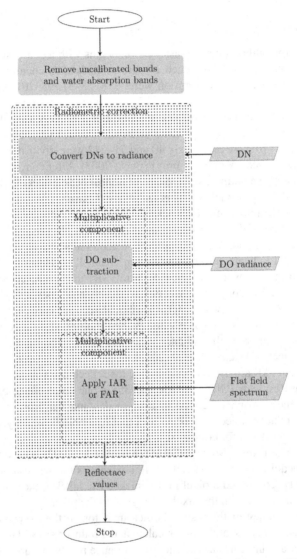

FIGURE 3.11 The overall methodology for an image-based reflectance calibration method.

Reflectance Calibration

3.6.3 Experimental Set up for Classification Experiments

Various reflectance calibration and classification experiments were performed as described in the overall procedure, to compare intra-class – inter-class confusion if any using SAM (Deshpande et al., 2017). As a strategy, reference signatures were extracted from small regions and tested on large areas. We selected test regions that are away from reference pixels by a large distance. All the experiments were performed on the April 22 image (USGS, 2013-1) (Table 3.4 and Table 3.5). The detail of the experimental set up is as explained below:

Experiment 1–2 (Baseline): We extracted reference signatures from the most representative materials of VIS classes. For example, vegetation signature is taken from trees, soil signatures from bare soil area with little or low grass cover, and impervious surface signatures are taken from a dense residential area with concrete roofs. In experiment 2, we changed the reference signature of the impervious surface to a concrete pavement and observed if all impervious surface targets are detected (Table 3.6 and Table 3.7).

Experiment 3–5: In experiment 3, we extracted reference signatures of the subclasses of impervious surfaces – concrete and industrial roofs and then further investigated (in experiment 4 and 5) if different income zones can be detected spectrally by extracting reference mixture signatures from respective zones (Table 3.6 and Table 3.7).

Experiment 6–7: Similar to the experiment for impervious surfaces, we successively extracted reference signatures of various subclasses in soil and performed VIS classification (Table 3.6 and Table 3.7).

Experiment 8: We divided vegetation reference signatures into two groups and performed VIS classification (Table 3.6 and Table 3.7).

Though the study area sub-image is 1000 rows and 256 columns, the average over the whole image (3200 × 256) is considered. This is required to reduce the dominance of any material within the scene. FAR experiments are designed to find the optimal Flat Field within the urban area.

The Flat Field codes along with their additional description are: (OGFC); (OGPA); (OGMO); (PLMU) – small, island-like open area in the Mutha River, enclosed by Mutha River on one side and by a small channel on the another; (PLCI) – a large open area near the Indrayani River channel; (Sector 5 – PLS5, and Sector 12 – PLS12); CONP – a smaller area used for parking – these bright pure pixels can be easily demarcated in the image; (IRTA, IRKI) – Provide a large very bright area; (QAMO) – irrespective of weathered rocks at places, it provides an excellent bright signature all across the bands (Table 3.6 and Table 3.7).

3.6.4 Results of Image-based Reflectance Calibration Experiments

3.6.4.1 General Observations

IAR and FAR calibration methods produce accurate signatures of VIS classes. Signatures extracted from EO-1-Hyperion images–with 10 nm spectral resolution–using these methods are sufficient to differentiate VIS types at different granularity. Preliminary results using image-derived signatures show particularly

TABLE 3.6
Reference Site Codes for Calibration and Classification Experiments

Expr. No.	Vegetation	Impervious	Soil
1	TRLP	REPG	PLMU
2	TRLP	CONP	PLMU
3	TRUN	CONP IRKI	PLMU
4	TRLP	REPG RUNS IRTA	PLMU
5	TRLP	REPG RUNS IRTA	PLCI
6	TRLP	CONP	PLMU QAMO DGKO
7	TRLP	REPG IRTA	PLMU OGFC DGKO BGKO
8	TRLP GGHB	CONP IRTA	PLMU DGKO

TABLE 3.7
Test Site Codes for Calibration and Classification Experiments

Expr. No.	Vegetation	Impervious	Soil
1	TRUN	RENS	PLCI
2	TRUN	REPG	PLCI
3	TRLP	REPG IRTA	PLCI
4	TRUN	RENS RUMO IRKI	PLCI
5	TRUN	RENS RUMO IRKI	PLCI
6	TRUN	REPG	PLCI QAMO DGKA
7	TRUN	RENS IRKI	PLCI OGPA DGKA BGKH
8	TRUN GGHI	REPG IRKI	PLCI DGKA

good classification accuracy for VIS classes (~90% best and ~80% worst overall accuracy over ~110 experiments using IAR and FAR). Broad-level VIS classes are separable spectrally whereas some of the subclasses show similar signatures. Road cement concrete and bitumen concrete show similar signatures with flat reflectance values without any diagnostic absorption. Concrete shows higher reflectance values than that from the bitumen road. Though the soil-covered bright spots in the image such as playgrounds have different spectral shapes, they are at times confused with built-up classes. A spectral matching technique such as NS3 (Nidamanuri & Zbell, 2011) that considers reflectance information as well would be helpful in such a scenario. Accuracy of classification is largely determined by the test region rather than the reference region: an average of all the pixels is taken as a reference, whereas each individual pixel similarity is calculated in the classification process. Further model generation that is creating a reference signature for classification, is not subjected to statistical variations as in the other machine learning techniques.

This is an added advantage of prototype-based learning methods such as SAM: the learning performance does not change with training sample size. Figure 3.12 and

Reflectance Calibration

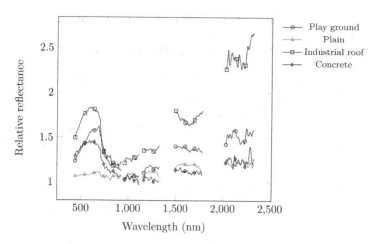

FIGURE 3.12 Signatures of the candidate Flat Fields in the study area.

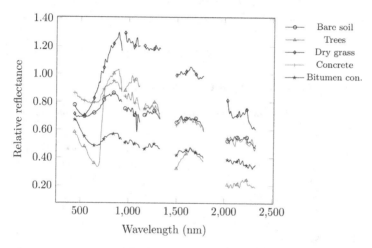

FIGURE 3.13 Sample reference signatures extracted from the image (extracted by FAR).

Figure 3.13 show extracted signatures of some of the candidate Flat Fields and the sample references for the VIS classes.

3.6.5 Calibration Performance for IAR and FAR

We compared the IAR and FAR results and analysed the performance of each calibration method for various urban land use and land cover classes at a broader level. We provide detailed discussion in the sections below (1, 2, 3, and 4).

3.6.5.1 Effect of Mixed Signature as a Reference

Reference signatures from pure pixels provide more accurate results (Baseline experiments 1 and 2). There is a jump of ~6% in accuracy, for IAR and for all the

TABLE 3.8
Per cent Overall Accuracy for Different Calibration Methods

No.	1	2	3	4	5	6	7	8	9	10	11	12	13	14
1	94	94	94	94	94	94	92	92	93	94	93	94	93	94
2	100	100	100	100	100	100	100	100	100	100	100	100	100	100
3	100	100	100	100	100	100	100	100	100	100	100	100	100	100
4	82	83	82	83	83	82	80	80	80	82	82	82	82	82
5	82	82	82	82	81	82	81	81	80	81	82	81	80	81
6	83	80	82	80	80	83	83	86	83	84	81	82	79	80
7	94	95	96	96	95	95	94	95	95	96	97	95	92	93
8	92	92	92	92	92	92	91	92	92	93	93	92	89	93
Avg.	91	91	91	91	91	91	90	91	90	91	91	91	90	90

Notes:
[1-IAR, 2-PLCI, 3-PLMU, 4-PLS12, 5-OGFC, 6-OGPA, 7-OGMO, 8-QAMO, 9-CONP, 10-IRTA, 11-IRKI, 12-Avg. of Flat Fields, 13–6SV, 14-EO-1-equation]

TABLE 3.9
Per cent Overall Accuracy for Different Flat Fields

No.	IAR	Plain	Dry grass	Playground	Concrete	Ind. roof	6SV	EO-1
1	94	94	94	94	92	93	93	94
2	100	100	100	100	100	100	100	100
3	100	100	100	100	100	100	100	100
4	82	83	83	82	80	82	82	82
5	82	82	82	82	81	82	80	81
6	82	81	82	82	84	83	79	80
7	96	95	96	95	95	97	92	93
8	92	92	92	92	92	93	89	93
Avg.	91	91	91	91	90	91	90	90

Flat Fields considered, when a concrete signature is used to identify a residential area dominated by concrete roofs (Table 3.8). Further, if the reference signature is taken from a carefully chosen large area having some mixed pixels, accuracy is not affected significantly (baseline experiments 1 and 2 for both the test regions). Producer's accuracy is also remarkably high for impervious surfaces (above 90%) except in cases where two bright signatures, namely stone quarry and concrete, are considered as references at the same time. Irrespective of a high Producer's accuracy for a stone quarry area (~100%), User's accuracy of 63–65% is achieved respectively (Experiment 6, Table 3.10). Similar confusion occurs when a large test region for soil contains some bright spots in it – they are confused with bright signatures of concrete.

Reflectance Calibration

TABLE 3.10
Per cent User's Accuracy for IAR and Best Flat Field Reflectance Calibration

No.	V		I			S			
	Trees	Hyacinth	Residential	Industrial	Upmarket residential	Plain	Quarry	Dry grass	Burnt grass
1	100		100			73/72			
2	100		100			100			
3	100		99	100		100			
4	25/24		98	94	97/98	69/71			
5	25/24		94/95	100	100	94			
6	100		98/100			100/98	63/65	98/100	
7	100		91/93	100		87/94	76/81	1	96/99
8	47/63	78/82	98	100		100/98		100	

Note:
* Only one number in the table cell **indicates** there is no difference between best IAR and Flat Field accuracies, grey cells indicate better FAR.

3.6.5.2 Vegetation Subclasses

Considering subclasses of vegetation, impervious surfaces and soil did not affect accuracy by a large margin. Results showed remarkably high overall accuracy with high User's and Producer's accuracy with occasional dips for vegetation and soil classes. Vegetation class confusion arises when thematic class like upmarket residential zone is considered which is mixture of concrete and tree pixels. Confusion between soil and impervious surface arises when bright pixels of a stone quarry are added to the reference signatures. Some of the bright pixels in the residential area comprising concrete roofs are misclassified as stone quarry areas.

Intra-class confusion is maximum for trees and resulted in poor Producer's and User's accuracy for the class. Water hyacinth is identified accurately with good User's and Producer's accuracy of 82% and 89% respectively (for FAR). Only two subclasses are considered at present: Trees and Water hyacinth (Experiment 8, Table 3.10, Table 3.11). Further investigation is required for other vegetation subtypes. The vegetation class is not confused with the soil or impervious surface class. However, in the case of a large test region for an upmarket residential class, some of the pure tree pixels are labelled correctly as trees instead of upmarket residential affecting User's accuracy for tree class (Experiment 5).

3.6.5.3 Impervious Surface Subclasses

Between the two impervious material classes considered, industrial roofs and residential roofs (concrete), there is little intra-class confusion. Industrial roofs are identified with high User's and Producer's accuracy all the time (Experiment 3, 4, 5, 7, and 8). Adding reference spectra of subclasses of vegetation and soil does not affect the accuracy of industrial roof detection (Experiment 5, 7, and 8) – the way soil classes

TABLE 3.11
Per cent Producer's Accuracy for IAR and Best Flat Field Reflectance Calibration

	V		I			S			
No.	Trees	Hyacinth	Residential	Industrial	Upmarket residential	Plain	Quarry	Dry grass	Burnt grass
1	100		92			100			
2	100		100			100			
3	100		100	96		100			
4	100		91/90	89	62/61	100			
5	100		100/99	89	62/61	100			
6	100		43/46			100	99/100	100/97	
7	100		96/99	83		100/97	80/85	100/97	93/96
8	38/50	84/89	100	89		100		97	

affect residential areas sometimes (Experiment 5, 6 and 7, Table 3.10, Table 3.11). In the presence of a bright quarry signature, a residential class's User's accuracy is marginally increased (by ~4–5%) but Producer's accuracy is decreased by a large margin (57%–53%) for IAR and FAR respectively because of mislabelling of concrete as a quarry (Experiment 5, 6 Table 3.10, Table 3.11).

3.6.5.4 Soil Subclasses

Overall soil classes show moderate intra-class and low inter-class confusion. Especially, bare soil (like playgrounds) creates intra-class confusion and inter-class confusion with impervious classes like concrete. The addition of a bare soil class lowered User's accuracy of a plain area for IAR and FAR by 13% and 4% respectively and lowered Producer's accuracy by ~3% for FAR (Experiment 6, 7; Table 3.10, Table 3.11). Intra-class confusion might be because of heterogeneous cover in plain test regions, which might have some sub-areas with less vegetation cover resembling bare soil. This confusion is less harmful and can be avoided by choosing test regions with a more uniform cover appropriately.

While residential class accuracy was unaffected in the presence of bare soil reference signatures; in the presence of stone quarry signatures, bright regions in the residential zones were mistaken for stone quarry. Hence, Producer's accuracy was decreased with respect to User's accuracy for residential class by 55%–54% for IAR and FAR respectively (Experiment 6; Table 3.10, Table 3.11). More detailed study is required to analyse the dominant effect of a gravel signature on a concrete signature. Irrespective of high Producer's accuracy for a stone quarry area User's accuracy of 65% and 63% is achieved by FAR and IAR (Experiment 6; Table 3.10, Table 3.11).

3.6.6 OVERALL PERFORMANCE OF PHYSICS-BASED METHODS

Performance of physics-based methods (Table 3.8, Table 3.9) agrees with image-based methods for broad VIS categories (Expr. 1, 2, and 3, Table 3.8) but is decreased relatively if subclasses of VIS are considered (Expr. 4 onwards, Table 3.8, Table 3.9). Especially, if soil subclasses are considered both the physics methods' overall accuracies are dropped by ~2–4% as compared to IAR and/or FAR. Among the physics-based methods, EO-1 equation (Equation 3.1) produced better results than 6SV (Table 3.8, Table 3.9). The overall drop in the accuracy could be attributed to the drop in Producer's accuracies for most of the subclasses (indicating increased omission errors) as compared with IAR and FAR. Overall physics-based methods provide no substantial advantages over image-based calibration methods. The accuracies of individual subclass experiments such as impervious surfaces, soil and so on are consistent with the image-based methods (see the section "Calibration performance for IAR and FAR").

3.6.7 OPTIMAL FLAT FIELD ASSESSMENT

We assessed the suitability of various Flat Fields for discriminating urban land use land cover and further compared the performance of various Flat Fields with physics-based methods.

3.6.7.1 Flat Fields Comparison

Of all the Flat Fields, PLCI and PLCT provide a marginal improvement in average overall accuracies over IAR. Industrial roofs also provide better average accuracy than IAR – increases by a per cent are observed for both surfaces. All the Flat Field reference signatures provide remarkably high accuracy for broad-level VIS classification (expr. 1, 2) and there is no single Flat Field that stands out. The scenario changes slightly for subclasses of impervious surfaces and soil. When impervious subclasses are considered, PLCI provides slightly better accuracy. Parking lot concrete (CONP) also matches these results. IRTA – Industrial roof – provides the best accuracies for soil subclass experiments (expr. 6, 7) and vegetation subclass experiments. Further, IRTA shows marginal improvement (1–2%) in accuracy over IAR. Interestingly – Plain regions are slightly better Flat Fields than playgrounds.

3.6.7.2 Best Flat Fields vs IAR, 6SV, EO-1 Equation

Industrial roof provides the best accuracies among all the Flat Fields – maybe due to its high radiance throughout the wavelength range. There is an average jump of ~3% in User's accuracies as compared with IAR. Accuracies are improved for soil classes for all the experiments–especially, experiments wherein soil subclasses are considered. Accuracies of most of the other classes and subclasses of vegetation and impervious surfaces remain unchanged (Table 3.12 to Table 3.17). Producer's accuracies also show an average jump of ~1. An average increase in User's accuracy and Producer's accuracy of FAR with respect to 6SV is ~8% and 5% and with respect

TABLE 3.12
Per cent User's Accuracy for 6SV/ Best Flat Field Reflectance Calibration

	V		I			S			
No.	Trees	Hyacinth	Residential	Industrial	Upmarket residential	Plain	Quarry	Dry grass	Burnt grass
1	100		100			71/72			
2	100		100			100			
3	100		99	100		100			
4	24		98	89/94	99/98	69/71			
5	24		95	84/100	100	97/94			
6	100		100			94/98	63/65	78/100	
7	100		96/93	89/100		76/94	63/81	74/100	100/99
8	57/63	82	98	100		94/98		78/100	

TABLE 3.13
Per cent Producer's Accuracy for 6SV/Best Flat Field Reflectance Calibration

	V		I			S			
No.	Trees	Hyacinth	Residential	Industrial	Upmarket residential	Plain	Quarry	Dry grass	Burnt grass
1	100		91/92			100			
2	100		100			100			
3	100		100	96		100			
4	100		9	89	61	100			
5	100		98/99	89	61	100			
6	100		41/46			85/100	100	91/97	
7	100		92/99	89/83		85/97	85	91/97	94/96
8	50	86/89	100	89		85/100		91/97	

to EO-1 equation (Equation 3.1) is ~4% and ~2% respectively. Figure 3.14 shows a comparative assessment of overall accuracies by different reflectance calibration methods. Overall, industrial roof covers provide near ideal Flat Fields in urban settings and provide marginal to substantial improvements over all the other reflectance calibration methods (Figure 3.15). Physics-based methods exhibit residual absorption errors whereas FAR (and IAR) relative reflectance signatures are without such errors (Figure 3.16). Figure 3.17 displays the classified image.

3.6.8 SUMMARY OF THE CASE STUDY FOR IMAGE-BASED METHODS

The urban VIS signatures extracted from a Hyperion image show distinct spectral curves at a broader level. Signatures extracted from EO-1-Hyperion images having ~10 nm spectral resolution are sufficient to differentiate VIS types at different

TABLE 3.14
Per cent User's Accuracy for EO-1/Best Flat Field Reflectance Calibration

	V		I			S			
No.	Trees	Hyacinth	Residential	Industrial	Upmarket residential	Plain	Quarry	Dry grass	Burnt grass
1	100		92			73/100			
2	100		100			100			
3	100		99/100	100		100			
4	24/100		98/90	94/89	97/61	71/100			
5	24/100		95/99	100/89	100/61	94/100			
6	100		100/46			100	60/100	100/97	
7	100		93/99	100/83		85/97	57/85	100/97	97/96
8	57/50	82/89	98/100	100/89		100		100/97	

TABLE 3.15
Per cent Producer's Accuracy for EO-1/Best Flat Field Reflectance Calibration

	V		I			S			
No.	Trees	Hyacinth	Residential	Industrial	Upmarket residential	Plain	Quarry	Dry grass	Burnt grass
1	100		92			100			
2	100		100			100			
3	100		100	96		100			
4	100		90	89	62/61	100			
5	100		99	89	61	100			
6	100		33/46			100	100	100/97	
7	100		90/99	78/83		93/97	85	100/97	94/96
8	50	86/89	100	89		100		97/97	

granularity. Preliminary results using image-derived signatures show particularly good classification accuracy for VIS classes (~90% best and ~80% worst overall accuracy over ~110 experiments using IAR and FAR). Inter-class confusion between bare soil and concrete or residential land use class is evident from the results.

Image-based calibration methods such as FAR and IAR provide many advantages over physics-based models such as 6SV. They are simple to use, and all the information required is available within the image itself. Both these methods are inherently explorative and sound assessment of the merits and demerits of both are required before wide application. In the case of IAR it is important to know if the dominant material in the scene has an adverse effect on the calibration and in the case of FAR, it is important to know which Flat Field might be the best in a given constraints of the study area. Some of the specific conclusions and recommendations of our study are:

TABLE 3.16
Per cent User's Accuracy for IAR/6SV Reflectance Calibration

	V		I			S			
No.	Trees	Hyacinth	Residential	Industrial	Upmarket residential	Plain	Quarry	Dry grass	Burnt grass
1	100		100			73/71			
2	100		100			100			
3	100		99	100		100			
4	24		98	94/89	97/99	69			
5	24		94/95	100/84	100	94/97			
6	100		98/100			100/94	63	98/78	
7	100		91/96	100		87/76	76/63	100/74	96/100
8	47/57	82	98	100		100/94		100/78	

TABLE 3.17
Per cent Producer's Accuracy for IAR/ 6SV Reflectance Calibration

	V		I			S			
No.	Trees	Hyacinth	Residential	Industrial	Upmarket residential	Plain	Quarry	Dry grass	Burnt grass
1	100		92/91			100			
2	100		100			100			
3	100		100	96		100			
4	100		91/90	89	62/61	100			
5	100		100/98	89	62/61	100			
6	100		43/41			100/85	99/100	100/91	
7	100		96/92	83/89		100/85	80/85	100/91	93/94
8	38/50	84/86	100	89		100/85		97/91	

1. Preliminary results using image-derived signatures by IAR and FAR show ~90% best and ~80% worst overall accuracy over ~110 experiments (Table 3.8 and Table 3.9). Producer's accuracy is also remarkably high for impervious surfaces (above 90%) except in cases where two bright signatures namely stone quarry and concrete are considered as references at the same time. Irrespective of high Producer's accuracy for stone quarry area (~100%), User's accuracy of 63–65% is achieved respectively achieved by FAR and IAR (Experiment 6, Table 3.10).
2. IAR and FAR produce sufficiently accurate signatures to detect several types of vegetation covers (for example, in experiment 8, related to water hyacinth and tree), but further experiments with more vegetation types in urban areas are required to understand the possibilities. EO-1 equation shows absorption

Reflectance Calibration

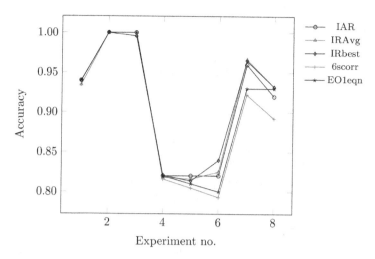

FIGURE 3.14 Overall classification accuracies for different reflectance calibration methods.

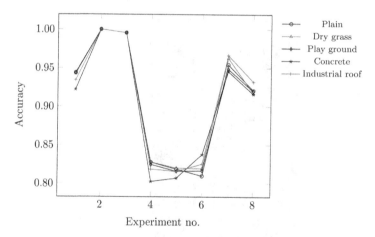

FIGURE 3.15 Overall classification accuracies for Flat Fields reflectance calibration.

artefacts and 6SV also show some reflection artefacts in the signatures. 6SV signatures appear to be overcorrected in a visible range (Figure 3.16).

3. FAR, on average, shows marginal improvements in overall classification accuracies as compared to IAR and other physics-based methods such as 6SV. The improvement in User's accuracy and Producer's accuracy over other methods is evident for many of the Flat Fields. Industrial roof covers provide one of the best classification results. Based on the availability of a sufficient number of pixels any of these three regions can be effectively used for calibrating a hyperspectral image of an urban area. Further, Flat Fields reduced confusion between bright signatures of soil and concrete.

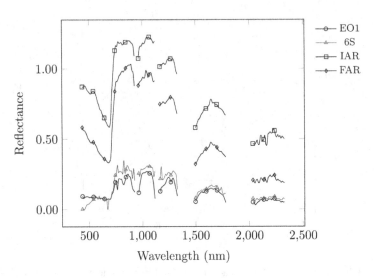

FIGURE 3.16 Comparison of the vegetation signatures derived by the different reflectance calibration methods.

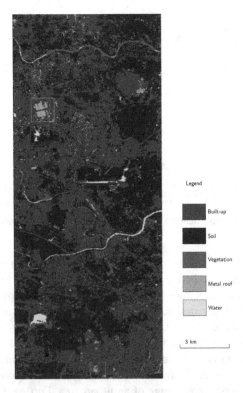

FIGURE 3.17 VIS classification results for the EO-1 Hyperion image of Pune city dated April 22, 2013.

4. Of all the Flat Fields, PLCI and PLCT provide a marginal improvement in average overall accuracies over IAR. Industrial roofs also provide better average accuracy than IAR – an increase by a per cent is observed for both surfaces. Broadlevel VIS classification (expr. 1, 2) does not have any favourites and all the Flat Fields provide remarkably high accuracy and are the same as IAR. Further, IRTA shows marginal improvement (1–2%) in accuracy over IAR. Interestingly, plain regions are slightly better Flat Fields than playgrounds. An average increase in User's accuracy and Producer's accuracy of FAR with respect to 6SV is ~8% and 5% and with respect to EO-1 equation (Equation 3.1) is ~4% and ~2% respectively.
5. In the presence of bare soil reference spectra, inter-class confusion between concrete and soil is reduced because bright spots in plain regions are classified as open grounds instead of concrete roofs. Similarly, some of the bright spots in the residential area, such as concrete roofs are mistaken to be surfaces like stone quarry areas.
6. Amplitude information is important for impervious surfaces as spectral shapes of many impervious surfaces are similar to each other and they differ only in amplitude information. Techniques like NS3 (Nidamanuri & Zbell, 2011) that consider amplitude information would be helpful to discriminate intra-impervious classes further.
7. In the absence of a pure material signature, all the reflectance calibration methods can handle a little mixture in the reference signatures for impervious surfaces. It reduces confusion because of bright pure signatures at times. Further, mixed signatures of VIS fractions in a certain proportion can represent land use classes, for example, upmarket residential vs low economy residential.

3.7 SUMMARY

Atmospheric correction is critical in the processing of hyperspectral data. It is one of the main steps in converting digital numbers to true reflectance values. The conversion to true reflectance is necessary to analyse the spectral signature of the target material, more specifically to determine the diagnostic absorption and reflection of the target material. The effect of the atmosphere on the radiance recorded by satellite is multiplicative and additive. Back-scattering by the atmosphere is an additive and the absorption and scattering in the direction other than the sensor are multiplicative.

There are two major methods for converting digital numbers or radiance values to reflectance. The first one is a physics-based method and the second is an image-based or empirical method. The physics-based methods model the atmosphere and light interaction using first principles and estimate the absorption and scattering by the atmosphere. The image-based methods are simpler than physics-based methods. They use all the data available in the image to extract the required multiplicative and additive correcting factors. Physics-based methods are complex and require measurements of atmospheric properties which are difficult. Image-based methods do not require any data. Atmospheric correction of the hyperspectral image needs careful attention because of its high spectral resolution.

WORKS CITED

Aspinall, R., Marcus, W., & Boardman, J. (2002). Considerations in collecting, processing, and analysing high spatial resolution hyperspectral data for environmental investigations. *Journal of Geographical Systems, 4*, 15–29. doi:10.1007/s101090100071

Atmospheric radiative transfer codes. (2018, June 7). Retrieved June 25, 2018, from Wikipedia: https://en.wikipedia.org/wiki/Atmospheric_radiative_transfer_codes

Beck, R. A. (2003, July). *EO-1 user guide v. 2.3*. Tech. rep., USGS Earth Resources Observation Systems Data Center (EDC), 47914–252nd Street, Sioux Falls, S.D., 57198–0001. Retrieved 10–19, 2022, from https://d9-wret.s3.us-west-2.amazonaws.com/assets/palladium/production/s3fs-public/atoms/files/EO1userguidev2pt320030715UC.pdf

Berk, A., Conforti, P., Kennett, R., Perkins, T., Hawes, F., & van den Bosch, J. (2014). MODTRAN6: A Major upgrade of the MODTRAN Radiative Transfer Code. *IEEE-WHISPERS – 6th Workshop on Hyperspectral Image and Signal Processing: Evolution in Remote Sensing*. Lausanne, Switzerland, June 24–27.

Biggar, S. F., Thome, K. J., & Wisniewski, W. (2003). Vicarious radiometric calibration of EO-1 sensors by reference to high-reflectance ground targets. *IEEE Transactions on Geoscience and Remote Sensing, 41*(6), 1174–1179.

Chavez, Jr. P. S. (1988, April). An Improved Dark-Object Subtraction Technique for Atmospheric Scattering Correction of Multispectral Data. *Remote Sensing of Environment, 24*(3), 459–479.

Chavez, Jr. P. S. (1996, September). Image-Based Atmospheric Corrections – Revisited and Improved. *Photogrammetric Engineering and Remote Sensing, 62*(9), 1025–1036.

Chen, J., & Hepner, G. F. (2001). Investigation of imaging spectroscopy for discriminating urban land covers and surface materials. *AVIRIS Earth Science and Applications Workshop, Palo Alto, California. JPL Publication 02–1.*, Pasadena, CA: JPL Publications. Retrieved from https://aviris.jpl.nasa.gov/proceedings/workshops/01_docs/2001Chen_web.pdf

Clark, R. N., Swayze, G. A., Livo, K. E., Kokaly, R. F., King, T. V., Dalton, J. B., ... McDougal, R. R. (2002). Surface Reflectance Calibration of Terrestrial Imaging Spectroscopy Data: a Tutorial Using AVIRIS. *Proceedings of the 10th Airborne Earth Science Workshop. 02–1*. JPL Publication.

Conel, J. E., Green, R. O., Vane, G. W., Bruegge, C. J., Alley, R. E., & Curtiss, B. (1987). AIS-2 radiometry and a comparison of methods for the recovery of ground reflectance. *Proceedings of the 3rd Airborne Imaging Spectrometer Data Analysis Workshop*, Pasadena, CA: JPL Publications.

Congalton, R. G. (1991). A review of assessing the accuracy of classifications of remotely sensed data. *Remote Sensing of Environment, 37*, 35–46.

Deshpande, S. S., Inamdar, A. B., & Buddhiraju, K. M. (2018). Spectral Overlap Correction Using Weighted Deconvolution for Improved Dark Object Technique for Correction of EO1-Hyperion Data. *Journal of the Indian Society of Remote Sensing, 46*(3), 337–344.

Deshpande, S. S., Inamdar, A. B., & Vin, H. M. (2017). Urban land use land cover discrimination using image based reflectance calibration methods for hyperspectral data. *Photogrammatric Engineering and Remote Sensing, 83*(5), 365–376.

Deshpande, S., Inamdar, A., & Vin, H. (2013). Discrimination of Vegetation-Impervious Surface-Soil Classes in Urban Environment Using Hyperspectral Data. *Proceedings of Asian Conference on Remote Sensing (ACRS 2013), Bali, Indonesia, Oct. 20–24*. Bali.

Gao, B.-C., Montes, M. J., Davis, O. C., & Goetz, A. F. (2009). Atmospheric correction algorithms for hyperspectral remote sensing data of land and ocean. *Remote sensing of environment, 113*, s17-s24.

Goetz, A. F., Vane, G., Solomon, J. E., & Rock, B. N. (1985, June). Imaging Spectrometry for Earth Remote Sensing. *Science, 228*(4704), 1147–1153.

Green, A. A., & Craig, M. D. (1985, April). Analysis of Aircraft Spectrometer Data with Logarithmic Residuals. In G. Vane, & A. F. Goetz (Eds.), *Proceedings of the Airborne Imaging Spectrometer Data Analysis Workshop. JPL Publications 85–41*. Pasadena, CA: JPL Publications.

Griffin, M. K., & Burke, H.-h. K. (2003). Comparison of hyperspectral data for atmospheric. *Lincoln Laboratory Journal, 14*, 29–54.

Ham, J., Chen, Y., Crawford, M. M., & Ghosh, J. (2005). Investigation of the random forest framework for classification of hyperspectral data. *IEEE Transactions on Geoscience and Remote Sensing, 43*(3), 492–501.

Harsanyi, J., & Chang, C.-I. (1994). Hyperspectral image classification and imensionality reduction: an orthogonal subspace projection approach. *IEEE Transaction on Geosciernce and Remote Sensing, 32*(4), 770–785.

Hepner, G. F., Houshmand, B., Kulikov, I., & Bryant, N. (1998, August). Investigation of the integration of AVIRIS and IFSAR for urban analysis. *Photogrammetric Engineering & Remote Sensing, 64*(8), 813–820.

Hoffbeck, J. P., & Landgrebe, D. A. (1996). Classification of remote sensing having high spectral resolution images. *Remote Sensing of Environment, 57*, 119–126.

Kruse, F. A. (1994). *Imaging spectrometer data analysis – a tutorial*. Retrieved December 16, 2015, from http://w.hgimaging.com/PDF/Kruse_isssr94tut.pdf

Kruse, F. A. (2004). Comparison of ATREM, ACORN, AND FLAASH atmospheric corrections using low-altitude data of Boulder CO. *13th JPL Airborne Geoscience Workshop, 05–3*. Pasadena, CA: JPL Publications. Retrieved December 16, 2015, from www.hgimaging.com/PDF/Kruse-JPL2004_ATM_Compare.pdf

Kruse, F. A., Lefkoff, A. B., Boardman, J. W., Barloon, K. B., & Goetz, A. F. (1993). The Spectral Image Processing System (SIPS) – Interactive Visualization and Analysis of Imaging Spectrometer Data. *Remote Sensing of Environment, 44*, 145–163.

Lee, C. S., Yeom, J. M., Lee, H. L., Kim, J.-J., & Han, K.-S. (2015). Sensitivity analysis of 6S-based look-up table for surface reflectance retrieval. *Asia-pacific Journal of Atmospheric Sciences, 51*(1), 91–101.

Levoni, C., Cervino, M., Guzzi, R., & Torricella, F. (1997). Atmospheric aerosol optical properties: a database of radiative characteristics for different components and classes. *Applied Pptics, 36*(30), 8031–8041.

Lillesand, T. M., & Kiefer, R. W. (1999). Digital image processing. In T. K. Lillesand & W. Ralph (Eds.), *Remote Sensing and Image Interpretation, Fourth Edition* (p. 480). John Wiley & Sons Inc.

Melgani, F., & Bruzzone, L. (2004). Classification of hyperspectral remote sensing images with support vector machines. *IEEE Transactions on Geoscience and Remote Sensing, 42*(8), 1778–1790.

Moran, M. S., Jackson, R. D., Slater, P. N., & Teillet, P. M. (1992). Evaluation of simplified procedures for retrieval of land surface reflectance factors from satellite sensor output. *Remote Sensing of Environment, 41*(2–3), 169–184.

Myint, S. W., Gober, P., Brazel, A., Grossman-Clarke, S., & Weng, Q. (2011). Per-pixel vs. object-based classification of urban land cover extraction using high spatial resolution imagery. *Remote Sensing of Environment, 115*, 1145–1161.

NCAVEO. (2007, August 21). *Direct methods*. Retrieved March 18, 2016, from www.ncaveo.ac.uk/calibration/radiometry/in-flight/

Nidamanuri, R. R., & Zbell, B. (2011, March). Normalized Spectral Similarity Score (NS3) as an Efficient Spectral Library Searching Method for Hyperspectral Image Classification.

IEEE Journal of Selected Topics in Applied Earth Observations and Remote Sensing, 4(1), 226–240.

Penn, B. S. (2002). Applying band ratios to hyperspectral data. *2002 Denver Annual Meeting*. Denver. Retrieved January 6, 2016, from http://aviris.jpl.nasa.gov/proceedings/workshops/02_docs/2002_Penn_Band_Ratios_web.pdf

Ridd, M. K. (1995). Exploring V-I-S (vegetation–impervious surface–soil) model for urban ecosystem analysis through remote sensing: comparative anatomy of the cities. *International Journal of Remote Sensing, 16*(12), 2165–2185.

Ridd, M. K., Ritter, N. D., Bryant, N. A., & Green, R. O. (April 1992). Neural network classification of AVIRIS data in the urban ecosystem. *Association of American Geographers Annual Meeting*. San Diego, CA.

Roberts, D. A., & Yamaguchi, Y. (1986, May). Comparison of various techniques for calibration of AIS data. In G. Vane, & A. F. Goetz (Eds.), *Second Airborne Imaging Spectrometer Data Analysis Workshop. 86–35*, pp. 21–30. Pasadena, CA: JPL Publications.

Schläpfer, D., Boerner, A., & Schaepman, M. (1999). The potential of spectral resampling techniques for the simulation of APEX imagery based on AVIRIS data. *Summaries of the Eighth JPL Airborne Earth Science Workshop, 99*(17).

Schowengerdt, R. A. (1997). *Remote sensing, models, and methods for image processing* (2 ed.). Academic Press.

Secker, J., Staenz, K., Gauthier, R. P., & Budkewitschb, P. (2001). Vicarious calibration of airborne hyperspectral sensors in operational environments. *Remote Sensing of Environment, 76*(1), 81–92.

Senthil Kumar, A., Radhika, T., Keerthi, V., Jain, D. S., Dadhwal, V. K., & Kiran Kumar, A. S. (2013, December 29). Spectral deconvolution and non-overlap bands sampling for IMS-1 hyperspectral imager. *Journal of Indian Society of Remote Sensing*. doi:DOI 10.1007/s12524-013-0345-5

Shettle, E., & Fenn, R. (1976, April). Models of the atmospheric aerosols and their optical properties. *AGARD Conf Proc*.

Slater, P. N., Biggar, S. F., Holm, R. G., Jackson, R. D., Mao, Y., Moran, M. S., … Yuan, B. (1987). Reflectance–and radiance-based methods for the inflight absolute calibration of multispectral sensors. *Remote Sensing of Environment, 22*, 11–37.

Slater, P. N., Biggar, S. F., Thome, K. J., Gellman, D. L., & Spyak, P. R. (1996). Vicarious radiometric calibration of EOS sensors. *Journal of Atmospheric and Oceanic Technology, 13*, 349–359.

Song, C., Woodcock, C. E., Seto, K. C., Lenney, M. P., & Macomber, S. A. (2001). Classification and change detection using Landsat TM data: when and how to correct atmospheric effects? *Remote Sensing of Environment, 75*, 230–244.

Spectral Sciences Incorporated. (2016-1). MODTRAN. Retrieved June 24, 2018, from MODTRAN: http://modtran.spectral.com/modtran_index

Spectral Sciences Incorporated. (2016-2). About MODTRAN. Retrieved June 24, 2018, from MODTRAN: http://modtran.spectral.com/modtran_about

Stamnes, K., Tsay, S.-C., Wiscombe, W., & Jayaweera, K. (1988, June 15). Numerically stable algorithm for discrete-ordinate-method radiative transfer in multiple scattering and emitting layered media. *Applied Optics, 27*(12), 2502–2509.

Thome, K. J. (2004). In-flight intersensor radiometric calibration using vicarious approaches. In S. A. Morain, & A. M. Budge (Eds.), *Post-launch calibration of satellite sensors* (pp. 95–102). London: Taylor and Francis.

USGS. (2011-1). Earth Observing 1 (EO-1)-FAQs. Retrieved December 16, 2015, from www.usgs.gov/centers/eros/earth-observing-1-general-questions-and-answers

USGS. (2011-2). Earth Observing 1 (EO-1)-Sensors. Retrieved 10–19, 2022, from www.usgs.gov/centers/eros/eo-1-sensors

USGS. (2013-1, April 22). EO1 Hyperion Scene: EO1H1470472013112110KZ_PF1_01, for Target Path 147, and Target Row 47. Retrieved from http://earthexplorer.usgs.gov

Vane, G., & Goetz, A. F. (1985, April). Introduction to the Proceedings of the Airborne Imaging Spectrometer (AIS) Data Analysis Workshop. In G. Vane, & A. F. Goetz (Ed.), *Airborne Imaging Spectrometer Data Analysis. 85–41.* Pasadena, CA: JPL Publications.

Vermote, E., Tanré, D., Deuzé, J. L., Herman, M., Morcrette, J. J., & Kotchenova, S. Y. (2006). Second simulation of a satellite signal in the solar spectrum–vector (6SV), 6S user guide, Version 3. Retrieved January 15, 2023, from https://salsa.umd.edu/6spage.html

Vincent, R. K. (1972). An ERTS multispectral scanner experiment for mapping iron compounds. *Eighth International Symposium on Remote Sensing of Environment* (pp. 1239–1247). Ann Arbor, MI. Retrieved February 17, 2016, from http://ntrs.nasa.gov/archive/nasa/casi.ntrs.nasa.gov/19730001633.pdf

Vincent, R. K., Quin, X., McKay, R. L., Miner, J., Czajkowski, K., Savino, J., & Bridgeman, T. (2004). Phycocyanin detection from LANDSAT TM data for mapping cyanobacterial blooms in Lake Erie. *Remote Sensing of Environment, 89*, 381–392.

Weisstein, E. W. (2017). *"Gaussian Function", from MathWorld – A Wolfram Web Resource.* Retrieved August 28, 2017, from http://mathworld.wolfram.com/GaussianFunction.html

Weng, Q., Hu, X., & Lu, D. (2008, June). Extracting impervious surfaces from medium spatial resolution multispectral and hyperspectral imagery: a comparison. *International Journal of Remote Sensing, 29*(11), 3209–3232.

Wilson, R. T. (2012). Py6S: A Python interface to the 6S radiative transfer model. *Computers and Geosciences, 51*, 166–171.

Xu, B., & Gong, P. (2007, August). Land-use/land-cover classification with multispectral and hyperspectral EO-1 data. *Photogrammetric engineering & remote sensing, 73*(8), 955–965.

Yang, L., Xian, G., Klaver, J. M., & Deal, B. (2003, September). Urban land-cover Change detection through sub-Pixel imperviousness mapping using remotely sensed data. *Photogrammetric Engineering & Remote Sensing, 69*(9), 1003–1010.

Zhang, X., Zhang, B., Geng, X., Tong, Q., & Zheng, L. (2003). Automatic flat field algorithm for hyperspectral image calibration. *Proc. SPIE 5286, Third International Symposium on Multispectral Image Processing and Pattern Recognition, 636 (September 29, 2003).* Beijing.

Zhao, W., Tamura, M., & Takahashi, H. (2000). Atmospheric and spectral corrections for estimating surface albedo from satellite data using 6S code. *Remote Sensing of Environment, 76*, 202–212.

4 Spectral Resources

This chapter will discuss:
- Importance of spectral resources
- Spectral libraries of urban materials
  - International
  - Indian
  - Availability
- Analysis methods/how to use these resources
  - Basic statistics
  - Chromatic properties
  - Optimal bands
- Analysis of Tarang
- Spectral characteristics of urban materials
  - Other notable studies by Herold and Hepner
  - Indian urban materials

4.1 SIGNIFICANCE OF SPECTRAL RESOURCES

A comprehensive spectral library of various natural and manmade materials is a vital resource for hyperspectral data analysis. One of the biggest advantages of hyperspectral data is its ability to detect the target material without any additional information, ideally.[1] This is because the chemical structure of the target material completely determines the spectral signature of the material. Thus, in an ideal situation when the spectral resolution is sufficient for measuring the diagnostic features, the spectral signature is used for identifying the target material in an entirely unsupervised manner. The spectral library is such a resource which stores the useful reference spectra. In the absence of the spectral library or a good reference spectrum, unsupervised target detection is not feasible, and the unique advantage of hyperspectral data is wasted.

The intention of developing a spectral library is the following:

a. The spectral library is used for matching the target pixels with the reference spectra in the library to resolve the identity of the target material or its chemical composition. The target spectrum can come from any source such as a non-imaging field device, or an imaging device on any of the mobile platforms such as airborne or spaceborne. In the case of the imaging source, the target spectrum is the spectrum of a given pixel or all the pixels in the image.
b. The spectral library references provide input endmembers to the unmixing algorithms. Thus, instead of finding endmembers from the image, the algorithm

takes the endmembers from the library and finds fractions (abundances) of each pixel in the image (Deshpande et al., 2015).
   c. The analysis of the spectral library is helpful in assessing the discrimination potential of the target materials using spectral properties alone. This is particularly important as many of the urban materials show remarkably similar signatures. The concrete signatures are confused with the bright soil signatures as well.
      i. In-depth analysis of the spectral library is further useful in identifying optimal bands or optimal spectral range for better discrimination of the urban materials.
      ii. The analysis is helpful in determining the limits of the classification algorithm and in designing a classifier with the right characteristics (Fukunaga, 1990).
   d. The references in the spectral library are useful in creating synthetic mixture signatures. This is especially helpful in creating mixtures that are difficult to recreate in the field. For example, oil spills.
   e. The spectral library constitutes a knowledge base for the spectral properties of the common urban (or non-urban) materials. The detailed spectral signatures in the library are critical resources to design a specific payload. For example, if we want to design a payload that will detect the age of concrete.

Many scientists have devoted a substantial amount of their time to building spectral libraries of natural and manmade materials for this reason. The United States Geological Survey (USGS) Digital Spectral Library Version 7.0 (Kokaly et al., 2017) is one such comprehensive library. Another notable library includes Advanced Spaceborne Thermal Emission and Reflection Radiometer (ASTER) Version 2.0 (Baldridge et al., 2009) and the Jet Propulsion Laboratory (JPL) library. These libraries are not dedicated to urban materials. However, urban materials are important constituents of them. Most of the urban materials are covered under the "man-made" materials category.

The Indian spectral libraries developed recently are rice (Shwetank et al., 2011), The Himalayan plant species (Manjunath et al., 2014), and Indian soils (Das et al., 2015). Besides, none of these libraries is available in the public domain.

These are a few examples of some of the spectral libraries available. Each library is region specific and may not contain the signatures of local materials specific to the study area. Many diverse spectral resources are required for two main reasons:

   a. the public domain libraries might not have signatures of regional materials.
   b. even if the signatures are available, the signatures of local material may vary because of changes in the ingredients and proportions thereof. Especially, the urban area is dominated by composite materials and hence signatures may vary because of local manufacturing practices.

Though spectral libraries are regarded as one of the most important resources, there are limited efforts in building spectral libraries. Understandably so as it is a very effort-intensive exercise: it took fifteen person-years to develop the USGS spectral library. The equipment required for the exercises is costly too. Systematic documentation of the samples and their chemical constituents is required. Even if it is built, it may not be available in the public domain in many cases, and hence its usage is restricted. The ideal spectral library has the following essential characteristics:

a. It has wide spectral coverage such as 400 nm to 2500 nm or beyond, maybe 14000 nm.
b. It has extremely high spectral resolution and sampling so that the spectra can be resampled to match the resolution of the study spectra.
c. It covers most of the common urban materials and represents the regional variety as well.
d. Enough samples are measured for each type of material so that the spectral properties are representative of the class.
e. The library spectra are available online for matching, downloading etc. and the library spectra should be accessible to any device using OGC standards or any such similar standards. The easy online access would enable real-time or near-real applications.

4.2 SPECTRAL LIBRARIES

4.2.1 USGS Digital Spectral Library – Version 7.0

The USGS Digital Spectral Library Splib06a (Clark et al., 2007) was a result of fifteen person-years of dedicated efforts to creating a knowledge base which can be used for remote detection of similar materials. It is one of the most comprehensive and well-documented libraries available in the public domain. The current version of the library stands at 7.0 (Kokaly et al., 2017).

The USGS Digital Spectral Library Version 7.0 includes the spectra for a wide variety of materials. Apart from minerals, it covers trees, shrubs, grasses, flowers, leaves, lichens and other microorganisms, manmade materials such as plastics, roofing materials, processed wood paint, and so on. All these spectra are grouped under various categories: Minerals, Soils, Rocks, and Mixtures (except those with vegetation), Coatings, Liquids, Liquid Mixtures, Water, Other Volatiles, Frozen Volatiles, Artificial (Manmade) Including Manufactured Chemicals, Plants, Vegetation Communities, Mixtures with Vegetation, and Microorganisms.

The spectra in the library were measured by a variety of instruments such as a laboratory spectrometer, a field spectrometer, and airborne spectrometers. The spectral coverage is from 0.2 to 200 microns. Samples were purified so that the chemical composition can be related to the spectral features. This library has synthetic mixture signatures as well. For the majority of the categories, the measurements are made using field and laboratory spectrometers. Some of the measurements that were difficult for the field spectrometers such as tall trees were made by the airborne spectrometers.

The spectra were collected using four different spectrometers:

a. Beckman™ 5270 with a spectral range of 0.2 to 3 μm,
b. The standard, high-resolution (hi-res), and high-resolution Next Generation (hi-resNG) models of Analytical Spectral Devices (ASD) field portable spectrometers with a spectral range of 0.35 to 2.5 μm,

TABLE 4.1
Record Description of Asphalt GDS376 Blck_Road Old ASDFRa_AREF

TITLE: Asphalt GDS376 Blck_Road old DESCRIPT
DOCUMENTATION_FORMAT: Man_Made
SAMPLE_ID: GDS376
MATERIAL_TYPE: Asphalt
MATERIAL: Road Asphalt
FORMULA: Unknown
FORMULA_HTML: Unknown
COLLECTION_LOCALITY: Denver, Colorado, USA, ORIGINAL_DONOR: Gregg Swayze
CURRENT_SAMPLE_LOCATION: USGS Denver Spectroscopy Lab
ULTIMATE_SAMPLE_LOCATION: USGS Denver Spectroscopy Lab
SAMPLE_DESCRIPTION:
    Weathered surface of a chunk of black road asphalt. Gravel is composed of quartz and feldspars.
      Surface colour of asphalt is lighter than interior of sample.
END_SAMPLE_DESCRIPTION.
XRD_ANALYSIS:
END_XRD_ANALYSIS.
COMPOSITIONAL_ANALYSIS_TYPE: NONE # XRF, EPMA, ICP(Trace), WChem
COMPOSITION_TRACE: None
COMPOSITION_DISCUSSION:
END_COMPOSITION_DISCUSSION.
MICROSCOPIC_EXAMINATION:
END_MICROSCOPIC_EXAMINATION.
SPECTROSCOPIC_DISCUSSION:
    Spectrum taken on weathered surface. Spectrum curves upward to higher reflectance at longer
      wavelengths. There are weak absorptions at 1.41, 1.73, 2.20, 2.31, and 2.35 microns. The
      absorption at 2.20 microns is probably due to trace amounts of kaolinite or another clay or mica.
      Bands at 1.73, 2.31, and 2.35 are probably C-H related absorptions.
END_SPECTROSCOPIC_DISCUSSION.
SPECTRAL_PURITY: 1a2_3_4_ # 1= 0.2–3, 2= 1.5–6, 3= 6–25, 4= 20–150 microns

    c. Nicolet™ Fourier Transform Infra-Red (FTIR) interferometer spectrometers with spectral coverage from 1.12 to 216 µm, and
    d. The NASA Airborne Visible/Infra-Red Imaging Spectrometer AVIRIS, with spectral coverage from 0.37 to 2.5 µm.

Each record in the USGS library, for example, "Asphalt GDS376 Blck_Road old ASDFRa AREF" (Table 4.1 and Figure 4.1), provides useful information such as wavelength range, spectral resolution, and spectral purity flags: a = spectrally pure and verified by the other analysis, b=spectrally pure but other analysis shows contamination, c= presence of some weak features because of contamination, d = significant contamination, "?" = insufficient knowledge for any comment. The description of the record also includes other supporting information such as sample source, location, chemical formula if any and so on. Thus, the demonstrated sample is an asphalt sample, with the code GDS376; it is black in colour and it is an old road sample. The spectral measurements are recorded by an ASD full-range standard resolution spectrometer having a spectral

Spectral Resources

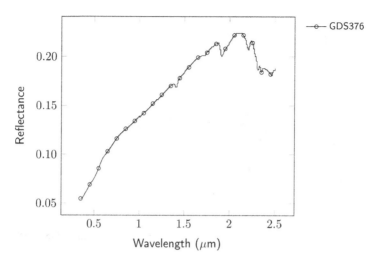

FIGURE 4.1 Spectral signature of the asphalt sample GDS376 from USGS library splib071.

range of 350 nm to 2500 nm. The sample is spectrally pure and is verified with the other analysis as well (ASDFRa). The codes for the other spectrometers are BECK, ASDHR, ASDNG, AVIRIS, NIC4 for Beckman 5270 (BECK), ASD FS hi-resolution (ASDHR), ASD FS4 hi-resolution Next Generation (ASDNG), AVIRIS-classic (AVIRIS), Nicolet 740, Magna-IR 760, and 6700 (NIC4) respectively. The last keyword AREF indicates that it is absolute reflectance. Other values are RREF for relative reflectance, RTGC for radiative transfer ground calibration, and TRAN for transmission spectra. The sample description for the code "Asphalt GDS376 Blck_Road old ASDFRa AREF" and its plot is in Table 4.1 and Figure 4.1 respectively.

(https://crustal.usgs.gov/speclab/data/HTMLmetadata/Asphalt_GDS376_Blck_R oad_old_ASDFRa_AREF.html)

The spectral data is available in the SPECtrum Processing Routines (SPECPR) data format. The reference spectra in the library are available for download using an interactive user interface at https://crustal.usgs.gov/speclab/QueryAll07a.php. The data is downloaded in the zip file format, and it contains all the records in separate files. For example, the reflectance values are available in an American Standard Code for Information Interchange (ASCII) text file, the plots are available in Graphics Interchange Format (GIF) images, sample meta data is available in Hypertext Markup Language (HTML) format and so on. For plotting the signature of a sample, the wavelength file and the signature files are used. For example, Figure 4.1 shows the plot for the sample "Asphalt_GDS376_Blck_Road_old_ASDFRa_AREF". The plot of wavelength against absolute reflectance was prepared by combining the channel file "splib07a_ Wavelengths_ASD_0.35–2.5_microns_2151_ch.txt" with the file "splib07a_Asphalt_ GDS376_Blck_Road_old_ASDFRa_AREF.txt" and then plotting. For additional details of the files and their format please refer to the report (Kokaly et al., 2017).

Figure 4.2 shows some of the urban construction materials in the USGS splib07a. These are categorized as "Artificial materials". These are roof and pavement materials. The code names of the signature files are: splib07a_Asphalt_ Shingle_GDS367_DkGry_ASDFRa_AREF, splib07a_Brick_GDS350_Dk_Red_

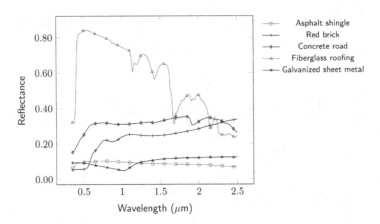

FIGURE 4.2 Spectral signatures of some of the urban materials from USGS splib07a.

Building_ASDFRa_AREF, splib07a_Concrete_GDS375_Lt_Gry_Road_ASDFRa_AREF, splib07a_Fiberglass_GDS335_Wh_Roofing_ASDFRa_AREF, splib07a_GalvanizedSheetMetal_GDS334_ASDFRa_AREF.

4.2.2 ECOSTRESS Spectral Library – Version 1.0

The ECOSTRESS Spectral Library – Version 1.0 (henceforth referred to as ECOSTRESS) is an updated version of Advance Spaceborne Thermal Emission and Reflection Radiometer (ASTER) V 2.0. The first version of ECOSTRESS was released on February 2, 2018. ECOSTRESS has 1100 new spectra of vegetation and non-photosynthetic vegetation. With the addition of these spectra to the original count of 2300 in ASTER 2.0, the number of the spectra of natural and manmade materials in ECOSTRESS is 3400. ASTER Spectral Library V 2.0 was published on Dec. 3, 2008.

ECOSTRESS is a compilation of spectral libraries from three diverse sources: the Johns Hopkins University (JHU) Spectral Library, the JPL Spectral Library, and the United States Geological Survey (USGS – Reston) Spectral Library. The library is developed as a part of the ASTER platform mission to support geological and other science studies using the ASTER terra platform by NASA. It provides a collection of 3400 spectral signatures of a variety of natural and manmade materials: rocks, minerals, soils, vegetation, snow and so on – including lunar soil and meteorites too! The library is available online and on order (Jet Propulsion Laboratory, 1999; Baldridge et al., 2009).

The spectra were measured using different spectrometers: Beckman 5270 (spectral range 0.2 to 3 μm, Analytical Spectral Devices (ASD) portable field spectrometer (spectral range 0.35 to 2.5 μm, Nicolet Fourier Transform Infra-Red (FTIR) Interferometer Spectrometer (spectral range 1.3 to 150 μm, and NASA Airborne Visible/Infra-Red Imaging Spectrometer AVIRIS (spectral range 0.4 to 2.5 μm). The spectrum was acquired by using at least one of the spectrometers as and when applicable. For example, AVIRIS spectra were used for samples such as tall trees.

The ECOSTRESS spectra are available for viewing and downloading using the interactive user interface at https://speclib.jpl.nasa.gov/library. The user can select the wavelength range and zoom in on that part for a detailed analysis of the signature. The

TABLE 4.2
Record Description of Sample No VH350–Bromus Diandrus

Name: Bromus diandrus
Type: vegetation
Class: grass
Genus: Bromus
Species: diandrus
Sample No.: VH350
Owner: UCSB
Wavelength Range: VSWIR
Origin: 34.5143; –119.798367; WGS84
Collection Date: 3/18/2015
Description: Samples were collected as part of the HyspIRI Airborne Campaign proposal titled: HyspIRI discrimination of plant species and functional types along a strong environmental temperature gradient. The same materials were processed in the Nicolet and then measured using the ASD.
Measurement: Bidirectional reflectance

Source: Reproduced from the ECOSTRESS Spectral Library through the courtesy of the Jet Propulsion Laboratory, California Institute of Technology, Pasadena, California. Copyright © 2017, California Institute of Technology. ALL RIGHTS RESERVED

---

x and y values of any point on the signature can be observed too by moving the mouse pointer over the graph. This is a useful feature for manually inspecting the diagnostic absorption if any (https://speclib.jpl.nasa.gov/library/ecoviewplot). The spectra are organized in the following categories: lunar, manmade, meteorites, minerals, non-photosynthetic vegetation, rock, soil, vegetation, and water.

The record description of the ECOSTRESS library spectrum is similar to the USGS library. The spectrum text file available for download encodes the description in the file name. For example, "vegetation.grass.bromus.diandrus.vswir.vh350.ucsb.asd.spectrum. txt" indicated the class, type, genus, species, wavelength range, sample code, owner, and the measuring instrument in that order. Table 4.2 shows the sample description.

Figure 4.3 shows some of the urban construction materials in the ECOSTRESS 1.0 spectral library. These are categorized as manmade materials. These are roof and pavement materials. The code names of the signature files are Asphalt-Shingle 0490UUUASP, Bare Red Brick 0413UUUBRK, Construction-Asphalt 0674UUUASP, Construction-Concrete 0598UUUCNC, Galvanized-Steel-Metal 0525UUUSTLa, Terra-Cotta tiles 0484UUUPOT, White fiberglass unspecified rubber 0834UUURBR.

4.2.3 JPL Library

The original JPL library was a collection of 160 minerals (Jet Propulsion Laboratory, 1990). The causes of the spectral features observed were not analysed. Most of the clay minerals were sourced from Source Clay Mineral Repository, University of Missouri, Columbia, Missouri and non-clay minerals from Ward's Natural Science Establishment, Rochester, New York; the Burnham Mineral Company (Burminco), Monrovia, California; or from an in-house collection.

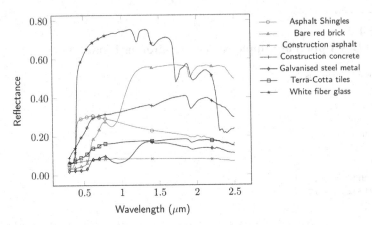

FIGURE 4.3 Spectral signatures of some of the urban materials from the ECOSTRESS 1.0 library.

Reproduced from the ECOSTRESS Spectral Library through the courtesy of the Jet Propulsion Laboratory, California Institute of Technology, Pasadena, California. Copyright © 2017, California Institute of Technology. ALL RIGHTS RESERVED.

135 out of the 160 minerals were studied for analysing the effects of particle size on the spectral signatures of the minerals. The samples were analysed for purity using X-ray Diffraction (XRD) analysis and additional chemical composition data was acquired by Camera CAMEBAX electron microprobe analysis. An elaborate chemical process was carried out to prepare the samples and powders of different grain sizes such as 125–500 µm, 45–125 µm, and less than 45 µm. The samples prepared after pulverization and grinding to the required size were poured into the small aluminium cylindrical containers of 3.2 cm in diameter and 0.5 cm in depth. The spectra in the range 0.4–2.5 µm were acquired using Beckman UV5240 Spectrophotometer with an interval of 0.001 µm from 0.4–0.8 µm and 0.004 µm from 0.8–2.5 µm. The spectra in the range of 2–15.4 µm were acquired using Nicolet 520 FT-IR spectrometer. The mineral classes include arsenates, borates, carbonates, elements, halides, hydroxides, oxides, phosphates, silicates, sulphates, sulphides, and tungstates.

4.2.4 Tarang – Spectral Library of Indian Urban Materials

The objective of the efforts (Deshpande et al., 2015; Deshpande et al., 2019) was to create an important spectral resource of urban materials and make it available for the researchers. An additional primary objective was to analyse the signatures systematically so that the specific insights can be leveraged for the development of indices for urban materials. In this section, we provide a detailed account of the efforts for the development of a spectral library of field signatures for Indian urban materials.

Two spectrometers were used for the collection of field signatures of the urban materials:

a. Portable spectrometer GER 1500 by Spectra Vista Corporation (SVC, 2016)
b. Analytical Spectral Devices (ASD) FieldSpec 3 (ASD, 2016) (Table 4.3).

The reason for choosing both spectrometers was to create a comprehensive collection of signatures of urban materials including urban vegetation. The SVC spectrometer is a portable device and provides practical advantages for field measurements over the ASD spectrometer though it lacks the wavelength coverage of the ASD spectrometer. A visible and near-infrared range is sufficient for studying vegetation signatures and hence the SVC spectrometer was appropriately used to record signatures of urban vegetation and some commercial crops cultivated on urban fringe areas which were difficult to record using the ASD spectrometer.

The SVC spectrometer was used in two modes 1) Handheld mode and 2) Tripod mode. The specification of each spectrometer is provided in Table 4.3. A makeshift platform using a wooden table was created to hold the samples. The location of the sample placement on the platform was carefully marked using the SVC target illuminator beam for the correct recording of the signature. Both spectrometers provide support for downloading the data to a personal computer or a laptop with the help of propriety software.

The locations for field study – in addition to individual in-situ field measurements – were carefully chosen. The SVC field measurements were recorded at a large playground ($\sim$30,000 m^2, 200 × 150 m). There were no high-rise buildings within the vicinity of the ground and the signature measurement location had no interference from the reflected light. Similarly, the bulk of the ASD measurements were taken at a large terrace of a multi-storied commercial building and had no interference from reflected light from other buildings or any other large objects such as trees. The target sample area, on average, was approximately four to five times of field of view of the respective instruments.

All the field measurements were taken within the two-hour window of local noon. The instrument or pistol grip of the instrument as applicable was held at $\sim$1 m distance away from the body to avoid the effect of scattering from clothing. The clothing was carefully chosen to avoid any interference because of the reflection of specific wavelengths from the clothes, if any. We used muted-coloured clothing for all the measurements. A direction perpendicular to azimuth (Goetz, 2012) was chosen for recording all the measurements. New, white-plate reference readings were taken at the beginning of every set of scans and on minor changes in the atmospheric

TABLE 4.3
Instrument Specifications for Tarang Measurements

	GER 1500 SVC	ASD Fieldspec 3
Wavelength range	350–1050 nm	350–2500 nm
Resolution (FWHM)	3 nm	3 nm (700 *nm*), 10 nm (2100)
Resampling interval	~1.4 (512 channels)	1.4 (350–1000) to 2 (1000–2500)
Field of View (FOV)	4°	25°
Integration time/Averaging	20/16	34/100

FIGURE 4.4 Field measurements.

conditions. All the field measurements were scheduled on cloud-free days with a clear atmosphere. The temperature and the humidity were also recorded after regular intervals (avg. temperature 37.2°C, avg. relative humidity of 60.0%) for the ASD measurements. Photographs (Figure 4.4) show the field measurement process for both spectrometers.

A set of 5 to 20 readings was taken for each sample or field location. The number of readings for each sample or field location was determined by the uniformity of the target surface. For example, for a colour-coated TataBluscope metal roof (Tata BlueScope Steel Ltd., 2016), five readings of one sample were considered sufficient, whereas twenty readings at various locations within the same sample were taken for a heavily corroded galvanized iron roof. This is over and above the multiple scans performed in a single reading by the instrument as per the settings in Table 4.3. The average of multiple readings for a sample was taken as a reference signature. Signatures recorded on the field were then downloaded to a personal computer (PC) using the data acquisition software provided by respective manufacturers. The signature files were converted into ASCI files for further processing.

Overall, ~1150 signatures of 105 unique samples were recorded using the portable spectrometer and ~200 signatures were recorded using the ASD spectrometer. The signatures of twenty-four unique samples of quite common and prominent urban materials were recorded by both the ASD and SVC spectrometers.

4.2.4.1 Sample Description and Specifications

Table 4.4 provides a description of the material samples and their specifications. Figure 4.5 shows photographs of the samples. The photographs were taken by a

TABLE 4.4
Description of Some of the Tarang Samples

Name/Trade name/(code)	Use	Description (Picture code[1])
Gravel (*Agr*)	Aggregates	Fine aggregates (P1), coarse aggregates (P2)
Rubble (*Bas*)	Soling	Basalt rock, construction grade bloc (P3)
Sand (*Snd*)	Aggregates	Fine sand light brown (P4)
Bitumen (*Bit*)	Pavement	Common bitumen used for roads (P5)
Bitumen concrete (*BCn*)	Pavement	Bitumen concrete pavement of helipad (P6)
Bitumen concrete	Pavement	Nominal mix bitumen concrete (P7)
Bitumen concrete	Pavement	Bitumen concrete surface coat, new road (P8)
Bitumen bound macadam/(*BBM*)	Pavement	BBM for the road of residential colony (P9)
Coloured Pavement blocks (*PB-Clr*[2])	Pavement	Coloured pavement blocks (P10-11)
Plain cement concrete (*PCC*)	Pavement	Plain cement concrete at a factory (12)
Rectangular block (*PB-Rt*)	Pavement	Rectangular common pavement block (13)
Reinforced Concrete (*CCn*)	Pavement	Road grade RCC (14)
Residential block (*PB-Pb*)	Pavement	Residential grade square pavement block (15)
Square block (*PB-Sq*)	Pavement	Square common pavement block (P16)
Zigzag block (*PB-Z*)	Pavement	Zig zag shape common pavement block (P17)

(*Continued*)

TABLE 4.4 (Continued)
Description of Some of the Tarang Samples

Name/Trade name/(code)	Use	Description (Picture code[1])
Asbestos/Swastik (*Asb-C*)	Roof	Common corrugated asbestos sheet (P18)
Asbestos/Swastik (*Asb-P*)	Roof	Plain asbestos sheet (P19)
RCC concrete	Roof	Pink cement coloured concrete Column (P20)
Reinforced Concrete/RCC	Roof	A roof concrete (P21)
Cement concrete (*CCn*)	Roof/Pavement	Nominal mix cement concrete cast (P22)
Clay tiles/Mangalore tiles (*CT-N*)	Roof	Common Mangalore tiles (big, small P23–24)
Clay tiles/ Mangalore tiles (*CT-O*)	Roof	Common Mangalore tiles used/weathered (P25)
Fibre (*Fbr*)	Roof	Square corrugated grey fibre sheet (P26)
Corrugated GI sheet (*GI-O-C*)	Roof	Corroded old GI sheet (P27)
Plain GI sheet (*GI-O-P*)	Roof	Heavily corroded old plain GI sheet (P28)
Plain GI sheet-new (*GI-N*)	Roof	New plain GI sheet (P29)
Metal roofs (*MR-R/G/B/W*)	Roof	Colour-coated industrial metal roofs (P30–33)
Brown soil (*Sol*)	Foundation	Brown inorganic silty clay soil (P34)
Red soil	Foundation	Field sample of red soil (P35)
Brown-grey soil	Parking lot	Silt inorganic from a Playground (P36)
Dark brown soil	Playground	a little organic content from a playground (P37)
Taiwan Lawn Grass (*Lwn-T*)	Lawn	Dense Taiwan lawn at a club house (P38)
American Lawn Grass (*Lwn-A*)	Lawn	Dense American lawn at the club house (P39)
Lawn Grass (2 samples) (*Lwn-Ws*)	Lawn	Playground lawn, water stressed (P40–41)
Acacia (Acacia arabica) (*Bbl*)	Arboriculture	Babhal trees near canal (P42)
Banyan (Ficus bengalensis) (*Bny*))	Arboriculture	Banyan tree (P43)
Gulmohar (Delonix regia) (*Gul*)	Arboriculture	Young Gulmohar tree (P44)
Almond (Terminalia catappa) (*Bdm*)	Arboriculture	Indian-Badam tree at an institute (P45)
Mango (Mangifera indica) (*Mng*)	Arboriculture	Common mango tree (P46)
Mast tree (Polyalthia longifolia) (*Ask*)	Arboriculture	Ashok tree at an institute (P47)
Peepal (Ficus religiosa) (*Ppl*)	Arboriculture	Young Peepal tree, Silver Oak, Pune (P48)
Raintree (Albizia julibrissin) (*Ssr*)	Arboriculture	Mature leaves of Shirish (P49)
Raintree (Albizia saman) (*Ssr*)	Arboriculture	Young fresh leaves of Shirish (P50)

[1] The picture code refers to the sequence number of the corresponding picture tile in Figure 4.5; [2] Depending upon the goal, all colour-coated materials are grouped together for analysis and referred to as ClC.

handheld digital camera. They were captured at the same time of spectrometer measurements. Additional details about samples are provided below. The Indian standard codes as per the Bureau of Indian Standards are recorded, referred to as IS code. The code specifies the construction materials and the ingredients of them. This is helpful to compare the signatures with materials with similar material composition.

Steel roofs (IS 277 (2003), November 2003): These are standard square corrugated steel roofs in the market from brands like Tata and Essar (Tata BlueScope Steel Ltd., 2016; Essar Steel, 2013). The thickness varies from 0.45 mm to 0.65 mm. Often the colour-coated versions of steel sheets are used. The common colours are brick red, blue, green, and white. The other grades include galvanized sheets from Uttam (Uttam Galva Steels Ltd., 2016) without any colour coating. The samples were acquired from local vendors.

Asbestos roof sheets (IS 459 (1992), March 1992): Two types of asbestos sheets are used – predominantly for low-cost industrial and residential construction – plain and corrugated. The samples from Sahyadri Industries (Sahyadri Industries Ltd., 2015) were selected for the measurements. The samples were acquired from local vendors.

Fibre Reinforced Plastic (FRP) sheets: FRP is one of the important materials for roofs as its usage in residential and industrial construction has increased in the recent past. The usage might vary from small-scale roofs for balconies and temporary shades to large roofs for terraced buildings and farmhouses. The large varieties of FRP-products are available from local and national level manufacturers. We chose the FRP sheets from one of the leading manufacturers, Vijay Agency Pvt. Ltd. (Vijay Agency Pvt. Ltd., 2015).

FIGURE 4.5 Photographs of some of the Tarang samples.

FIGURE 4.5 (Continued)

Spectral Resources

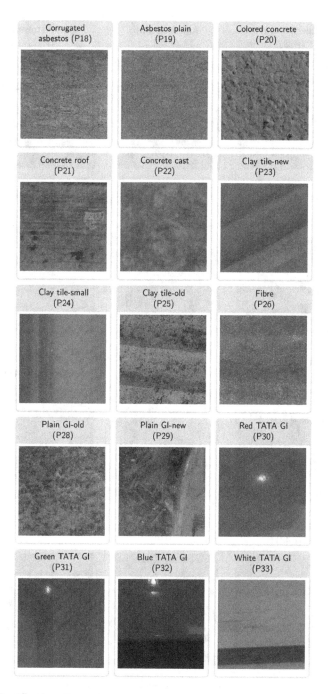

FIGURE 4.5 (Continued)

FIGURE 4.5 (Continued)

Concrete (IS 10262 (2009), July 2009): Cement concrete block with nominal mix (1:1.5:3 by volume) conforming to M30 grade was cast and cured for 21 days. Bitumen concrete (1:2:2 by weight) was also cast.

Pavement blocks (IS 15658 (2006), June 2006; IS 10262 (2009), July 2009): Pavement blocks by various local traders with 1:4 mix and confirming to M30 to M50 grade. These blocks are manufactured in many sizes and shapes. For example, the zigzag-shaped pavement block is quite common for all roadworks, whereas square and rectangular blocks are common for residential pavements (Ronak Tiles Pvt. Ltd., 2014). The samples were collected from local vendors.

Clay tiles (IS 654 (1992), March 1992): These are common clay tiles and go by the trade name Mangalore Tiles.

Aggregates (IS 10262 (2009), July 2009): Samples comprise common aggregates of varied sizes made by crushing basalt rock. Soil names follow IS code IS-1498–1970 (IS 1498 (1970), March 1972).

4.3 ANALYSIS METHODS

The main objective of the spectral library is to provide reference spectra for the spectral matching methods. To enhance this process, it is necessary to understand the spectral characteristics of the urban targets for their discrimination. The detailed spectral analysis of the spectra in the reference library is necessary to design the target detection and discrimination methods to be deployed later. The analysis is motivated by the following questions: Which spectral range VNIR or SWIR is more effective in discriminating urban target materials? Which bands are optimal for discriminating the target materials? Are there any diagnostic absorption or reflections – is it possible to create a material-specific index for the detection of the target? What is the best degree of discrimination possible for various urban targets? The following sections describe common methods used for getting the answers to these questions.

4.3.1 Basic Statistics

Simple statistics such as mean (μ) over reflectance values of all the bands for a spectrum and standard deviation (σ) are particularly good indicators of the spectral properties of a given material. For example, bitumen concrete has extremely low reflectance values over VNIR-SWIR bands and is relatively flat, thus, it has very low μ and σ. If we compare μ and σ of the spectrum over VNIR and VNIR-SWIR range of the spectrum for a given material, then the variation in μ and σ reveals the spectral range that is significant to the spectrum. For example, the higher σ values of a VNIR-SWIR spectrum of the material as compared to its VNIR spectrum indicate that the SWIR range is more significant to the spectrum. The comparison of μ and σ of the VNIR and VNIR-SWIR spectra from a single spectrometer or different spectrometer is possible as well.

Here is the simple procedure to perform this basic analysis.

1. Divide the spectrum into two spectra; one VNIR and one compete that is VNIR-SWIR.
2. Calculate the mean and standard deviation of the two spectra generated above.
3. Plot a point representing the material in the μ-σ space.
4. The materials will form natural clusters according to their spectral properties that is if the VNIR region is more important or SWIR.
5. The result of the plot is two points for the same material in the same space. The distance between these two points is distinct enough to enable discrimination. The distance can be further converted into a movement vector, that is, in which direction the point has moved with respect to its VNIR spectrum. The movement vectors are visually more intuitive, and the chart can easily explain the spectral characteristics of the target materials (Figure 4.6 and Figure 4.7).

We showcase the analysis done using the method for the spectral library Tarang. For convenience, we used the SVC and ASD spectra directly to represent VNIR and VNIR-SWIR spectra of the same material. Figure 4.6 and Figure 4.7 show these plots. As can be seen, most of the urban materials are grouped into three main categories based on the mean (μ) and standard deviation (σ):

a. The materials that move South-West (SW) with respect to the VNIR-SWIR spectrum co-ordinates in μ and σ (μ-σ) space (group A materials).
b. The materials that move South-East (SE) in μ-σ space with respect to the VNIR-SWIR spectrum (group B materials).

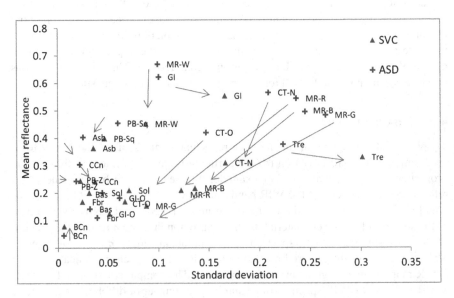

FIGURE 4.6 μ and σ of reflectance values for a given spectrum in the range VINIR and VNIR-SWIR.

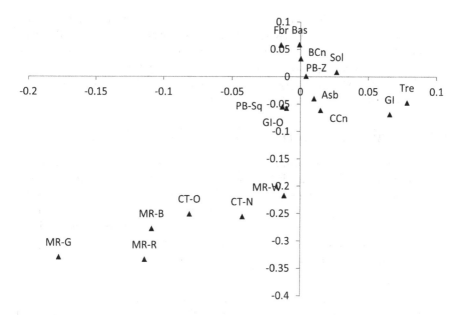

FIGURE 4.7 The deviation between the VNIR and VNIR-SWIR spectra in μ-σ space.

c. The materials that shift a little along the x or y-axis in μ-σ space, or with minor angles with x- or y-axis (group C materials).
d. The movement vector of a spectrum in the μ-σ space is distinct enough to enable discrimination/identification (Figure 4.6).

All the colour-coated materials (ClrC), new clay tiles (CT-N) and so on (group A) have limited variation in the VNIR spectrum. The VNIR range for these materials exhibits below-average (for a given spectrum of this group) reflectance values and they correspond to the respective colouring agent whereas the rest of the energy in the VNIR range is absorbed. In addition, the SWIR region exhibits an above-average reflectance for group A materials. This results in reduced group μ reflectance values for the VNIR spectrum with respect to the corresponding VNIR-SWIR spectrum. Further, the spectral signatures show more variation in the SWIR region and hence the σ of the VNIR spectrum is reduced with respect to the corresponding VNIR-SWIR spectrum.

The cement composites such as cement concrete (CCn), Asbestos (Asb), and pavement blocks (PB) (group B), on the other hand, shift SE as the μ of the VNIR spectrum is decreased and σ of the VNIR spectrum is increased for the corresponding VNIR-SWIR spectrum. The shift in position in the μ-σ space for the group b materials is not as prominent as the group A materials. Spectra of most of the cement composites are relatively flat over VNIR and SWIR range and thus exhibit minor differences in the μ and σ of the corresponding VNIR-SWIR and VNIR co-ordinates in the μ-σ space. A minor variation shown by the group B materials is in the visible range.

The group *C* materials such as bitumen composites are characterized by a nearly flat spectrum with extremely low reflectance values over VNIR and SWIR range. Only mean shift is observed in the μ-σ space for the corresponding VNIR-SWIR and VNIR spectra. The soil spectrum is shifted toward the right-hand side on the x-axis (with a slight positive angle) as the σ is increased for the VNIR spectrum as compared with the corresponding VNIR-SWIR spectrum (Figure 4.7).

4.3.2 THE BHATTACHARYYA DISTANCE ANALYSIS FOR URBAN CLASSES

One of the most important questions before deploying any classifier for hyperspectral data is to analyse the separability of the target materials using spectral properties. This is essential to estimate the bounds on classification error for the best classifier. The error indicates, in general, the natural limits of the best classifier in separating the target materials. Bhattacharyya distance is one such important measure that indicates class separability. It provides the upper bound on the Bayes error rate for the given data (Fukunaga, 1990). The B-distance B is given by:

$$B = \frac{1}{8}(m_i - m_j)^t \left\{ \frac{\Sigma_i + \Sigma_j}{2} \right\}^{-1} (m_i - m_j) + \frac{1}{2} \ln \frac{\{|(\Sigma_i + \Sigma_j)/2|\}}{\sqrt{|\Sigma_i||\Sigma_j|}} \quad \text{Equation 4.1}$$

where, m_i, m_j = Mean vector for class i and j, Σ_i, Σ_j = Covariance matrices of class i and j. Generally, the B-distance using all the wavelength features is used for the initial assessment.

The first term of the B-distance indicates the class separability because of the mean-difference, and the second term indicates the separability because of covariance-difference (Fukunaga, 1990). Thus, these two components help in determining the nature of the classifier as well. If the B-distance is mainly because of mean-difference, then the classifier using the mean vector would perform better. In case, the covariance-difference is the major component, then the classifier using the covariance structure of the data would perform better.

Furthermore, the covariance matrix in Equation 4.1 is not well conditioned. This is because of the high dimensionality of hyperspectral signatures and relatively small number of samples as compared to the dimensions. The covariance matrix in such conditions may be singular, that is, it is not invertible (Ledoit & Wolf, 2004; Manolakis, Lockwood, Cooley, & Jacobson, 2009). In such cases, the inverse of the covariance matrix is to be estimated using the existing singular covariance matrix. The Leodoit-Wolf estimator for covariances (Σ_i, and Σ_j) (Ledoit & Wolf, 2004) provides a satisfactory solution to this problem. Estimating the covariance using the Leodoit-Wolf estimator is a better solution as compared to calculating the pseudo inverse of the covariance matrix. The Leodoit-Wolf estimator ensures more accurate results than each individual estimator, namely the sample covariance matrix and the structured estimator, by taking a weighted average. The simplest but inaccurate alternative to calculating the B-distance in the low sample scenario is to consider only the mean difference or assume some constant covariance-difference. The latter is used by commonly used software MultiSpec for feature selection of hyperspectral data – the distance of 0.872 to be specific (Biehl & Landgrebe, 2002). The other alternative is to use a Moore-Penrose pseudoinverse.

4.3.3 Most Significant Wavelength Span: Fine Granularity

The most significant span or *Optimalspan* is defined as a wavelength range that is most important for discriminating a given class pair. Intuition is; if wavelengths within the given span produce maximum separation of the classes, then the span is an *Optimalspan* (objectives b and c). The first step to do this is to find n wavelengths that maximize the B-distance between a given class pair. If a particular span provides wavelengths that maximize the B-distance most of the time, then we can assume that the span is *Optimalspan*. This can be decided by conducting a considerable number of B-distance experiments and counting the number of times a span of say ~100 nm resulted in maximum B-distance for the class pair. The topmost bin provides *Optimalspan* and counts for each bin reveals a complete distribution of significant wavelengths over a complete range (350–2500 *nm*) *Spandistribution*.

Finding n that maximizes the B-distance between two classes is not possible analytically (Fukunaga, 1990). Hence, we need to explore the space by choosing some n bands randomly and repeating the experiment i number of times, where i is sufficiently large. A sufficiently large i value assures that the most discriminating features are selected for the calculation of B-distance.

The algorithm to calculate Optimalspan and Spandistribution sorts the results of 2000 experiments and then counts c to further generate the results (Algorithm 4.1). For example, for k=5% and n=3, there would be 300 (k × n × 2000) wavelengths/bands to calculate c in each bin within a 350–2500 nm range. The c values were normalized further to indicate percentage and are used to calculate Optimalspan and Spandistribution as defined. Optimalspan metric can be visualized as an array; each cell in this array represents a value indicating the starting wavelength of a bin within the range of 350–2500 nm for a class pair. For example, for a c1 and c2 class pair, a cell value of 350 indicates that a span of 350–450 nm is the most important span for separating c1 and c2. Additionally, Optimalspans for different n values indicate changes in important span (if any) with an increase in the number of features (n). For example, if for a given class pair for n=3 optimal span is 350 and for n=15 optimal span is 950 nm; this indicates with more features, the optimal span shifted from a visible range to an infrared range (Algorithm 4.1).

n wavelengths/bands are selected using any random number generator function. For example, if the number of bands is 1000 then a random number between 1 to 1000 is selected. Different n values such as 2, 5, 10, 15, and 20 are useful to understand the optimal number of 3 n bands for the best possible discrimination. Once the wavelengths are selected, the reflectance values from these bands are used for calculating the B-distance between the class pairs in consideration. One random selection of any n wavelengths constitutes a single experiment. Each experiment provides the B-distance for a given n features for a given class pair. Any number of iterations can be performed; for example, 2000 iterations (i) of an experiment were performed for each n (Deshpande et al., 2019).

ALGORITHM 4.1
Algorithm for Selecting Most Significant Wavelength Span

```
Input: i = 2000, n = 3,5,7,10,15, k= 5%, lc = list of class labels
Output: Optimalspan array/list, Spandistribution

OptimalBands(i, n, k, lc)
{
        lcp = Create class pairs from a list of class labels
        FOR every i starting from 1
        Choose n wavelengths (bands) randomly from 350-2500 nm range
            FOR every class pair in lcp
                Calculate the B-distance
        Sort the results in descending order (Largest values first)
        FOR 100 nm bins within 350-2500 nm (starting at 350, 450 and so
on) range
                Count number of times a particular wavelength appears
                in the top k results
        RETURN top bin (Optimalspan); RETURN complete distribution
        (Spandistribution)
        }
```

Spandistribution is depicted in a stacked bar chart. Each stacked bar for a given class pair indicates how c is distributed over the complete range. The height of the sub-bar indicates score c for a particular wavelength range/span in terms of the number of times they appear in top k experiments in 2000 iterations for each n. Each sub-column in a stacked bar is arranged in sequence starting from 350 and increased by step size 100 nm. They are shown in half tones with increasingly darker tones for the larger wavelengths. Thus, a predominant white bar indicates that the difference in each class pair is because of the visible range of electromagnetic spectrum and the darker shades indicate the dominance of an SWIR range (Algorithm 4.1). At present, there is no comparable algorithm to calculate the significant span for hyperspectral data. This new algorithm is especially useful in identifying features for the materials without any diagnostic features (Algorithm 4.1).

4.3.4 MOST SIGNIFICANT SPAN: COARSE GRANULARITY (VNIR VS SWIR)

The mean μ and standard deviation σ of each spectrum over the VNIR and VNIR-SWIR range are useful to assess the utility of the VNIR and SWIR range. The variation in μ and σ values in the VNIR range as compared to the VNIR-SWIR range reveals whether the SWIR spectrum provides additional information, if any. For example, the higher σ values of a VNIR-SWIR spectrum of a given material than a corresponding SVC spectrum indicate that variations in the reflectance values over the VNIR range are less as compared with a VNIR-SWIR spectrum. Hence, the SWIR region is more useful to identify the material.

4.3.5 CHROMATIC PROPERTIES OF THE URBAN MATERIALS

Extending colour vision principles to display hyperspectral data has been investigated by researchers in the recent past. However, the colour obtained by this approach (artificial colour) is not used comprehensively for discriminating target materials. Especially, some of the interesting artificial colour properties could be used effectively in the analysis and discrimination of urban targets. Most of the composite urban materials for pavements and roofs are cement composites and display shades of grey in a visible range. The visible observation is true for the entire spectrum too as the spectrum over 350 nm to 2500 nm is relatively flat for most urban materials. Thus, the "colour" for a hyperspectral of an urban material should appear grey even if the VNIR-SWIR range is considered.

Alternatively, the spectrum can be coloured by sampling the spectrum with stretched CIE 1964 colour-matching functions. Analysis shows, indeed, that the urban materials appear in shades of grey after colouring them by stretched CIE matching functions. Further, it shows the potential of artificial colour to discriminate subclasses of impervious surfaces. The procedure for colouring the spectrum or creating an artificial colour for the material is given in the sections below.

4.3.5.1 Colour Conversion

Sampled values of CIE 1964 Standard observer colour matching functions at every 5 nm are available within 380–780 nm range (lEclariage, 2015). The matching functions are stretched according to the spectral range of the sensor providing the spectrum. For example, EON-1-Hyperion spectral coverage is from ~350–2500 nm with a spectral resolution of ~10 nm, and the ADS spectral range is 350–2500 nm with ~1 nm spectral resolution. The stretch on the standard colour-matching function is performed so that new colour-matching functions (spectral sensitivity curves) are spread over the desired 350–2500 nm range in both cases, that is EO-1 and ASD. It could be a simple linear stretch by equating starting values to 380 nm, and end values (780 nm) to a new end of 2380 nm. EO-1-Hyperion spectra and ADS field spectra are also sampled accordingly. For possible accuracy improvement, CIE function values after every one nm could have been interpolated from the values available at five nm. Further, any other sensitivity curves could have been used. CIE colour-matching functions are useful as they are de facto standards for chromatic discrimination studies for natural colours. Once the colour matching functions are available over new range, XYZ tristimulus values are calculated by following integrations:

$$X = \frac{1}{N}\int_\lambda \bar{x}(\lambda)S(\lambda)I(\lambda)P(\lambda)d\lambda \qquad \text{Equation 4.2}$$

$$Y = \frac{1}{N}\int_\lambda \bar{y}(\lambda)S(\lambda)I(\lambda)P(\lambda)d\lambda \qquad \text{Equation 4.3}$$

$$Z = \frac{1}{N}\int_\lambda \bar{z}(\lambda)S(\lambda)I(\lambda)P(\lambda)d\lambda \qquad \text{Equation 4.4}$$

$$N = \int_{\lambda} \overline{y}(\lambda) I(\lambda) d\lambda \qquad \text{Equation 4.5}$$

Where $\overline{x}$, $\overline{y}$, $\overline{z}$ are CIE 1964 Standard observer colour mapping functions stretched for the range 380–2380 nm range, $I(\lambda)$ is the standard illuminant function (D65 for present work), and S (λ) is a sample reflectance spectrum. In actual practice, the integrations are replaced by summations over a wavelength range of 380–2380 nm range. XYZ values are further converted to AdobeRGB-1998 (Lindbloom, 2015) values for display purposes:

$$\begin{bmatrix} R \\ G \\ B \end{bmatrix} = \begin{bmatrix} 2.0413690 & -0.5649464 & -0.3446944 \\ -0.9692660 & 1.8760108 & 0.04155560 \\ 0.0134474 & -0.1183897 & 1.0154096 \end{bmatrix} \begin{bmatrix} X \\ Y \\ Z \end{bmatrix} \qquad \text{Equation 4.6}$$

Similarly, XYZ values are converted to CIE Lab co-ordinates (Lindbloom, 2015). CIE Lab colour space is based on the opponent theory of human perception. According to the CIE Lab model, human visual sensation is divided into three parts: light intensity, colour sensation based on Red-Green, and Blue-Yellow opponents (represented by a and b axis). Positive a value indicates redness and positive b values indicate yellowness of the colour.

4.4 SPECTRAL ANALYSIS

4.4.1 Spectral Properties of Various Urban Materials

Urban places often show a remarkably high degree of variation in the land covers; developing countries show even more variety. Large variations of socio-economic activities in the cities of developing countries result in large variations in land use and land covers. Manmade materials such as cement and bitumen composites, colour-coated materials, metal roofs, and so on, dominate the urban area. Bare soil cover such as playgrounds and open areas covered with different degrees and vegetation such as parks and golf courses, and trees used for road arboriculture constitute natural materials. Cultivation on the fringes of the city completes the plethora of materials present in and around urban areas in developing countries. We discuss below, the spectral properties of some of the important materials commonly occurring in Indian cities.

Cement composites: Cement composites (CC) are one of the most important classes of urban materials. The cement composites such as cast-in-situ road-grade concrete, pre-cast pavement blocks, plain cement concrete pavements, asbestos roofs, and concrete roofs of residential areas dominate urban surfaces. The spectral properties of all these materials show similar characteristics. The cement composites do not show any diagnostic absorption in the VNIR-SWIR range. Their spectra are nearly flat with a little concavity in the visible range. The average brightness of the cement composites is higher than the soil brightness though the signature of the soil is similar to the cement composite signatures. Discrimination of soil, especially bare soil, and cement composites spectrally is a challenging problem. As can be seen from Figure 4.8 to Figure 4.13, the spectral shape of soil and cement composites differ

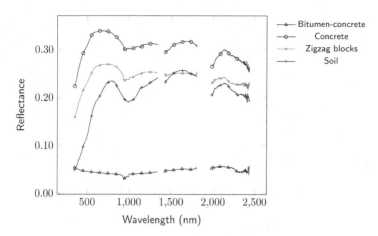

FIGURE 4.8 Spectral signatures of the pavement samples in Tarang.

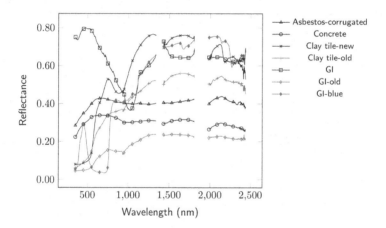

FIGURE 4.9 Spectral signatures of the roof materials in Tarang.

very slightly from each other – the difference is noticeable only in the visible range. The slope of the soil signature in the blue range is slightly higher than the cement composites. This is the only possible spectral feature that can be used for spectral discrimination (Figure 4.8 to Figure 4.11).

Bitumen Composites: Bitumen composites (BC) have the lowest average brightness of all urban materials. The signature is very flat without any prominent diagnostic absorption and BCs can easily be identified by visual inspection of the spectral signature. The addition of aggregates with different proportions does not make much difference to the shape of the spectral signature and low average brightness. The old and new bituminous pavements and the bitumen-bound macadam soling show very similar signatures (Figure 4.8 to Figure 4.11).

Metal roofs and clay tile roofs: Metal roofs (MR) show a large degree of variation in their surface coatings and hence the variations in the signatures. Overall, high average brightness values in the SWIR region are exhibited by all the metal roofs irrespective of their surface coating (Figure 4.9). The majority of the metal roofs

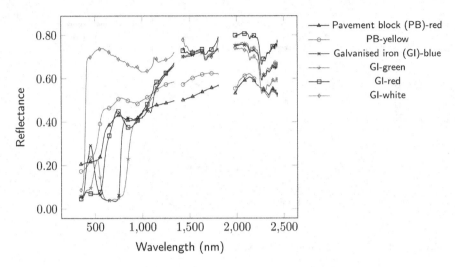

FIGURE 4.10 Spectral signatures of the colour-coated materials in Tarang.

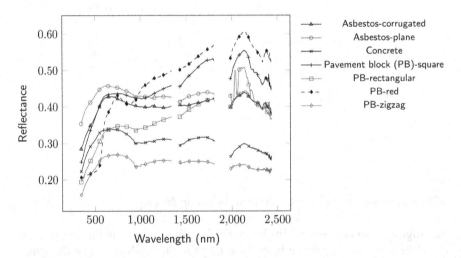

FIGURE 4.11 Spectral signatures of the cement composites in Tarang.

are colour-coated, and they show a diagnostic reflection respective to their colour coating. Galvanized iron (GI) sheets have extremely high reflectance throughout the spectral range but ~1000 nm range. The GI sheets and the colour-coated metal roofs such as blue and white show extremely high average brightness and can be easily detected in hyperspectral images as well (Deshpande et al., 2017).

The corroded GI shows signatures similar to soil signatures although corroded GI shows slightly higher average brightness, especially in the SWIR region. Thus, the characteristic high reflection of the GI sheet in the blue region is diminished with a degree of weathering and hence can be used to detect ageing of the GI roofs. New clay tiles show steep slopes in the visible range and high average brightness in the

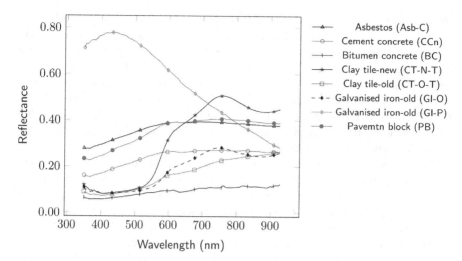

FIGURE 4.12 Spectral signatures of the urban materials in Tarang, recorded using an SVC spectrometer.

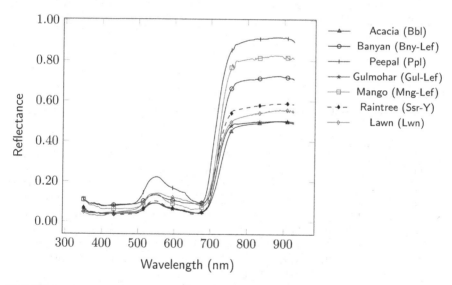

FIGURE 4.13 Spectral signatures of the urban trees in Tarang.

SWIR and are easy to discriminate. On the other hand, old tiles show signatures similar to soil and are difficult to discriminate from soil classes.

4.4.2 THE BHATTACHARYYA DISTANCE ANALYSIS FOR URBAN CLASSES

4.4.3 B-DISTANCES FOR BROADER URBAN CLASSES

The materials showing low or extremely high reflectance over all the bands (VNIR-SWIR), without any dominant absorption or reflection at a specific wavelength, show above-average separability. The spectral signature curve of these materials is flat relatively. Most of these materials belong to groups b and c in the μ-σ space (GI and bitumen composites, new clay tiles, and other colour-coated metal roofs). The GI, and bitumen composites are the brightest and dullest (darkest) materials respectively among all the urban materials. The distance between GI and bitumen composites is predominantly because of mean differences (Table 4.5; Figure 4.8, Figure 4.9). The new clay tiles also have high reflectance over all the NIR-SWIR range and very low reflectance in the blue range. Soil and trees are well separable from other classes and are marked with higher B-distances.

Cement composites, on the other hand, show a low to extremely short distance between some of the class pairs. The trees are well separable from cement composites. The most troublesome confusion for cement composites occurs with bare soil. The other classes where the B-distance is relatively low are thematic classes such as metal roofs (the new GI and the old GI grouped together), colour-coated/colour-coated metals (all the colours are considered in one group). The material class confusion, if any, is hard to overcome rather than the thematic class. As for the present study, the thematic class contains signatures of different materials; the mean and covariance of such a group do not represent a single material. Cement composites and aggregates are close to each other with respect to the B-distance (Table 4.5).

TABLE 4.5
Absolute Bhattacharyya Distances for Broader Urban Classes in Tarang

	1	2	3	4	5	6	7	8	9	10	11	12	13	14	15
1	0	13	5	32	1529	40	61	202	191	85	1080	47	5	67	623
2	13	0	170	103	2.6E4	198	192	3203	585	469	3.4E5	147	12	953	2989
3	5	170	0	56	2453	75	16	62	182	140	1221	58	7	79	696
4	32	103	56	0	0	0	33	217	93	163	294	26	11	104	407
5	1529	2.6E4	2453	0	0	477	194	4706	261	5730	1.9E4	1440	94	7223	3493
6	40	198	75	0	477	0	38	403	118	265	600	41	12	203	570
7	61	192	16	33	194	38	0	0	0	132	154	67	0	77	328
8	202	3203	62	217	4706	403	0	0	207	2265	4158	627	58	675	1754
9	191	585	182	93	261	118	0	207	0	290	205	159	0	245	472
10	85	469	140	163	5730	265	132	2265	290	0	1.0E5	140	16	408	1808
11	1080	3.4E5	1221	294	1.9E4	600	154	4158	205	1.0E5	0	2090	3	1.2E4	9835
12	47	147	58	26	1440	41	67	627	159	140	2090	0	0	176	1032
13	5	12	7	11	94	12	0	58	0	16	3	0	0	11	207
14	67	953	79	104	7223	203	77	675	245	408	1.2E4	176	11	0	1685
15	623	2989	696	407	3493	570	328	1754	472	1808	9835	1032	207	1685	0

Notes:
1. Agr, 2. BC, 3. CC, 4. ClC, 5. CT-N, 6. CT-O, 7. ClrC, 8. PB-Clr, 9. MT-Clr, 10. Fbr, 11. GI-N, 12. GI-O, 13. MR, 14. Sol, 15. Tre, Grey cells show lower B-distances indicating poor class separability.

4.4.4 BHATTACHARYYA DISTANCES FOR CEMENT COMPOSITES AND METAL ROOFS

Cement composites, overall, show extremely low intra-class separability; the B-distances between class pairs of various cement composites are low to very low (Table 4.6). The primary results indicate that the B-distances are moderately affected by the proportions of cement content or aggregate content/size of aggregates. For example, the B-distance between cement composites with a relatively high proportion of cement is extremely low, whereas the distance increases with proportions aggregates (Table 4.6). This further requires elaborate investigation to study the correlation between cement proportion in cement composites and B-distance. The distance between asbestos and cement concrete is the highest among all the cement composite pairs. The cement concrete thus is well separated from other cement composites in the Bhattacharya distance space. The pavement blocks and asbestos sheets show relatively lower B-distances and are difficult to separate from each other by a statistical classifier (Table 4.6).

Metal roofs, on the other hand, show very high B-distances between intra-class pairs (Table 4.7). As the metal roofs wear surface coatings of distinct colours, the class pairs are marked by large B-distances. Exceptions are the old GI sheets. The Old GI sheets are close to red/green colour-coated metal roofs because of the corrosion of the GI surface. Most of the class pairs show higher B-distances. The B-distance between the blue- and white-surface-coated metal roofs is the highest. New GI and old GI sheets also exhibit a remarkably high B-distance (Table 4.7).

TABLE 4.6
Absolute Bhattacharyya Distances for Cement Composites in Tarang

	Asb-C	Asb-P	PB-Z	PB-Rt	CCn
Asb-C	0	102.79	281.87	445.66	162.49
Asb-P	102.79	0	337.25	4595.92	4078.57
PB-Z	281.87	337.25	0	2010.14	166.39
PB-Rt	445.66	4595.92	2010.14	0	15738.04
CCn	162.49	4078.57	166.39	15738.04	0

TABLE 4.7
Absolute Bhattacharyya Distances for Metal Roofs in Tarang

	MR-B	MR-G	MR-R	MR-W	GI-O	GI-N
MR-B	0	2673.15	43216.01	243465	3800.61	152211.95
MR-G	2673.15	0	6406.74	5529.29	657.07	9697.99
MR-R	43216.01	6406.74	0	87946.7	1291.63	24417.61
MR-W	243465.41	5529.29	87946.65	0	19546.64	142005.91
GI-O	3800.61	657.07	1291.63	19546.6	0	2089.67
GI-N	152211.95	9697.99	24417.61	142006	2089.67	0

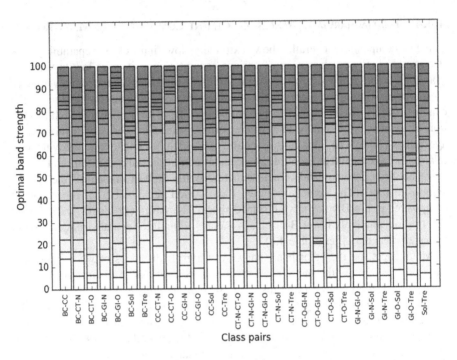

FIGURE 4.14 Spandistribution of the urban materials.

4.4.5 Most Significant Wavelengths for Discrimination of Urban Classes

4.4.5.1 Topmost Optimal Span/Important Wavelengths for Urban Materials at Class Level

The topmost wavelength range/span for the higher B-distances, mostly (considering 106 class pairs), is the visible range for most of the urban classes (Figures 4.14 to Figure 4.17). This is counterintuitive as the SWIR range is supposed to provide more information about the target material, generally speaking. However, close inspection reveals that there are many urban classes with coated surfaces, which give them unique features in the visible range. The classes with a flatter signature through the VNIR-SWIR region, such as cement composites, bitumen composites, fibre (roofs), and GI exhibits an SWIR range as the topmost region for the higher B-distance. The dominance of visible range as a topmost discriminating span is valid at both material and thematic levels (Figure 4.14 to Figure 4.17).

Though the visible range is the topmost discriminating span, other spans (bins) count c is close to the count of the topmost span/bin. Thus indicating a unique spectral behaviour: the important bands for spectral separation are distributed all over the range with a slight bias for the visible range (Figure 4.17). It is interesting to note that an increase in the number of bands chosen for a B-distance experiment (k) does not affect much the optimal span for most of the class pairs but trees. For trees, the topmost span for the higher B-distance shifts from the NIR region to the visible (red)

Spectral Resources

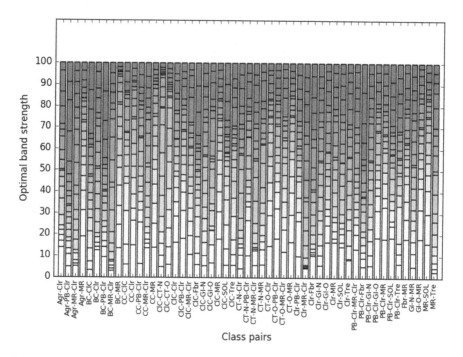

FIGURE 4.15 Spandistribution of the urban themes or the land use.

region for k=10, 15, 20, which is the diagnostic absorption region for vegetation. For GI, experiments with k = 10, 15, 20 indicate the most important span which could be treated as diagnostic absorption. GI cover has very bright signatures, but reflectance is lowest at a ~1000 nm range. Thus, the procedure also indicates its utility to detect diagnostic absorption if any (Figure 4.17).

Table 4.8 and Table 4.9 show optimal span results for urban themes. The colour coding of the cells helps in quick visual interpretation of the results. Successive bins are shaded and the tone of the shading increases as the wavelength of the span increases. For example, a 350 nm bin is marked with 5% grey, a 450 nm bin with 25% grey and so on. The dominance of a particular grey level quickly indicates the dominance of a particular range. For example, further, the analysis also helps in identifying diagnostic absorption, if any. For example, for the tree (Tre) class one can observe the dominance of a 650 nm bin. The tabular results for urban themes and materials for k=3 to 5 are further used to create three-dimensional charts.

4.4.6 Most Optimal Span and Important Wavelengths for Urban Material at Intra-class Level for a Few Important Classes Such as Cement Composites and Metal Roofs

We further analysed the two most important urban classes for intra-class separation: Cement composites and metal roofs. Cement composites show a slight bias for the SWIR range, although the spans for high B-distances are distributed over VIS

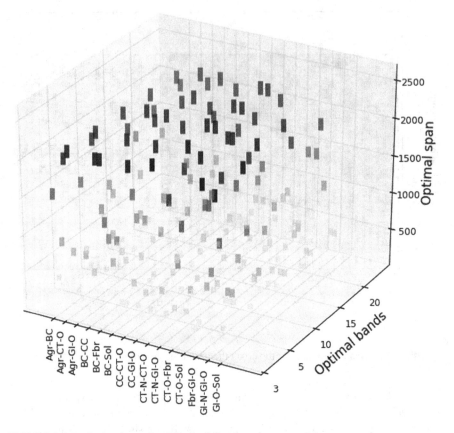

FIGURE 4.16 Optimalspan for different k for the urban materials.

and SWIR ranges. The spans within the range of 950–1500 nm range score very low on the optimal span metric. This behaviour is reflected in Figure 4.14 to Figure 4.17 appropriately. The bars are short and closely spaced within the range. There is no specific preference for a particular wavelength range and thus cement composites do not show any diagnostic wavelength or span. The topmost span does not vary much for different ks.

Metal roofs intra-class pairs and metal roofs inter-class pairs show behaviour similar to broader urban classes, that is, there is a slight bias towards the visible range for the most favourable span/bin, though the number of times a particular span/bin is selected is distributed all over the spectrum (VNIR-SWIR) with very close scores. This is because most of the metal roofs are colour-coated. On the other hand, there is no specific preference for a particular wavelength span for GI sheets because of their high reflectance through the VNIR-SWIR range (Figure 4.14 to Figure 4.17).

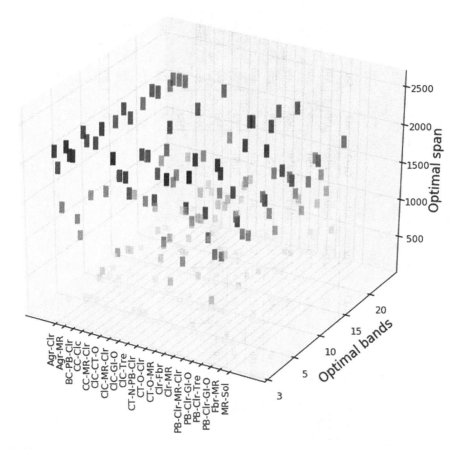

FIGURE 4.17 Optimalspan for different k for the urban themes.

4.4.6.1 Whether the NIR Range Provides Any Advantages over the SWIR Range

The distribution of c indicates that there is no preference for any specific wavelength range. The most favourable bands are distributed across the range. Colour-coated materials show a preference for the visible range (Figure 4.14 to Figure 4.17).

4.4.7 Chromatic Properties

The hypothesis that most of the impervious surfaces might appear grey because of their uniform (mostly) reflectance across the 350–2500 nm range is supported by the experiment results. Concrete has a concave shape in the 350–1000 nm range. However, it does not affect the greyness of the colour as the concavity is subtle, and overall reflectance is flat without any dominant spectral feature. Bitumen concrete retains its black colour and is closest to zero in CIE Lab space. Roof materials display observations similar to pavements. Asbestos sheets and GI sheets both appear

TABLE 4.8
Optimalspan for Urban Themes for k = 5

	Agr	BC	CC	ClC	CT-N	CT-O	ClrC	PB-Clr	MR-Clr	Fbr	Gl	Gl-O	MR	Sol	Tre
Agr	0	1550	650	2150	450	2150	2250	2150	1450	2250	1250	450	350	350	950
BC	0	0	350	2150	750	650	350	2350	1550	2250	1550	1250	450	750	1250
CC	0	0	0	550	450	550	350	450	350	2250	1150	450	350	350	650
ClC	0	0	0	0	850	850	350	350	350	2150	1550	2150	350	550	1950
CT-N	0	0	0	0	0	850	1050	450	1150	450	2050	2350	450	450	650
CT-O	0	0	0	0	0	0	350	350	350	2250	1650	1650	350	550	650
ClrC	0	0	0	0	0	0	0	450	1450	2250	750	350	350	350	350
PB-Clr	0	0	0	0	0	0	0	0	350	2250	2150	350	350	350	1950
MT-Clr	0	0	0	0	0	0	0	0	0	1650	750	1450	350	1450	1450
Fbr	0	0	0	0	0	0	0	0	0	0	2050	450	350	450	450
Gl	0	0	0	0	0	0	0	0	0	0	0	1050	450	2250	1450
Gl-O	0	0	0	0	0	0	0	0	0	0	0	0	350	450	850
MT	0	0	0	0	0	0	0	0	0	0	0	0	0	350	1950
Sol	0	0	0	0	0	0	0	0	0	0	0	0	0	0	950
Tre	0	0	0	0	0	0	0	0	0	0	0	0	0	0	0

TABLE 4.9
Optimalspan for Urban Themes for k = 20

	Agr	BC	CC	ClC	CT-N	CT-O	ClrC	PB-Clr	MR-Clr	Fbre	Gl	Gl-O	MR	Sol	Tre
Agr	0	1450	650	2350	450	550	350	2150	2250	2250	2150	450	350	350	350
BC	0	0	350	2250	2350	650	350	2250	1250	2250	1050	1250	550	2350	350
CC	0	0	0	550	450	550	350	450	350	2250	550	450	350	450	650
ClC	0	0	0	0	950	850	350	350	450	2250	750	1650	350	650	650
CT-N	0	0	0	0	0	850	1050	450	1150	2350	850	2350	350	450	650
CT-O	0	0	0	0	0	0	350	350	350	2250	950	1050	350	550	650
ClrC	0	0	0	0	0	0	0	450	350	350	2150	350	350	350	350
PB-Clr	0	0	0	0	0	0	0	0	350	2250	1050	350	350	350	650
MT-Clr	0	0	0	0	0	0	0	0	0	1650	1050	1450	350	1250	1450
Fbre	0	0	0	0	0	0	0	0	0	0	650	450	350	1650	450
Gl	0	0	0	0	0	0	0	0	0	0	0	1650	750	2050	1250
Gl-O	0	0	0	0	0	0	0	0	0	0	0	0	450	450	650
MT	0	0	0	0	0	0	0	0	0	0	0	0	0	350	1950
Sol	0	0	0	0	0	0	0	0	0	0	0	0	0	0	650
Tre	0	0	0	0	0	0	0	0	0	0	0	0	0	0	0

160 Hyperspectral Remote Sensing in Urban Environments

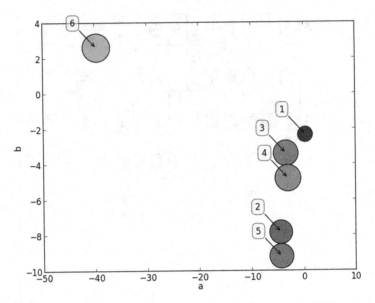

FIGURE 4.18 Artificial colours of common pavements (ASD spectra): 1-bitumen concrete, 2-irregular shaped PB, 3-rectangular PB, 4-residential PB, 5-concrete, 6-tree.

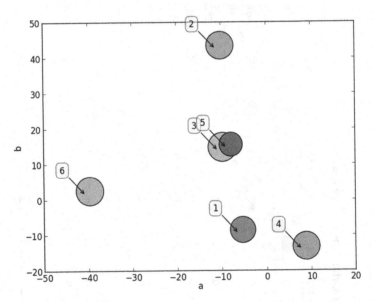

FIGURE 4.19 Artificial colours of common roof materials (ASD spectra): 1-asbestos sheet, 2–blue Tata BlueScope, 3-new clay tile, 4-new GC, 5-old GC, 6-tree.

Spectral Resources

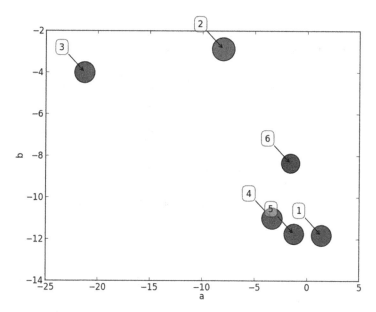

FIGURE 4.20 Artificial colours of spectra extracted from EO-1-Hyperion: 1-basalt from stone quarry, 2-dry grass, 3-trees, 4-road concrete, 5-residential concrete roofs, 6-bitumen concrete.

grey and are close to zero in CIE Lab space. The age has a negligible effect on the GI sheet: because of corrosion there is a slight shift in the grey with added yellow. Colour-coated (blue) roof metal sheet shows a yellow shade but is close to zero on the a axis. A key observation is that the blue colour in the visible range does not have a dominant effect on artificial colour. Other roof materials such as clay roof tiles and old (corroded) GI sheets appear very close to each other in CIE Lab space and show shades of grey with a slight shift towards yellow (Figure 4.18 to Figure 4.20).

Thus, chromatic discrimination of a variety of impervious surfaces using hyperspectral data seems a viable option. Dimensionality reduction is an added advantage of this approach. The advantages of artificial colour features over other features in a generic classification framework need to be investigated further. In addition, the chromatic discrimination function is not dependent only on diagnostic absorption or reflection and hence has the potential to be a suitable alternative to spectroscopic methods (Sodickson & Block, 1994). Further, the method is useful in situations where coated materials might disguise the identity in a narrow visible range but not beyond it.

4.5 TARANG – THE SPECTRAL LIBRARY AND WEB SERVICES

The spectral knowledge base ("Tarang") built so far (see Section 4.2.4 Tarang – spectral library of indian urban materials") is available through the portal "Tarang" at

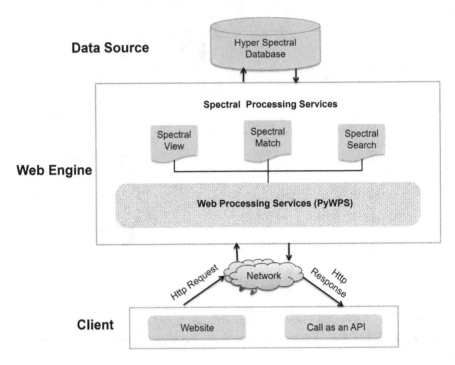

FIGURE 4.21 Tarang architecture.

http://splibtarang.com/index.php "Tarang" would provide user services to "search", and "match":

a. Search – the user would provide a search string for the material or class and the web page would show the signatures in the library matching the string,
b. Match – the user can provide the stream of a wavelength and a corresponding reflectance value of the unknown signature file and find the identity of the target.

The system is built using the Open Geo-consortium (OGC) Web Processing Service (WPS) standards (OGC, 2007). This makes the service interoperable in complex distributed environments. Various mobile sensing platforms capable of hyperspectral sensing can record spectra of a variety of land covers. This sensory data can then be analysed by fetching Tarang's web services (Deshpande et al., 2015). The data will be processed by the server as per the user input (Figure 4.21).

4.5.1 Tarang Components

Tarang is based on Service Oriented Architecture (SOA), implementing OGC compliant WPS. SOA and WPS make Tarang more flexible, interoperable, and reusable as it can be accessed by another third-party application as a service to get the desired output. Tarang architecture is divided into three core components (Figure 4.21): Database, Web engine, and Web interface.

TABLE 4.10
Tarang Table Attributes

Attributes	Specifications
Material Class	Class of material
Material Name	Common name of the material
Material Usage	Land use of the material, e.g., roofing, pavement
Sig	Spectrum–reflectance values
Date	Date of collection in DD-MM-YY format
Time	Time of collection in HH:MM:SS PM/AM format
Temp	Ambient air temperature during signature collection (in Celsius)
Humidity	Specific humidity of surrounding during signature collection

4.5.1.1 Database

Tarang's data source is a relational database based on PostgreSQL, an object-relational database management system (POSTGRESQL, 2018). The data is indexed in tables in such a way that users can easily search the key terms to get the relevant information. Meta information is referenced using foreign keys. The spectra are stored as a comma-separated list of reflectance values. This simple format is helpful in interoperability. Pyscopg2, a PostgresSQL adapter for python language has been used to connect with the database. The data is accessed by executing SQL queries, to give the desired output to the user.

The table consists of various attributes such as Material Class, Material Name, and Material Usage; for example, roofing, pavement, Spectrum – reflectance values, Date of collection in DD-MM-YY format, Time of collection in HH:MM:SS PM/AM format, Ambient air temperature during signature collection in Celsius, Specific humidity during signature collection (Table 4.10).

4.5.1.2 Web Engine

The web engine is the main component of Tarang's architecture as it takes input from the user, processes the spectral data with the help of various system functions, and provides specified outputs. Further, the web engine is divided into two parts namely the OGC Web Processing Services and Spectral Processing Services.

a. **Web Processing Services (WPS)** – This component defines the three standard operations that are GetCapabilities, which returns service level metadata; DescribeProcess, which returns process description; Execute, which returns an output. For example, the GetCapabilities query to the services returns two functions: "Search Spectra" and "Match Spectra". That is, the system is conveying information to the reader about the services that Tarang provides. A further query DescribeProcess for "Match Spectra" would return an input and output of the "Match Spectra" function.

b. **The web engine** implements PyWPS a python implementation of the Web Processing Services (PyWPS, 2009). It takes the HTTP Request from the client in the GET or POST form using a key-value pair or XML respectively and gives an HTTP Response in the form of data like text, GeoTIFF or WPS Response in the form of XML. In PyWPS various python classes (such as "Match Spectra") can be defined in the form of processes, which execute the data accordingly. Python provides certain advantages over other development environments because of strong array processing support in Python-Numpy: each spectrum can be treated as m dimensional array where m = number of bands, and then can be easily processed further. Each algorithm is modelled as a process in the process library under the PyWPS process and thus makes the system easily extensible. Some other implementations of OGC WPS are:

   i. **deegree 3 WPS**: deegree web services implement the Geospatial Web Service Specifications of the OGC. In addition, it implements INSPIRE Network Services (INSPIRE, n.d.) too. "deegree" is a comprehensive framework that provides support for other hosts of services such as Web Map Services (WMS), Web Map Tile Services (WMTS) and so on.
   ii. **ZOO WPS**: The ZOO API service is a Java Script library which has the capability to add user logic in chaining ZOO services. The ZOO API kernel is made in C. Zoo also supports numerous services which can be written in python, Perl, Java etc. The project is released under MIT/X-11 style licence (ZOO, n.d.).
   iii. **52º North WPS**: The 52°North Web Processing Service is a full java-based open-source implementation with pluggable architecture. Other features include XML data handling, support for exception handling, storing execution results and so on (52°North WPS, n.d.).

c. Spectral Processing Services – The initial processing of data takes place in the Spectral Processing Services of the web engine module. These are the python classes/functions under the PyWPS context, which defines various algorithms to process the spectrum and return the output to the calling services (PyWPS, 2009). Presently, three processes have been defined in the system: Spectral View – for browsing the data, Spectral Search – for searching the spectrum of a specific material, and Spectral Match – for identifying the target spectrum using SAM (Figure 4.21).

   a. **Spectral View**: This function enlists all the spectra in the library with their specifications such as class, subclass, sample location, time of recording etc.
   b. **Spectral Search**: This function provides the search functionality to get the information related to specific material. The user can provide the specific keywords related to name, class, and usage of material and the function will retrieve the related data from the spectral data source.
   c. **Spectral Match**: This is the most important feature of Tarang. User (any device or other software system) recorded spectrum can be uploaded (send) on the Tarang web page. The Spectral match function then performs matching to give the top N materials matched within the spectral library for a given target spectrum. Presently we are using Spectral Angle Mapper (SAM) and NS3 algorithms to determine the closeness of two spectra. The spectral match function processes the values by applying the above

algorithms and executing various SQL queries on the spectral database. In future, we are planning to incorporate other algorithms like Bhattacharya Distance etc. to make spectral discrimination more accurate.

4.5.1.3 Web Interface (Client)

The primary client consists of the Tarang website, which provides options to users such as view, search, and match. The website interacts with the web engine and fetches data through HTTP Request – response protocol. All the services available on the website are available to any external user (device or software system which can communicate with the internet) as web services.

4.5.2 Using a Spectral Library

A spectral library can be used as a dictionary of reference signatures. The reference signatures can be exploited in two ways:

a. by searching a match for an unknown signature and to identify it,
b. by using the reference signatures of known materials within the library as end members for unmixing analysis (Bioucas-Dias et al., 2012; Keshava, 2003).

The first use case can be exploited using the web services; for example http://splibtarang.com/index.php. Alternatively, web services can be consumed by multiple connected devices using http protocol for communication. The in-situ sensors often have a constraint computing machinery. Therefore, they may not have all the hardware and software for complex computation. A common solution to such problems is to offload the computation to a central computing facility. Figure 4.22 shows such a

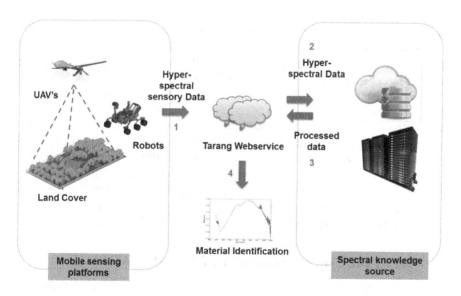

FIGURE 4.22 Tarang as a service.

TABLE 4.11
Comparison of No. of Pixels Using Hard Classification and Fractions Calculated Using Tarang Signatures

Class	Hard classification	Abundances using Tarang
Vegetation	16289	10892
Impervious surfaces	57887	57570
Soil	77566	83250

scenario. Imagine a fleet of robots, say for precision farming. These mobile devices are designed to measure soil nutrients, plant health, and so on, using hyperspectral sensors. Based on the measurements, the device can take corrective action as necessary. For example, the pesticides/fertilizers are applied as per the requirements. The devices can capture the spectral signatures and then can send them to the cloud. Then the "Tarang" web services can perform the required computation by fetching a match (or other service) and then send the results back to the devices. The results could be the target identity, or it could be the concentration chemicals requested, or it could be the actuation commands, according to the system design and need.

The second use case provides a practical solution to unmixing analysis difficulties: In the absence of reference data, when end members are extracted from a given image, the identities of the end members are not known afront and need to be resolved after the abundances are calculated. Additional difficulties arise because of multiple end members (Roberts et al., 1998; Keshava, 2003; Bioucas-Dias et al., 2012; Khopkar & Zare, July 2013; Heylen et al., 2016). Thus, the automatic endmember detection from an image might provide 4–5 endmembers of "vegetation" class and so on. Instead, to overcome these difficulties, researchers can choose the endmembers directly from a library and then perform unmixing analysi s.

We discuss an unmixing exercise performed using the signatures of known materials in Tarang. Three representative signatures of the VIS classes were chosen from Tarang, and abundances were calculated of these members for a subsection of the EO1-Hyperion image of Pune city (USGS, 2013-1). The section starting from row number 1550 to 2050 was selected. The region was selected as mixed pixels had a good distribution of VIS classes in it. Water pixels in the image were masked while calculating the VIS abundances. The Fully Constrained Least-Square Based Linear unmixing (FCLS) algorithm was used to calculate the abundances (Heinz et al., 1999: Heinz & Chang, 2001). The abundances were then compared with the total area of each VIS class with the results for the hard classification using SAM. In the present case study, hard classification resulted in underestimation of plain area (soil) by 7%, overestimation of vegetation by 33%, and overestimation of 1% of impervious surfaces (Table 4.11).

4.6 SPECTRAL CHARACTERISTICS OF URBAN MATERIALS

Most of the impervious surface detection studies begin with a spectral signature analysis of urban materials. These efforts are critical as it motivates building a spectral library for future use in target detection, and discrimination analysis. A separate study for a different region may be required as signatures of composite urban materials vary with their raw materials.

Chen and Hepner (2001) study hyperspectral signatures of urban materials to assess their discriminating power and utility of reference spectra for image classification. Field spectra of twenty roofing materials, twelve paving materials, twenty-three vegetation samples, and twenty-five miscellaneous urban materials were collected using an ASD field spectrometer. Along with the field studies, AVIRIS data was acquired for the study area – Park City, Utah. The field spectra were rescaled to magnitude 0 to 1 and were resampled to match AVIRIS field data.

Herold et al. (2004) provide a comprehensive analysis of the spectra of specific types of materials in urban areas. Over 5000 (5500) spectra for 147 unique materials in Santa Barbara and Goleta, CA, USA, were recorded and analysed for the signatures using statistical measures. The spectra were measured using ASD, and Full-Range (FR) spectrometer in the range of 350–2400 nm with a spectral resolution of 1.4 nm for the Visible and Near Infrared (VNIR), and 2.0 nm for the Short Wavelength Infrared (SWIR). The spectra were recorded during two hrs of solar noon. The spectra were resampled with a 2 nm width to create 1075 bands. Resampled 108 spectra were further analysed for separability using Bhattacharya distance (Fukunaga, 1990).

We summarize below the important spectral characteristics of impervious surfaces as recorded by these studies:

4.6.1 PARK CITY AND SANTA BARBARA/GOLETA OBSERVATIONS

a. Sand, gravel, and concrete are not distinguishable in broadband data (Chen & Hepner, 2001). Spectra of asphalt roads, parking lots, and some roofs show constant reflectance without a specific absorption. The bare soil spectrum also shows similar behaviour (Herold et al., 2004).

b. New paving asphalt has an overall extremely low reflectance (15% lower than its medium-weathered counterpart) with a relatively high increase towards the 2100 nm range (Chen & Hepner, 2001). Ageing asphalt shows increased reflectance in all parts of the spectrum – the increased reflectance can be attributed to a loss of oil material in sealing. Ageing also changes the nature of the curve – an ageing spectral curve shows concavity in the range 1000–1600 nm while new asphalt surfaces show convexity (Chen & Hepner, 2001; Herold et al., 2004)

c. Concrete surfaces have the highest reflectance (among the surface types studied) and with ageing the reflectance decreases. This behaviour is exactly the opposite of asphalt surfaces. Concrete roads have high separability. Concrete roads and some roof materials show similar signatures. Concrete shows low separability with bare soil and beach (Herold et al., 2004).

d. Urban materials lack diagnostic absorption and full spectrum matching techniques should be considered. In the cases of two materials with the same spectral curve with a difference in reflectance magnitude, SAM would not be able to detect the difference – such as asphalt, and paving concrete (Chen & Hepner, 2001).

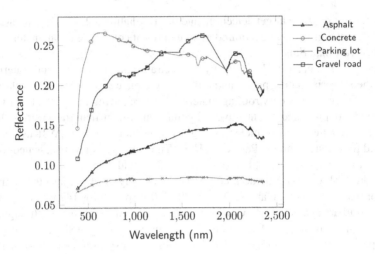

FIGURE 4.23 Spectral signatures of different road surfaces, recorded by Herold et al. (2004).

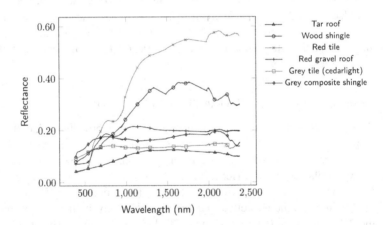

FIGURE 4.24 Spectral signatures of different roofs, recorded by Herold et al. (2004).

e. Though the spectral reflectance of roof materials (mainly shingles) changes with illumination angle, the shape of the spectrum does not change – indicating the utility of spectra to discriminate objects under different (direction-wise) illumination conditions (Herold et al., 2004).

Figure 4.23 and Figure 4.24 show the spectral signatures of the most common road and roof materials used in the study by Herold et al. (2004). Some of the spectra are slightly affected by high-frequency noise because of water absorption at 934 nm.

4.6.2 Tarang Observations

The Tarang spectral library records similar observations as earlier. Urban places commonly show a remarkably high degree of variation in the land covers; developing countries show even more variety. A large variation of socio-economic activities in the cities of developing countries results in a large variation in land use and land covers. Manmade materials such as cement and bitumen composites, colour-coated materials, metal roofs and so on dominate the urban area. Bare soil cover such as playgrounds, and open areas covered with different degrees and vegetation such as parks and golf courses and trees used for road arboriculture constitute natural materials. Cultivation on the fringes of the city completes the plethora of materials present in and around urban areas in developing countries. We discuss below the spectral properties of some of the important materials commonly occurring in Indian cities.

Cement composites (CC) are one of the most important classes of urban materials. The cement composites such as cast-in-situ road-grade concrete, pre-cast pavement blocks, plain cement concrete pavements, asbestos roofs, and concrete roofs of residential areas dominate urban surfaces. The spectral properties of all these materials show similar characteristics. The cement composites do not show any diagnostic absorption in the VNIR-SWIR range. Their spectra are nearly flat with a little concavity in the visible range. The average brightness of the cement composites is higher than the soil brightness, though the signature of the soil is similar to the cement composite signatures. Discrimination of soil, especially bare soil, and cement composites spectrally is a challenging problem. As can be seen from figures (Figure 4.8 to Figure 4.11), the spectral shape of soil and cement composites differ very slightly from each other – the difference is noticeable only in the visible range. The slope of soil signature in the blue range is slightly higher than cement composites. This is the only possible spectral feature that can be used for spectral discrimination (Figure 4.14 to Figure 4.17).

Bitumen composites (BC) have the lowest average brightness of all the urban materials. The signature is very flat without any prominent diagnostic absorption and BCs can easily be identified by visual inspection of the spectral signature. The addition of aggregates with different proportions does not make much difference to the shape of the spectral signature and low average brightness. The old and new bituminous pavements, and the bitumen-bound macadam soling show very similar signatures (Figure 4.8 to Figure 4.11).

Metal roofs (MR) show a large degree of variation in their surface coatings and hence the variations in the signatures. Overall, high average brightness values in the SWIR region are exhibited by all the metal roofs irrespective of their surface coating (Figure 4.14 to Figure 4.17). The majority of the metal roofs are colour-coated, and they show a diagnostic reflection respective to their colour coating. Galvanized iron (GI) sheets have very high reflectance throughout the spectral range but ~1000 nm range. The GI sheets and the colour-coated metal roofs such as blue and white show very high average brightness and can be easily detected in hyperspectral images as well (Deshpande et al., 2017). The corroded GI shows signatures similar to soil signatures although corroded GI shows slightly higher average brightness, especially in the SWIR region. Thus, the characteristic extremely high reflection of the GI sheet in the blue region is diminished with a degree of weathering and hence can be used to detect the ageing of the GI roofs. New clay tiles show a steep slope in the visible range and high average brightness in SWIR and are easy to discriminate. On the other

TABLE 4.12
Significant Bands for Urban Materials Reported in Literature

	430–450	450–510	530–590	640–670	850–880			1570–1650		2110–2290
Landsat 8 range[1] (nm)										
Landsat 8 band centres[2] (nm)	455	480	560	655	865			1610		
Landsat 8 band names[3]	Aerosol	B	G	R	NIR			SWIR1		SWIR2
(Herold, Roberts, Gardner, & Dennison, 2004) (nm)	420	440	570	640	750	1050	1315		1990	2004
(Deshpande, Inamdar, & Vin, Spectral library and discrimination analysis of Indian urban materials, 2019) (nm)	400	500	600	700	1100	1105	1300	1700	2300	2400
(Herold, Roberts, Gardner, & Dennison, 2004; Heiden, Segl, Roessner, & Kaufmann, 2007)[4] (nm)	445	576	638	759	1100		1316		1989	
Built-up index (Zha, Gao, & Ni, 2003)					NIR			SWIR		
Concrete condition index (Samsudin, Shafri, & Hamedianfar, 2016) (nm)			526		770					
Metal roof condition index ((Samsudin, Shafri, & Hamedianfar, 2016) (nm)					743,753					
Concrete quality (Shaban, 2013) (nm)		450					1380		1850	

Notes:
[1,2,3] www.usgs.gov/faqs/what-are-band-designations-landsat-satellites

[4] The bands reported by Herold et al. (2004) were used by Heiden, Segl, Roessner, & Kaufmann (2007); The bands mentioned in this row were selected using the plot from Herold, Roberts, Gardner, & Dennison (2004) and by Heiden, Segl, Roessner, & Kaufmann (2007) for comparative assessment of robust features.

hand, old tiles show signatures similar to soil and are difficult to discriminate from soil classes.

4.6.3 SIGNIFICANT BANDS FOR URBAN MATERIALS

Table 4.12 shows a landscape of significant wavelengths for discriminating urban materials. The Landsat 8 bands and its nomenclature is provided for benchmarking.

4.7 SUMMARY

A spectral library is a critical resource for any target detection characterization. Understanding the spectral properties of the target material is essential for designing analysis methods. It is a preliminary step for payload design as well. The USGS and ECOSTRESS are the two main libraries in the public domain. However, more such libraries are required for local studies of urban materials. This is because urban composites show a large variation in the raw materials.

Tarang is one such library of Indian urban materials. It has ~1150 signatures of 105 unique samples of Indian urban materials including pavements, roofs, urban trees, and so on. Tarang is available at http://splibtarang.com/index.php. Search and match web services are available for any device communicating using http protocol.

Spectral signatures are analysed the diagnostic features if any that uniquely identify the target material and so on. In addition to this, choosing bands that are optimal in separating the target material from others is essential too. Bhattacharyya distance is used to decide how separable the two target materials are. Larger values of the distance indicate that they are easily separable. Another interesting way to characterize the urban material is by using colour as a discriminator.

NOTE

1 The instrument for measuring the light-matter interaction is the limitation as it may not be able to resolve fine spectral lines in some of the cases and hence the target detection may need additional information. The resolution limitation may further lead to uniqueness assumption of light-matter interaction as at the coarse resolution two materials may show similar signature and cannot be resolved spectrally. In addition, in the spaceborne or the airborne imaging spectroscopy the spectral and spatial resolution are the conflicting goals and hence the design parameter may limit the spectral resolution.

WORKS CITED

52°*North WPS*. (n.d.). Retrieved March 11, 2015, from http://52north.org/communities/geoprocessing/wps/

ASD. (2016). *FieldSpec Portable Spectroradiometer*. Retrieved March 11, 2015, from www.asdi.com/products/fieldspec-spectroradiometers

Baldridge, A. M., Hook, S. J., Grove, C. I., & Rivera, G. (2009). The ASTER Spectral Library Version 2.0. *Remote Sensing of Environment, 113*, 711–715.

Biehl, L., & Landgrebe, D. (2002, December). MultiSpec-a tool for multispectral-hyperspectral image data analysis. *Computers and Geosciences, 28*(10), 1153–1159. Retrieved from https://engineering.purdue.edu/~biehl/MultiSpec/

Bioucas-Dias, J., Plaza, A., Dobigeon, N., Parente, M., Du, Q., Gader, P., & Chanussot, J. (2012, April). *Hyperspectral unmixing overview: geometrical, statistical, and sparse regression-based approaches.* Retrieved from http://arxiv.org/pdf/1202.6294v2.pdf

Chen, J., & Hepner, G. F. (2001). Investigation of imaging spectroscopy for discriminating urban land covers and surface materials. *AVIRIS Earth Science and Applications Workshop, Palo Alto, California. JPL Publication 02-1.* Jet Propulsion Laboratory, Pasadena, CA. Retrieved from https://aviris.jpl.nasa.gov/proceedings/workshops/01_d ocs/2001Chen_web.pdf

Clark, R. N., Swayze, G. A., Wise, R., Livo, E., Hoefen, T., Kokaly, R., & Sutley, S. J. (2007). USGS Digital Spectral Library splib06a: U.S. Geological Survey, Digital Data Series 231. *USGS Digital Spectral Library splib06a: U.S. Geological Survey, Digital Data Series 231.* Retrieved from http://speclab.cr.usgs.gov/spectral.lib06

Das, B. S., Sarathjith, M. C., Santra, P., Sahoo, R. N., Srivastava, R., Routray, A., & Ray, S. S. (2015, March 10). Hyperspectral remote sensing: opportunities, status and challenges for rapid soil assessment in India. *Current Science, 108*(5), 860–868.

Deshpande, S. S., Gupta, M., Yadav, P., Inamdar, A. B., & Vin, H. M. (2015). Spectral library search Tarang for Indian urban materials: OGC compatible web services for connected devices. *TACTiCs, TCS Technical Architect Conference.* Bangalore.

Deshpande, S. S., Inamdar, A. B., & Vin, H. M. (2017). Urban land use land cover discrimination using image based reflectance calibration methods for hyperspectral data. *Photogrammatric Engineering and Remote Sensing, 83*(5), 365–376.

Deshpande, S. S., Inamdar, A. B., & Vin, H. M. (2019). Spectral library and discrimination analysis of Indian urban materials. *Journal of the Indian Society of Remote Sensing, 47,* 867–877.

Essar Steel. (2013). *Colour coated.* Retrieved July 5, 2016, from www.essarsteel.com/section _level2.aspx?cont_id=iGXYH/GHeTQ=&path=Products_and_processes_%3e_Produc ts_%3e_Colour_coated

Fukunaga, K. (1990). *Introduction to pattern recognition* (2 ed., Vol. Computer Science and Scientific Computing). (R. Werner, Ed.) Academic Press, San Diego, CA, USA.

Goetz, A. F. (2012, October). *Making Accurate Field Spectral Reflectance Measurements.* Retrieved July 15, 2016, from Boulder, CO, 80301, USA: http://discover.asdi.com/Port als/45853/docs/Measurements-paper-10-26-12.pdf

Heiden, U., Segl, K., Roessner, S., & Kaufmann, H. (2007, December). Determination of robust spectral features for identification of urban surface materials in hyperspectral remote sensing data. *Remote Sensing of Environment, 111*(4), 537–552.

Heinz, D. C., & Chang, C.-I. (2001, March). Fully constrained least squares linear spectral mixture analysis method for material quantification in hyperspectral imagery. *IEEE Transactions on Geoscience and Remote Sensing, 39*(3), 529–545.

Heinz, D., Chang, C.-I., & Althouse, M. G. (1999). Fully constrained least-squares based linear unmixing. *Geoscience and Remote Sensing Symposium* (pp. 1401–1403). Hamburg, Germany.

Herold, M., Roberts, D. A., Gardner, M. E., & Dennison, P. E. (2004). Spectrometry for Urban Area Remote Sensing–Development and Analysis of a Spectral Library from 350 to 2400 nm. *Remote Sensing of Environment, 91*(34), 304–319. Retrieved from www. sciencedirect.com/science/article/pii/S0034425704000768

Heylen, R., Zare, A., Gader, P., & Scheunders, P. (2016, August). Hyperspectral unmixing with endmember variability via alternating angle minimization. *IEEE Transaction on Geoscience and Remote Sensing, 54*(8), 4983–4993. doi:10.1109/TGRS.2016.2554160

INSPIRE. (n.d.). Retrieved March 11, 2015, from http://inspire.ec.europa.eu/index.cfm/ pageid/5

IS 10262 (2009). (July 2009). *Guidelines for concrete mix design proportioning, first revision.* New Delhi: Bureau of Indian Standards.
IS 1498 (1970). (March 1972). *Classification and identification of soils for general engineering purposes, first revision.* New Delhi: Bureau of Indian Standards.
IS 15658 (2006). (June 2006). *Precast concrete blocks for paving.* New Delhi: Bureau of Indian Standards.
IS 277 (2003). (November 2003). *Galvanized steel sheets (plain and corrugated)-specification, sixth revision.* New Delhi: Bureau of Indian Standards.
IS 459 (1992). (March 1992). *Corrugated and semi-corrugated asbestos cement sheets, third revision.* New Delhi: Bureau of Indian Standards.
IS 654 (1992). (March 1992). *Clay roofing tiles, mangalore pattern-specification, third revision.* New Delhi: Bureau of Indian Standards.
Jet Propulsion Laboratory. (1990). *JPL Spectral Library.* (Jet Propulsion Laboratory, California Institute of Technology) Retrieved August 23, 2013, from ASTER Spectral Library: http://speclib.jpl.nasa.gov/documents/jpl_desc
Jet Propulsion Laboratory. (1999). *ASTER Spectral Library–Version 2.0.* (Jet Propulsion Laboratory, California Institute of Technology) Retrieved August 23, 2013, from ASTER Spectral Library: http://speclib.jpl.nasa.gov/
Keshava, N. (2003). A survey of spectral unmixing algorithms. *Lincoln Laboratory Journal, 14*(1), 55–78.
Khopkar, P., & Zare, A. (July 2013). Simultaneous band-weighting and spectral unmixing for multiple endmember sets. *IEEE International Geoscience and Remote Sensing Symposium (IGARSS)* (pp. 2164–2167). doi:0.1109/IGARSS.2013.6723243
Kokaly, R. F., Clark, R. N., Swayze, G. A., Livo, K. E., Hoefen, T. M., Pearson, N. C., ... and Klein, A. J. (2017). USGS spectral library version 7: *U.S. Geological Survey data series* 1035. 61. doi:https://doi.org/10.3133/ds1035
lEclariage, I. C. (2015). *Selected colorimetric tables.* Retrieved January 28, 2015, from www.cie.co.at/index.php/LEFTMENUE/DOWNLOADS
Ledoit, O., & Wolf, M. (2004). A well-conditioned estimator for large-dimensional covariance matrices. *Journal of Multivariate Analysis, 88*, 365–411. Retrieved from http://perso.ens-lyon.fr/patrick.flandrin/LedoitWolf_JMA2004.pdf
Lindbloom, B. (2015). *Useful color equations.* Retrieved January 28, 2015, from www.brucelindbloom.com/
Manjunath, K. R., Kumar, A., Mehra, M., Renu, R., Uniyal, S. K., Singh, R. D., ... Panigrahy, S. (2014). Developing spectral library of major plant species of western Himalayas using ground observations. *Journal of Indian Society of Remote Sensing, 42*(1), 201–216.
Manolakis, D., Lockwood, R., Cooley, T., & Jacobson, J. (2009). Hyperspectral detection algorithms: Use covariances or subspaces? In S. S. Shen, & P. E. Lewis (Ed.), *SPIE. 74570Q,* pp. 1–8. San Diego: SPIE. Retrieved from http://hdl.handle.net/1721.1/52735
OGC. (2007). *OpenGIS® Web Processing Service: OGC 05–007r7.* Open Geospatial Consortium Inc. Retrieved January 5, 2015, from file:///C:/Users/114528/Downloads/05–007r7_Web_Processing_Service_WPS_v1.0.0.pdf
POSTGRESQL. (2018). *About.* Retrieved June 27, 2018, from POSTGRESQL: www.postgresql.org/about/
PyWPS. (2009). *PyWPS.* Retrieved July 7, 2016, from http://pywps.org/
Roberts, D. A., Gardner, M., Church, R., Ustin, S., & Scheer, G. (1998). Mapping chaparral in the Santa Monica mountains using multiple endmember spectral mixture models. *Remote Sensing of Environment, 65*, 267–279.

Ronak Tiles Pvt. Ltd. (2014). *Pavement blocks*. Retrieved July 5, 2016, from http://ronaktiles.in/

Sahyadri Industries Ltd. (2015). *Swstik, the roof of India*. Retrieved July 5, 2016, from www.silworld.in/Download/Products/SWASTIK_Brochure.pdf

Samsudin, S., Shafri, H., & Hamedianfar, A. (2016, 1). Development of spectral indices for roofing material condition status detection using field spectroscopy and WorldView-3 data. *Journal of Applied Remote Sensing, 10*, 025021.

Shaban, A. (2013). *Determination of Concrete Properties Using Hyperspectral Imaging Technology: A Review*, 1–11.

Shwetank, J. K., & Bhatia, K. (2011). Development of digital spectral library and supervised classification of rice crop varieties using hyperspectral image processing. *Asian Journal of Geoinformatics, 11*(3), 43–51.

Sodickson, L. A., & Block, M. J. (1994). Kromoscopic analysis: a possible alternative to spectroscopic anlaysis for noninvasive measurement of analytes in vivo. *Clinical Chemistry, 40*(9), 1838–1844.

SVC. (2016). *Field Portable Spectroradiometers*. Retrieved July 5, 2016, from www.spectravista.com/ground.html

Tata BlueScope Steel Ltd. (2016). *Durashine roof and wall sheets*. Retrieved July 5, 2016, from http://tatabluescopesteel.com/Roof-and-Wall#Durashine%20Roof%20&%20Wall

USGS. (2013-1, April 22). EO1 Hyperion Scene: EO1H1470472013112110KZ_PF1_01, for Target Path 147, and Target Row 47. *EO1 Hyperion Scene: EO1H1470472013112110KZ_PF1_01, for Target Path 147, and Target Row 47*. Sioux Falls, SD USA: U. S. Geological Survey. Retrieved from http://earthexplorer.usgs.gov

Uttam Galva Steels Ltd. (2016). *Gp and GC*. Retrieved July 5, 2016, from www.uttamgalva.com/products/gp_gc.html

Vijay Agency Pvt. Ltd. (2015). *Corrugated sheets*. Retrieved July 5, 2016, from www.vijayagency.co.in/corrugated-sheets.html#design-corrugated-sheets

Zha, Y., Gao, J., & Ni, S. (2003). Use of normalized difference built-up index in automatically mapping urban areas from TM imagery. *International journal of remote sensing, 24*(3), 583–594.

ZOO. (n.d.). Retrieved March 11, 2015, from www.zoo-project.org/

5 Classification of Urban Land Use and Land Cover

This chapter will discuss the following topics:

- Classification of urban land use and land cover
  - Image classification
  - Its significance to hyperspectral data
  - Supervised and unsupervised machine learning techniques overview
  - Challenges in land use land cover classification
  - Class codes/models USGS etc.
  - Mathematical preliminaries
- Index-based methods
  - Method intuition, method
  - Different urban indices
- Spectral matching methods
  - Structure – general algo
  - Similarity measures
  - Some of the case studies
- Conventional machine learning (ML, k-NN)
  - Simple image features
  - Proxy features
- Deep learning (with spectral convolutional focus)
  - Spectral, spatial, spectral-spatial features
  - Soe common deep learning networks
  - Spectral capsules

5.1 INTRODUCTION

Image classification is one of the first steps[1] in the analysis of remotely sensed data. It is a process of identifying large areas in an image which are coherent with respect to some predefined properties – for example, land cover or land use (Egmont-Petersen et al, 2002). The classification results are used for further analysis that generates more actionable insights from the image. For example, urban growth is studied for urban planning, impervious surface classification is helpful for identifying flood-prone areas in the city, and so on.

The classification is broadly divided into two areas: supervised classification and unsupervised classification. Supervised classification is a process of learning from the labelled data also known as training data, whereas unsupervised classification is a process of learning naturally occurring groups or patterns in the data without any human-assigned labels. The former is useful when we have sufficient annotated pixels with the class labels by human experts, and the task is to assign one of the class labels to each unlabelled pixel. For example, we want to classify each pixel into vegetation, soil, and impervious surface using a few pixels in the image marked with either of the vegetation, soil, or impervious surface classes. The latter (unsupervised classification) is used when we are interested in understanding some interesting patterns within the data. For example, we may find after creating naturally occurring groups that a set of pixels represent an up-market residential zone, whereas the other groups indicate a mid-economy residential zone and so on. We may also find that the vegetation has two subgroups in it; one is tree canopy, and the other is lawn. Thus, the unsupervised classification is used for understanding the underlying structure of the data. It helps in understanding how cohesive the data is, what are the naturally occurring groups in the data and so on. Machine learning is a broad area, and the complete overview of the topic is not in the current scope of the chapter. However, we will provide a brief overview of the machine learning topics in the context of hyperspectral data processing. We will focus on unsupervised machine learning approaches and spectral features in the context of supervised learning approaches.

The term image classification in the context of the hyperspectral image refers to assigning a land cover or a material category to a pixel in the image (Chen et al., 2015; Ghamisi et al., 2017; Li, et al., 2019). In that sense, it is a dense prediction problem. The terms "image segmentation" or "dense prediction" are common in computer vision researchers for such a task. In computer vision, image classification usually refers to the assignment of a label (or multiple labels) to each image in a collection of images, usually. This is possible because most of the images used for such tasks are taken from handheld cameras. The subject of the photographs for such imaging, most of the time, is a single well-defined object such as a group of people, person, animal, vehicle and so on. Thus, the image label may, in fact, represent an object category of the object the image. This task hence may be referred to as an object recognition task as well (Deng et al., 2009; Russakovsky et al., 2015; Li et al., 2022). In remotely sensed data an image rarely has a single object associated with it. Usually, it would be a group of objects or land cover land use patterns. The entity of classification usually is a pixel or a group of pixels within the image.

We will use the "image classification" term in the first sense, that is, the pixel/s in the image are assigned a label. It could be an object category label or a land cover label, depending upon the classification goal and resolution of the image. With the recent advances in machine learning, especially learning with neural networks, the amount of work in hyperspectral image classification has seen rapid growth. We would discuss a few important classification methods that focus on the spectral properties of a pixel. That is, we will give more importance to the methods that exploit the spectral properties of a pixel in the image.

5.1.1 CLASSIFICATION

Figure 5.1 shows the broad-level steps for the general classification process. The objective of the image classification process is to label each pixel in the image with a land cover or land use class. The class names can be represented by a number or a string. In the case of regression, the output is a real value. For example, instead of just classifying pixels of a lake in distinct categories of turbidity, we may need a quantitative relation between the pixel and the turbidity in Nephelometric Turbidity Unit (NTU). A label is a qualitative output whereas real value is a quantitative output. We continue to use the word "label" to represent both the qualitative and qualitative output. Thus, a pixel is a sample or an object to be labelled. A pixel is represented by its band values. The classification procedure must use these features. Features are also known as attributes or predictors. They are independent variables, and the class label or real output value are dependent variables. The features are the quantitative or qualitative descriptors of the object to be classified, a pixel in this case. The classifier can be thought of as a function that outputs a class label when the function is evaluated for the pixel features. The features could be simple band values, or they could be complex features derived from band values. It can be represented informally as:

a. Label = f (band values)
b. Label = f (band values, and the association of the given pixel with the others in spectral or spatial features)

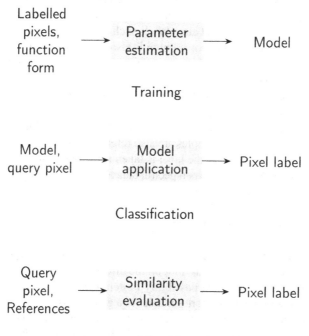

FIGURE 5.1 Overall classification process.

Or

c. Real valued output = f(band values)

It is worth discussing here the need for multiple features. If there is a single most important band which is helpful in assigning the class label to the pixel, we would not need the multiple features and a complex function for mapping that to a label. For example, water absorbs infrared radiation, and it would appear black. A simple threshold on the diagnostic band values can determine the class of the pixel. On many occasions that is not possible and hence requires consideration of multiple features. The ultimate pursuit in feature engineering or sensing is to find that single feature of the data which is good enough for labelling or identifying the object alone.

The functions that take features as an input and produce a pixel label as an output can take any form. It could be a probability distribution function; it could be a linear combination of the band values and so on. Informally, we can express this as:

a. Label probability = prior probability of the label and likelihood (given a mean and variance of the class
b. Label = weighted summation of band values
c. Label probability = weighted summation of band values.

The learning, given the above description of the process, would be estimating the mean and variance of the data for case A and the weights for the band values in B and C cases. The parameters of the functions are learned using training data. Training data is nothing but pixels with human-annotated class labels. For example, for a pixel, the label can be assigned from a set of class labels such as *tree, road, building, water, soil* and so on. The result or outcome of training is also referred to as a model. When we have a new pixel to label, also referred to as a query pixel, we can use the model to calculate the pixel label.

In some of the cases, especially of interest to the hyperspectral data, the process of training does not have to be as elaborate as explained earlier. It could be a simple measurement of similarity between a query pixel and a single or a few good examples of the class. These examples are also referred to as references or prototypes or instances. The decision about the label is taken if the query pixel is close to some known reference. In the case of hyperspectral data analysis, the references are precise laboratory or field spectra of known materials.

a. Label = *label of* (closest reference)

The popular spectral angle mapper (SAM) method belongs to this category. The effectiveness of such algorithms depends a lot on the effectiveness of the distance measure used in the algorithm. There is a possibility of defining a new distance or similarity measure as well, which may be more suitable for a given classification problem. These methods do not depend on the hyperspectral data collection mode: the techniques can be easily applied to the data from non-imaging equipment such as a field spectrometer, which is not possible for some of the classifiers using spatial features. These are the situations in which the target is required to be resolved entirely on spectral properties.

Classification of Urban Land Use and Land Cover

In addition to this, one more key point about the classification of remotely sensed data is the number of labels we assign to a single pixel. Usually, it is one for convenience. In that case, it is referred to as hard classification. In the case of multiple-label assignment to a single pixel, it is called soft classification. Soft classification is performed by unmixing analysis in remote sensing. Unmixing analysis is a process by which each pixel in the image is represented as a linear combination of a few pure pixels within the image. These pure pixels are called endmembers. Without going into the details of unmixing and various algorithms for unmixing, we provide details of the analysis of unmixing in later sections in the context of land use classification. Thus, the reader should be well equipped to use unmixing analysis as and when required for hyperspectral data analysis.

5.1.2 Hyperspectral Image Classification

Hyperspectral image classification is a challenging task on many accounts. Some of the challenges we have seen already in the second chapter. Here we review them briefly in the context of classification.

The high dimensionality of the data needs careful attention. In many of the commonly handled image-processing tasks, such as object detection or image classification, the images are RGB or black and white. Some of the modern neural networks may not pay much attention to the depth of the images in such cases. However, in the case of a hyperspectral image, the depth (number of bands) carries vital information. Ideally, the spectral information should be sufficient to detect the target material. Reducing the spectral dimensions of the data for ease of processing is a common strategy. However, the dimensionality reduction should not lose the important spectral information.

A small number of training samples is another problem requiring the careful formulation of the classification problem. The effects are two-fold – first, the covariance matrix of the training samples becomes singular and requires either covariance estimation or pseudo inverse in statistical formulation (see Section 3.3.2, "The Bhattacharyya distance analysis for urban classes", as well). Second, a small number of training samples may not be a true representative of the classes or may not have enough in number to extract enough statistics for precise classification. The small number of training samples is a problem for remote sensing in general. However, a commonly suggested number of training samples for hyperspectral images are higher as compared to the RGB images.

Because of the physical limitations, hyperspectral images have medium to coarse resolution as compared to their RGB counterparts. This leads to an increased number of mixed pixels in the image. Thus, hard classification, that is, assigning a single label to a pixel, results in the estimation error for the land covers. For example, a vegetation area in the image would be underestimated or overestimated than the actual area. Soft classification, that is, assigning multiple class labels to a pixel, is required for more accurate results.

For urban targets, the problem is further aggravated as various urban materials have similar spectral signatures, especially cement composites. Cement composites are confused with bright soils or bare soils as well. The confusion with the bright soil is well reported in the literature (Myint et al., 2011; Xu & Gong, 2007; Phinn et al., 2002). The confusion results in inaccurate results in the hard and soft classification of soil and built-up areas.

With this brief introduction to the classification, we define the classification more formally in Section 5.1.3, "Mathematical preliminaries".

5.1.3 Mathematical Preliminaries

Image X is a collection of x pixel vectors. If the spatial order of the pixels in the form of rows and columns is considered, then $X \in \mathbb{R}^{H \times W \times D}$ where, H and W are the number of rows and columns, and D is the number of dimensions or bands. For example, an RGB image of $H=200$ rows and $W=100$ columns is denoted by $X \in \mathbb{R}^{200 \times 100 \times 3}$. Similarly, the hyperspectral image of the same size with $D=200$ number of bands is denoted by $X \in \mathbb{R}^{200 \times 100 \times 200}$. Thus, the image X is represented as a three-dimensional array, rows and columns indicate the spatial extent and the depth indicates the spectral extent or number of bands. It is equally easy to consider image X as a list of D-dimensional ($H \times W$) pixel vectors. This is because many of the classifiers work on a single pixel and assign a label to it during the classification. The list of pixels vector is prepared by stacking each pixel vector below the previous one, starting at the upper left corner of the image. Image in this format (we will call it a list format) is denoted by $X \in \mathbb{R}^{(H \times W) \times D}$. Thus, RGB image and the hyperspectral image mentioned in the example earlier are denoted by $X \in \mathbb{R}^{20000 \times 3}$ and $X \in \mathbb{R}^{20000 \times 200}$ respectively. $X = \{x_{t=1}, x_{t=2}, x_{t=3}, \ldots, x_{t=T}\} \in \mathbb{R}^{(H \times W = T) \times D}$ is an image in the list format where $x_t = \{x_{d=1}, x_{d=2}, \ldots x_{d=D}\} \in \mathbb{R}^D$ is the tth pixel vector. The classification is a process of labelling each x_t in the image X using labelled examples from X^L, where X^L and X^U are the list of labelled samples and unlabelled samples from X respectively. Thus, $\mathbf{X} = \mathbf{X}^L \cup \mathbf{X}^U$. The labels are denoted by $\mathbf{Y} = \{y_{k=1}, y_{k=2}, \ldots y_{k=K}\}$ K classes. The pair $(x_{h,w}, y_k)$ is a pair of pixels indexed using a row and a column and a class label assigned to it. The pair (x_t, y_k) is a pair of a pixel indexed using list format and a class label assigned to it. For example, (x_{12}, y_4) denotes the 12th pixel belongs to the 4th class.

With this formal description we can formulate the classification problem as: Given the input – a set of labelled pixels = $\{(x_t, y_k)(x_t, y_k)(x_t, y_k) \ldots (x_t, y_k)\} \in X$, the learning task should create a model such that given a new pixel vector $x_t \in X$ it outputs the class label $y_k \in Y$ of the pixel. This is a supervised machine learning problem. We will call a new instance to be classified as a query pixel or a test pixel. The process of learning is also called training or model development and using the model for classifying a query pixel is called classification or testing or deployment, depending upon the context. Ideally, the learning process learns a function that maps the input to the output space, that is, it takes the pixel vectors and their labels and learns a mapping function between them such that $f: \{x_t\} \to y_k \in Y$. In the real world, however, we learn the approximation or function parameters assuming some functional form. For example, we may assume that the output value is a linear combination of the input predictors/features/attributes, and the learning task is to learn the weights of the independent variables (Please see Hastie et al., 2008 for more detailed discussion). The learning, for example, is equivalent to identifying the right parameters that minimize

Classification of Urban Land Use and Land Cover 181

the prediction error on the given example data (training data, labelled samples). It is the equivalent of choosing a hypothesis (a set of parameters that meet the criteria) from a set of possible hypotheses (all possible values of the parameters). (Readers are encouraged to see the detailed discussions in Mitchell, Introduction, 1997; Hastie et al., 2008.)

5.1.4 FEATURES FOR HYPERSPECTRAL IMAGE CLASSIFICATION

The classification has two important dimensions: Features or initial inputs given to the algorithm and, the form of the classification function. The function can be seen as a discriminant[2] function as well – a function evaluated for a data instance (a pixel in the case of image classification) results in values which help in deciding the class of the data instance. The function can be a simple ratio of two specific bands (index-based methods), it could be a similarity with a reference signature (k-Nearest Neighbour – k-NN, Spectral Angle Mapper – SAM), it could be a conditional probability of a class given a data instance (Naïve Bayes' – NB, Maximum likelihood – ML), it could be a linear or non-linear combination of band values (regression, Artificial Neural Network – ANN) and so on.

Similarly, there are multiple choices for the input features. Broadly there are two types – considering the band values of the single pixel, and the features which require a group of pixels for extracting them such as texture or edges, shapes, etc. within the local neighbourhood. The former is also called spectral features and the latter as spatial features. The spatial features are calculated using intensity variation over a group of pixels. Band values is the most common choice of features for hyperspectral data. The next pixel-level spectral features derived from the band values are the slope of the spectral signatures in various wavelength regions or average brightness and so on (Heiden et al., 2007). The indices as bands can be used as features too (Xu, 2007). The pixel-level features are commonly used for detecting the target material or detecting the land cover. However, if the ground pixel size is from medium to large then the pixel level land use can be identified using proportions of land covers in that pixel as well (more later in Section 5.5, "Economic zone classification").

The spatial features are required if the spectral features are not able to resolve the identity of the target pixel. The spatial features are extracted using the intensity variation over a group of pixels within the scope. For example, images can be segmented using a texture and the texture category of the pixel can be added as a feature in addition to the pixel band values. The geometric properties of the group of pixels delineated by a boundary such as shape area etc. are used to identify the objects in the image. The lack of 3D information unlike handheld imagery (as only the top view is available), and the medium-to-coarse resolution of the hyperspectral spaceborne imagery provides less opportunities for such object-based classification. However, it is possible with high-resolution imagery by airborne or UAV platforms. Though spaceborne hyperspectral imagery may not provide delineated object boundaries, it still can provide the geometric properties for the delineated irregular boundary over the group of pixels. Fractal information of the region boundary is particularly useful in identifying man-made and natural classes. The boundaries of the manmade land

TABLE 5.1
Common Classifiers and the Associated Features Used for Hyperspectral Image Classification

Classifier	Discrimination function	Features	Scope
Urban indices	Band ratio, normalized index	Specific band values	Target pixel
SAM	Similarity measure	Band values	Target pixel
KNN	Similarity with k neighbour	Band values	Target pixel
NB	Conditional probability of a class given a pixel	Band values	Target pixel
ML	Conditional probability of a class given a pixel	Band values	Target pixel, local neighbourhood
SVM	Equation of a hyperplane	Band values	Target pixel, local neighbourhood
ANN	Non-linear combination of learned parameters, logistic regression	Band values, band values of surrounding pixels	Target pixel, local neighbourhood
CNN	Non-linear combination of learned parameters, logistic regression	Band values, edge or region features in the local neighbourhood of the pixel	Target pixel, local neighbourhood

cover or land use class are more regular than the natural ones (Deshpande, Sowmya et al., 2017).

All the above techniques mentioned for classification use the handcrafted features, that is, a set of predictors are chosen by a human expert. In the case of spectral features, the expert determines which bands are to be used, if other derived features are used how to calculate them and so on. If they are the spatial features, then the expert may define a particular filter to extract the edges or shapes directly and so on. Modern machine learning techniques such as Convolutional Neural Networks (CNN) learns the spatial features or spectral features automatically. Note that due considerations are required in designing CNN to extract the spectral features. Without such considerations, the automatic features extracted from the pixel vector input are not interpretable in the spectral domain. This is discussed in more detail in Section 5.6.3, "Spectral, spatial, and spectral-spatial neural networks". Table 5.1 summarizes some of the common discriminant functions and the associated common features used. Note that this is not a strict mapping.

With this background, we will discuss some of the basic classifiers for hyperspectral data, beginning with k-Nearest Neighbour (k-NN) and Spectral Angle Mapper (SAM) and then more complex classifiers such as Maximum Likelihood (ML). Finally, we will discuss the convolutional neural network (CNN) for the classification of hyperspectral images.

5.1.5 Urban Land Cover and Land Use

Detecting land use and land cover are the two main broad-level goals for the classification of hyperspectral images with moderate resolution. The high spatial resolution hyperspectral imagery from drones or any other imaging device is helpful in detecting urban objects as well as, for example, buildings, cars, trees and so on. The dominant features in such cases are spatial features. However, spectral features are dominant features for moderate resolution, say with 20 m to 30 m, imagery from spaceborne or airborne platforms. Hence the challenge is to classify the image into different land covers or land use with different granularity. The problem of extracting the fine granular land use class information or land cover information using spectral properties is a nontrivial problem.

Land use and land cover are related to each other and have been used in the literature interchangeably. Most of the urban studies consider individual classes or modified USGS classes (Cadenasso et al., 2007; Anderson et al., 1976) and consider land use land cover in an integrated manner. However, remote sensing observations provide spectral and spatial information about the surface properties only. Apart from a few, land use classes are hence to be identified using supplementary data or using proxy features derived from primary spectral and spatial features (Gamba, 2013; Turner et al., 1995; Anderson et al., 1976). Inferring land use patterns, for example, economic zones, from remote sensing has been a long-standing problem (Comber et al., 2012).

The land cover classification is relatively easy as compared to the land use classification. Any classifier with the basic primary features such as band values performs the required task. Whereas any classifier can classify the land covers using any type of features, spectral matching methods, and band values of the pixel (D-dimensional pixel vector) are the combination used often. These methods are easy to implement, and various land covers are easily detected using a library of reference spectra. One of the most important objectives for analysing the urban image is to classify images into broad-level VIS classes, and subsequently.

5.1.5.1 Vegetation-Impervious Surfaces-Vegetation (VIS) Model

Vegetation-Impervious Surface-Soil (VIS) classes (Ridd, 1995) are three broad-level urban classes that interact with their surroundings in a distinct manner physically. To the extent that, VIS classes at a broader level are sufficient to characterize the urban environment; for example, VIS distribution in each land unit may help in determining its land use. Ridd (1995) understood the need for an ecological perspective to study urban areas and developed a model that helps in better understanding the functional interaction between urban units and their surroundings.

The VIS model is similar to the conventional diagram used for soil classification. The VIS model can easily characterize multiple dimensions of the urban environment under the study (Ridd, 1995). Urbanization, depending upon the initial conditions at the development site and the temporal stage of the development, reduces the soil component or vegetation component. Initial plot clearance activities change vegetation or soil cover to impervious surfaces as these plots would occupy the structure intended for public or private use (Figure 5.2). As we can see, some of the conventional USGS classes (Anderson et al., 1976) coincide with the VIS classes. Each of the conventional USGS classes may be remodelled using VIS. However, the motivation of the VIS model is to identify classes that reflect biophysical functions and not just LU/LC

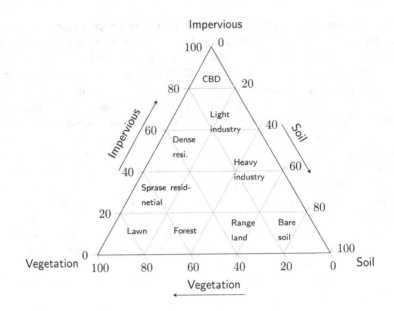

FIGURE 5.2 VIS model.

classes. The VIS model is used by researchers as a basic unit to interpret the urban structure and is a focus of many remote sensing studies (Weng et al., 2008; Xu & Gong, 2007; Phinn et al., 2002; Arnold & Gibbons, 1996).

As can be seen, the VIS model is useful for urban analysis on multiple levels. It not only provides the basic class schema but also a tool to analyse urban dynamics. The biggest advantage of the model is that VIS-based analysis is available for remote sensing investigations directly. This is because hyperspectral data can detect the broad-level and fine-level VIS classes more accurately and efficiently.

High Ecological Resolution Classification for Urban Landscapes and Environmental Systems (HERCULES) is another model focusing on VIS types and their structure for monitoring urban areas (Cadenasso et al., 2007). The objective of the HERCULES model is to complement existing approaches and enhance the ecological understanding of urban systems. HERCULES begins initially with three basic classes as VIS and then breaks each class further into six features: (1) coarse-textured vegetation (trees and shrubs); (2) fine-textured vegetation (herbs and grasses); (3) bare soil; (4) pavement; (5) buildings; and (6) the building typology. Each of these six components is allowed to vary independently of the other. Preliminary analysis suggests that HERCULES is better at predicting nitrate yield from urban watersheds than USGS classification (Cadenasso et al., 2007). Figure 5.3 shows the higher-level depiction of the model.

The VIS classification schema provides a unique flexibility to either use it for land cover or land use classification exclusively. Spatial configurations of land covers in a given area as a definition of land use class have been introduced by a few earlier studies (Phinn et al., 2002; Wu & Murray, 2003; Barnsley & Barr, 1996). VIS land

Classification of Urban Land Use and Land Cover 185

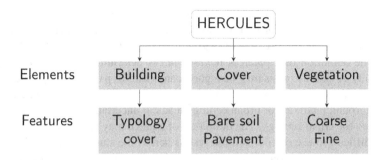

FIGURE 5.3 HERCULES model.

covers at broader class granularity are good candidates to begin this investigation. The potential of HS data to detect a large variety of urban materials provides an added advantage for such approaches. Furthermore, the VIS classification schema is more suitable for ecological studies as the land covers represent unique ecological functions.

VIS fractions can be used as proxies as well for land use. VIS fractions in a given area and its spatial configuration encode the information about its land use. Any land use can be seen as a collection of land covers/objects with some spectral properties and with a certain spatial configuration. The combination of VIS classes for land use would generally be unique. For example, sandy soil, trees, and green grass arranged in a certain manner and certain proportions represent a golf course visually (Deshpande et al., 2020). Of these two types, extracting fractions of VIS classes is relatively easy using hyperspectral data. Unmixing analysis is helpful. Before we see how VIS fractions are used for identifying economic zones, we briefly review some of the proxy features used for land use classifications.

5.1.5.2 VIS as Proxies

Deriving different proxy variables such as landscape metrics for understanding landscape ecology is very well studied in the past and by many researchers recently as well (O'Neill et al., 1996; Li & Reynolds, 1993; Herold et al., 2002; Wu, 2004; Ebert et al., 2009). The usefulness of comparative assessment of landscape metrics for different regions for ecological studies shows the possibility of extending similar approaches in understanding the urban landscape.

O'Neill et al. (1996) developed three indices to study landscape patterns, namely, Dominance Index (D1) – indicating dominant class, Contagion Index (D2) – indicating the probability of co-occurrence of the classes, and Fractal Index (D3) indicating irregularity of the class boundary. The metrics were modified by Li and Reynolds (1993) further and used by a number of researchers in later studies (O'Neill, et al., 1996; Li & Reynolds, 1993; Herold et al., 2002; Wu, 2004; McGarigal, 2014). Turner et al. (1989) and Wu (2004) study the effect of scale on landscape metrics, whereas Herold et al. (2002) use various landscape metrics to identify characteristics of land use classes, namely commercial zone, high-density residential zone, and low-density residential zones.

Three important concepts in landscape ecology are patches, corridors, and background matrixes. Each point in a landscape belongs to either a patch, a corridor, or a background matrix for any type of land mosaic, including suburban land (Forman, 1995). The definition of a patch is flexible and is determined by the phenomenon under study. A land use class may be a good candidate for patch, depending upon the scale of the analysis, However, a lack of systematic definitions of patch, corridors, and background matrixes for urban ecology leads to uncertainty in the calculation of the indices for urban ecology. The definition of a patch may vary with scale, making it more difficult to use or compare the results across the studies. Further, biotic and abiotic ecological functions interact with each other in a complex manner in an urban patch.

Barnsley and Barr (1996) used frequency and spatial configuration of land covers to infer land use from a SPOT-1 HRV image of London, England. In the first stage, different land covers within the images were identified and in the second stage, the frequency and spatial arrangement of land covers was aggregated using a square kernel. The system is called SPARK. The spatial arrangement of land covers was modelled by counting the adjacency of different classes in a kernel: for example, for three classes, the co-occurrence matrix would be 3×3 with a maximum possible adjacency score of 20. These concurrence metrics were compared with the template of a given land use. A matching score of 1 indicated a perfect match and 0 indicated otherwise. Classes considered were namely residential, low-density, medium-density, or industrial/commercial. A similarity of 0.83, 0.81, and 0.88 was found with the template respectively.

Barnsley and Barr (1996) did not account for the order of adjacency in deciding the co-occurrence score. Though regional properties are aggregated for the classification of land use, the label is assigned to a single pixel. Further, a pixel is assigned to a single class. This is a critical drawback as it distorts the true adjacency relations present in the kernel. The mixed pixels in the test or template kernel would exhibit complex adjacency relations and are not captured by a co-occurrence matrix, for example, two adjacent pixels having a 50% fraction of two land covers.

Veldkamp and Lambin (2001) provided an overview of land use models and discussed ideal attributes of the land use modelling in a detailed manner. Authors recognized the difficulty in incorporating social, political and economic factors because of a lack of spatially explicit data: the spatial units for each sub-process would be different. The authors suggest to:

1. Model proxies for driver parameters for a deeper understanding of the driving forces and land use changes; for example, the distance to a road is a proxy driver parameter for land use changes.
2. Incorporate existing known drivers of land use and investigate new ones. For example, factors affecting land use changes are global policies, microeconomics, local land use policies, and land management technologies.
3. Develop dynamic models of land use and biophysical parameters interaction. The biophysical response is feedback to the land use changes and land use models are often developed for predicting and assessing the impact of land use changes on the biophysical parameters.

4. Model the variability in the spatio-temporal behaviour of the system at different scales.

Thus, Veldkamp and Lambin (2001) highlighted the importance of proxy variables for understanding land use classes and land use changes. The suggestions are appropriate and there are limited efforts in that direction, generally speaking. Of all the suggestions, modelling spatio-temporal behaviour at different scales is a very challenging task and might not be an appropriate goal as the driver parameters change with scale.

Herold et al. (2002) used digitized colour infrared photographs of 3 m resolution and landscape metrics to study the land use structure of Santa Barbara, California. The classes considered were namely commercial zone, high-density residential zone, and low-density residential zones. Landscape metrics such as Fractal Dimensions, Contagion Index, Patch Density, and Percentage Land were calculated for the images covering different districts of Santa Barbara, California (1978, 1088, 1998). The patch was considered to be the unit with homogenous properties namely size, shape, tone, colour, and texture. These are called homogenous urban patches – HUPs. The HUPs were selected in such a manner that a given HUP would correspond to one type of land use; for example, residential.

The study by Herold et al. (2002) showed the possibility of using landscape metrics for the separation of land use classes. The right combination of various metrics can further be considered as a prototype of a given class. Four metrics were found more suitable in particular: Dominance of vegetation or built-up cover, housing density, mean plot size, and spatial aggregation of built-up area. The metrics considered were also found useful in temporal studies of urban growth. Occurrence of built-up area and vegetation in certain proportions were observed for each land use class considered. The ratio of built-up to vegetation for commercial, high-density residential, and low-density residential was 0.90, 0.65 and 0.30 respectively.

Rashed and Weeks (2003) used mixture analysis and developed a relation between wealth index and fractions of impervious surface, soil, vegetation, and additional landscape metrics. The wealth or net worth index indicated the average wealth of a household weighted by age, race, and income for a census tract. Further, the wealth index was treated as a proxy to access the resources. The VIS fractions were normalized and aggregated at a census tract level. Soft classification results by unmixing were converted to hard classification per pixel basis by majority rule. The pixel was assigned to a land cover class having maximum proportion in that pixel. They found a strong correlation between the patch density of impervious surfaces and vegetation with wealth indices. However, the definition of a patch was ambiguous. Census tract might not conform to the patch form ecological perspective and hence results are difficult to interpret. The fraction of land covers was used independently to correlate to the wealth index and the spatial association between these fractions within a given pixel was ignored. The study was limited to multispectral data (Landsat).

Other urban studies on vulnerability assessment and spatial organization highlight the utility of proxy variables in inferring important socio-economic properties. Cutter, Boruff, and Shirley (2003), in their study of urban properties influencing the vulnerability index, indicated a correlation between economic factors and vulnerability.

Cutter et al. (2003) studied various socio-economic factors of urban areas to arrive at a vulnerability index. The socio-economic data was collected for 3141 counties in the US around 1990, and 85 factors were selected for further analysis. Factor analysis using PCA was performed to identify eleven latent socio-economic variables explaining a maximum variance of 76\%. Among the factors, personal wealth, age, and density of built environment collectively explained 30\% of the variance with a correlation with social vulnerability, which is 0.87 and 0.98 for economic data and density of built environment respectively.

Bertaud (2004) characterizes the spatial structure of the city using built-up density. The spatial structure is mainly characterized using population density and its spatial profile. The study of 49 cities around the world showed a correlation between density and location on the continent; for example, US cities have the lowest population density whereas Asian cities have the largest. The increasing population density in an outward direction from the city centre constitutes a liability to the city.

5.2 CLASSIFICATION WITH URBAN INDICES

One of the simplest ways to classify a hyperspectral image pixel is to use the inherent spectral properties of the target material and define a discriminating function which is a ratio of the two or more significant bands. A more advanced discriminator is a ratio of differences and the addition of two or more bands. The ratio provides normalized output values that are proportional to a particular property of the target material, or it indicates the confidence in detection of the target material. For example, for a hypothetical concrete index whose values are ranging from 0 to 1, a value of 1 indicates concrete, a value of 0.5 indicates the mixture of concrete with some other material, and 0 indicates vegetation may be. Please note the vegetation is not the opposite of concrete in this example – the range of values for different land covers will fall within a certain range and the threshold is determined by performing validation studies accordingly. Thus, the index value is a degree of belief about the target material which is deduced from its spectral properties. This is a much more deterministic method as the unique spectral properties of the material are used to discriminate the target materials. The general set-up of these methods is like the following:

Two (or multiple) bands are chosen that uniquely identify the target material. This is a critical step as it determines the success of the index. These bands are generally around the region of the diagnostic absorption and/or reflection. The ratio or difference of these bands is supposed to be statistically significant to differentiate the target material. For example, the vegetation index uses red and near-infrared band differences and the addition of the same bands. The difference between the infrared band value and the red band value is diagnostic for vegetation, that is, the difference between these two bands is relatively high for vegetation as compared to the other materials. Moreover, the difference is proportional to the chlorophyll content of the leaves. The difference decreases as the chlorophyll content decreases.

The difference or ratio is further normalized between −1 to 1 or 0 to 1. It could be a ratio of the difference of two bands and the addition of the same two bands, as in the Normalized Difference Vegetation Index (NDVI). It could be a simple ratio as well, like the Ratio Vegetation Index (RVI) (Xue & Su, 2017). The ratio of differences

Classification of Urban Land Use and Land Cover 189

of two bands or ratio of difference and addition of two bands or simply a ratio, normalizes the illumination differences and to some extent atmospheric attenuation effects (Richards & Jia, "Multispectral transformations of image data", 2006; Xue & Su, 2017). The index is further modified by adding a constant or by other operations.

The index is generally related to the bio-physical characteristics of the target and hence it is useful in monitoring those parameters of the natural or anthropogenic system. For example, NDVI is directly proportional to the chlorophyll content and/or leaf area index of a tree. The concrete index (Samsudin et al., 2016) is proportional to the degradation of concrete because of age.

The range of values of an index for different materials is determined by extensive validation studies and/or the basics of spectral properties (Xue & Su, 2017). Based on the studies and analysis of the index function, a range of values are assigned to the given target material or the degree of it. For example, the values above 0.2 in NDVI are values for vegetation and the number is directly proportional to the health of the vegetation (Hashim et al., 2019). The set of values can be even used for intraclass classification. For example, grass and trees in urban areas can be identified by setting up the right location-specific threshold on NDVI. There are many advantages of index-based classification:

a. The index is related to a bio-physical parameter and hence the changes in the state of the target over time are easily monitored. It may require complex machine learning models otherwise to classify the target material state. The degree of confidence associated with the indices is very high as well. This is because of the inherent unique spectral properties used in the development of the index.
b. They are basic mathematical operations and hence are very fast. They can be easily deployed on a personal computer to process large spatio-temporal data quickly. This is essential in many earth observation scenarios. Processing multiple images over time and space is required rather than a single image. In all such occasions, index-based methods are the first choice. For example, understanding anthropogenic activities and their effects on the surrounding environment requires monitoring many parameters at a given instance over a large space and time. Simultaneous studies of increased human activity and its effects such as air, water and soil pollution are required too. Index-based methods can capture each physical state by an index and then they can be used for establishing causal relations between the human activities and the environmental effects. The low computational complexity of index-based methods enables such studies.
c. They can even be used to reduce the communication load between satellites and ground stations – the information about the target material such as identity or state can be sent directly to a ground station rather than the raw data. For example, the BIRD satellite by DLR uses onboard hardware implemented to classify the image to create a thematic map. The thematic map generation is processed by a neural network implemented in the neuromorphic "Recognition Accelerator NI1000" from Nestor Inc. The main idea was to reduce the size of the data sent to the ground station and provide the results without any time delays. The classifier was designed to detect a fire directly. In addition to the fire, the additional classes were *water, bare soil, cloud,* and

urbanization. Unsuccessful detection of these classes resulted in "no class" labels. An additional goal was to create a thematic map and send it to the ground station (Halle et al., 2002).

d. Similarly, the onboard computer can calculate the forest fire index onboard and send it to the ground station instead of complete information. These kinds of onboard computing systems are in an experimental stage currently (Lorenz et al., 2017; Lopez, et al., 2013; Bernabé et al., 2011). As indices use basic mathematical operations, they provide certain advantages in such onboard process chain architectures. They can be implemented at the hardware level and hence they can provide near real-time results. Even if they are implemented by onboard software, the computational load will be very minimal as compared to the other process chain results such as classification or unmixing.

5.2.1 Useful Properties of Urban Materials for Developing Indices

The spectral properties of urban materials are discussed in detail in Chapter 4. Here we take a quick review of them in the context of the urban material indices. Most of the urban materials don't show very strong diagnostic absorption within the spectral range of 400 nm to 2500 nm, with a 10 nm spectral resolution. However, the slope of the spectral signature plot – wavelength vs reflectance – for different wavelength ranges is distinct. Thus, the discriminating spectral information for urban materials is spread over the entire 400 nm to 2500 nm range. The blue reflectance of the concrete and metal roofs is very high as compared to the soil and vegetation. Additionally, the reflectance of these two materials is very high in the SWIR range as well. These bands provide the basis of spectral indices for individual urban materials, and the flat nature of spectral signature provides the basis of the built-up area index or impervious surfaces index.

5.2.2 Various Urban Indices

Xu (2007) used a soil-adjusted vegetation index, a water index, and a built-up index by Zha et al. (2003) as the features and used a maximum likelihood classifier for extracting the built-up area of two cities in south-eastern China. The indices were calculated using Thematic Mapper (TM) and Enhanced Thematic Mapper (ETM) bands. The classification accuracy of ~92% to ~98% was achieved. Zha et al. (2003) introduced one of the earliest built-up indices that use TM bands 4 and 5. The bands are referred to as near infrared (NIR) and shortwave infrared (SWIR) respectively (USGS, 2023). The spectral coverage of the NIR band is from 770 nm to 900 nm and for SWIR it is from 1550 nm to 1750 nm. The index is defined as the ratio of difference of band 5 and 4 and the sum of the same bands. The defined NDBI includes some of the spectrally similar classes like barren land, that is, some of the urban land covers other than built-up are considered as built-up. Nonetheless, this is one of the initial attempts to use spectral properties for developing the index for urban areas.

The developed normalized difference built-up index (NDBI) was used to classify the built-up area of the city of Nanjing, in eastern China. The difference of NDBI and normalized difference vegetation index (NDVI) was used for actual classification. The NDBI values were mapped to 0 to 254 first where negative values were mapped to 0 and positive values to 254. NDVI was subtracted from the outcome of the first step, that is, the mapped NDBI values to 0 to 254. The second step results in negative or 0 values for farmlands and woodlands, thus separating built-up with positive NDBI values. The accuracy of the classification of the built-up and the non-built-up was comparable with the manual classification and better than the maximum likelihood classifier. NDBI overestimated the built-up area by 2% as compared to the manual method.

The index-based built-up index (IBI) developed by Xu (2008) is an index defined over the indices, that is, it is developed using the other indices instead of using band values directly. The intuition is to enhance the NDBI further when the soil-adjusted vegetation index (SAVI) and modified normalized difference water index (MNDWI) values are low. Thus, it takes the difference between NDBI and the summation of SAVI and MNDWI. The values are normalized by dividing the difference by the sum of all the indices used in the numerator. The values thus arrived at are divided by 2 to avoid small values. The index ranges from –1 to 1. The IBI results for the Fuzhou area of China were compared to the built-up area extracted using SPOT imagery. The threshold of 0.013 was chosen manually for this study to classify built-up areas. An overall accuracy of ~98% was achieved.

The initial efforts were focused on developing indices for broad-level classes such as built-up instead of individual urban materials. This was because of the moderate resolution and broad band data collection of the Landsat imagery. Moderate resolution results in a mixed pixel and hence it is difficult to label a pixel with a single urban material such as concrete or bitumen or a metal roof and so on. Naturally, these were the attempts to develop the index for a built-up class. Attention was not given to the index development for individual built-up land cover categories. However, availability of the high-resolution multispectral data and/or hyperspectral data in recent years has helped in developing indices for individual members of the built-up class. For example, concrete index, roof index and so on. With high-resolution data, the chances of pixels with a single material increases, hence it becomes possible to detect the material using an index. With hyperspectral data or narrow band data collection detailed spectral properties are available which enable material-level detection.

Samsudin et al. (2016) developed two indices, namely the normalized concrete condition index (NDCCI) and the normalized difference metal condition index (NDMCI). These indices were developed not only to identify concrete and metal roofs but also to qualitatively assess their age. The indices were developed for the World View 3 (WV3) data. First, the spectral properties of the target materials were studied using field spectrometer measurements. Next, the significant wavelengths for the target materials were identified using the ratios that were providing the best classification results by SVM. The significant wavelengths identified for the concrete were 526 nm and 770 nm, corresponding to the green and NIR1 bands of WV3. The significant wavelengths identified for metal roofs were 743 nm and 753 nm, corresponding to the red edge and NIR1 WV3 bands. Thus, NDCCI and NDMCI is given by:

$$NDCCI = \frac{NIR1 - Green}{NIR1 + Green} \qquad \text{Equation 5.1}$$

$$NDMCI = \frac{NIR1 - Red\,edge}{NIR1 + Red\,edge} \qquad \text{Equation 5.2}$$

The indices for WV3 bands. As the indices are specified by the significant wavelengths they can be easily extended to any hyperspectral data. The last step was the validation of the results and selection of the ageing-related thresholds for both the target materials. The index values for the new concrete were found to be in the range of 0.3264 to 0.5775 and for the weathered concrete 0.1922 to 0.3264. Similarly, the index values for the new metal were found to be in the range of −1 to 0.0196 and for weathered metal, 0.2216 to 0.9588. The indices provided better accuracy in classification as compared to the SVM results. The NDCCI accuracy was ~84% and NDMCI accuracy was ~94 as compared to ~73%, and ~62.5% for the SVM, respectively.

Sameen et al. (2016) followed a similar approach to Samsudin et al. (2016), that is, they selected significant bands and then using these significant bands they developed the index. The significant bands were selected using particle-swarm optimization. Thus, particle swarm optimization is used as a tool for selecting the significant bands. The index is still based on the spectral properties of the target materials. The index was developed for WV3 bands. Note that the significant bands were selected using WV3 data unlike Samsudin et al. (2016) who used field spectrometer data and then applied it to WV3. The index WV3 BSI is given by:

$$BSI = \frac{band4 - (2 \times band7)}{band4 - (2 \times band7)} \qquad \text{Equation 5.3}$$

$$BSI = \frac{Yellow - (2 \times NIR1)}{Yellow - (2 \times NIR1)} \qquad \text{Equation 5.4}$$

There are many other minor variations of the basic built-up indices. Some of them add a temperature band to the basic index for better results (Liu et al., 2013; Assyakur et al., 2012; Xu, 2010; Bouzekri et al., 2015) and some define index using other indices (Xu, 2008). Such indices are not discussed in detail here. This is because currently most of the hyperspectral sensors, be they airborne or spaceborne, are confined to the spectral range of 400 nm to 2500 nm. An example of an index defined over index (Xu, 2008) is given below:

$$IBI = \frac{NDBI - \frac{SAVI + MNDW}{2}}{NDBI + \frac{SAVI + MNDW}{2}} \qquad \text{Equation 5.5}$$

Classification of Urban Land Use and Land Cover

$$IBI = \frac{\left(\frac{([1550,1750]-[770,900])}{([1550,1750]+[770,900])}\right) - \frac{\left(\frac{([770,900]-[630,690])}{([770,900]+[630,690])}\right) + \left(\frac{([520,600]-[1550,1750])}{([520,600]+[1550,1750])}\right)}{2}}{\left(\frac{([1550,1750]-[770,900])}{([1550,1750]+[770,900])}\right) + \frac{\left(\frac{([770,900]-[630,690])}{([770,900]+[630,690])}\right) + \left(\frac{([520,600]-[1550,1750])}{([520,600]+[1550,1750])}\right)}{2}}$$

Equation 5.6

where, IBI is an index-based built-up index, and the wavelength range is specified using TM bands as the index was validated using TM bands. The values above 0.013 were used for mapping built-up areas using this index.

Table 5.2 summarizes a few important urban indices. The indices are specified using wavelengths as well, along with their band numbers according to the sensors for which it was developed. It is difficult to name the bands of the hyperspectral sensors, like red, green, NIR etc. as there are many. It is easy to specify them with a number or the band centre wavelength. Describing an index using wavelength over the band number is preferred to avoid any ambiguity arising from the differences in the specification of the sensor bands for different sensors.

5.3 NEAREST NEIGHBOUR CLASSIFICATION

Nearest neighbour is one of the simplest of the classifiers. The decision is made based on the k-nearest training samples for a query pixel. The distance between any training pixel and query pixel is calculated by any distance or inverse of similarity measure, for example, Euclidian distance. There is no processing required during training, all the processing takes place during the time of classification or deployment or testing. Hence it is referred to as lazy learning as well. It also belongs to the category of instance-based learning (Mitchell, "Instance-based learning", 1997).

The classification procedure for a query pixel calculates the distance from each training pixel in the list of labelled pixels. Then, top k training pixels which are close to the query pixels are chosen. The majority class label among the top k training pixels is then assigned to the query pixel. Thus, it takes a constant amount of time for the classification of any pixel. The total time for classification of each pixel is aggregated time taken for calculating the distances, sorting the list, and taking the majority among top k labels. The procedure can be viewed as seeking a local approximation to a complex function defined over the population. Hence k-NN provides advantages in the case when a complex function can be defined as a collection of local approximation. The main disadvantage of the k-NN algorithm is the overall time

TABLE 5.2
Important Urban Indices along with Their Formulas

Index	Name	Inventors	Bands	Wavelengths[1]	Threshold
Normalized difference vegetation index	NDVI	(Rouse Jr J., Haas, Schell, & Deering, 1973; Rouse Jr J., Haas, Schell, & Deering, 1973)	MSS1: $\dfrac{(MSS7 - MSS53)}{(MSS7 + MSS5)}$	$\dfrac{([800,1100] - [600,700])}{([800,1100] + [600,700])}$	>0.2 (Hashim, Abd Latif, & Adnan, 2019)
Soil adjusted vegetation index	SAVI	(Huete, 1988)	TM:[2] $\dfrac{(NIR - Red)(1+L)}{(NIR + Red + L)}$ Where, L = 0 to 1 (from low to high vegetation density)	$\dfrac{([770,900] - [630,690])}{([770,900] + [630,690])}$	>0 default
Normalized difference water index	NDWI	(McFeeters, 1996)	TM: $\dfrac{(TM2 - TM4)}{(TM2 + TM4)}$	$\dfrac{([520,600] - [770,900])}{([520,600] + [770,900])}$	>0 default, >0.19 to >0.243[3]
Modified normalized difference water index	MNDWI	(Xu H., 2006)	TM, ETM: $\dfrac{(TM2 - TM5)}{(TM2 + TM5)}$	$\dfrac{([520,600] - [1550,1750])}{([520,600] + [1550,1750])}$	>0 default
Normalized difference built-up index	NDBI	(Zha, Gao, & Ni, 2003)	TM: $\dfrac{(TM5 - TM4)}{(TM5 + TM4)}$	$\dfrac{([1550,1750] - [770,900])}{([1550,1750] + [770,900])}$	NA[4]
Index-based built-up index	IBI	(Xu H., 2008)	TM: $\dfrac{NDBI - \dfrac{SAVI + MNDW}{2}}{NDBI + \dfrac{SAVI + MNDW}{2}}$	Each individual index is as defined earlier	>0.013

Classification of Urban Land Use and Land Cover

Name	Abbrev	Reference	Formula	Threshold/Range
Normalized difference concrete condition index	NDCCI	(Samsudin, Shafri, & Hamedianfar, 2016)	WV3: $\dfrac{(NIR1 - Green)}{(NIR1 + Green)}$; $\dfrac{(770-526)}{(770+526)}$	0.3264 to 0.5775 new concrete, 0.1922 to 0.3264 weathered
Normalized difference metal condition index	NDMCI	(Samsudin, Shafri, & Hamedianfar, 2016)	WV3: $\dfrac{(NIR1 - Rededge)}{(NIR1 + Rededge)}$; $\dfrac{(753-743)}{(753+743)}$	-1.00 to 0.0196 new, 0.2216 to 0.9588 weathered
Built-up spectral index	BSI	(Sameen & Pradhan, 2016)	WV3: $\dfrac{(Yellow - 2NIR1)}{(Yellow + 2NIR1)}$; $\dfrac{([585,625] - 2[770,895])}{[585,625] + 2[770,895]}$	Visually interpreted based on the results
Enhanced Built-Up and Bareness Index	EBBI	(As-syakur, Adnyana, Arthana, & Nuarsa, 2012)	ETM+: $\dfrac{(ETM5 - ETM4)}{(10\sqrt{ETM5 + ETM6})}$; $\dfrac{([1550,1750] - [770,900])}{(10\sqrt{[1550,1750] + [10400,12500]})}$	0.10–0.35 built-up, >0.35 soil
Normalized difference impervious surface index	NDISI	(Xu H., 2010)	ETM+: $\dfrac{ETM6 - \left(\dfrac{NDWI + ETM4 + ETM5}{3}\right)}{ETM6 + \left(\dfrac{NDWI + ETM4 + ETM5}{3}\right)}$; $\dfrac{[10400,12500] - \left(\dfrac{NDWI + [770,900] + [1550,1750]}{3}\right)}{[10400,12500] + \left(\dfrac{NDWI + [770,900] + [1550,1750]}{3}\right)}$	>0 (default), ~0.10–0.12 (manually adjusted)

Notes:

[1] The wavelength range used in the table is based on the band number mapping designated at www.usgs.gov/faqs/what-are-best-landsat-spectral-bands-use-my-research. The indices are specified using the wavelength range for easy implementation of them using hyperspectral data. The band names such as red, green, NIR etc. and the associated wavelengths are not used consistently in the literature (for example Xu, 2008, uses MIR for TM band 5, the band 5 is referred to as SWIR in Landsat documentation www.usgs.gov/faqs/what-are-best-landsat-spectral-bands-use-my-research). To avoid this confusion the index is given as the band arithmetic used in developing them and the corresponding wavelengths.

[2] The original index is defined with the band names such as NIR etc., but the IBI uses TM bands hence the wavelengths are given as per the TM bands range.

[3] for achieving same accuracy as MNDWI.

[4] for achieving same accuracy as MNDWI.

taken for classifying the whole image as the training procedure is repeated for every pixel. Additionally, all the bands are used in calculating the similarity with the target training classes. Some band values would be more important than others for a class.

Algorithm 5.1 is the simplest nearest neighbour implementation. The improvement over the basic version assigns different weightage to the vote by each training pixel. Training samples close to the query pixel are given more importance than the distant ones. The weightage calculation step in Algorithm 5.1 changes to:

$$y = \max_{y \in Y} \sum_{i=0}^{k} w_i \delta(y, Ld_i)$$

Equation 5.7

where, $w_i = \dfrac{1}{d(x, x_t)^p}$. The value of p is usually 1. The values greater than 1 would penalize the distant training pixels more (Cunningham & Delany, 2020).

The popular classifier for hyperspectral data – spectral angle mapper (SAM) – is one implementation of k-NN, where k is 1, there is only one training pixel for each class (generally speaking), and the distance measure is cosine similarity or spectral angle (as it is referred to). In SAM, we often choose the references, that is, the training pixels, from the library of laboratory spectra. Another way is to choose an average over a very homogenous region of pixels within the image. In both cases, the strong assumption about representatives of the single example for the class may be a

---

ALGORITHM 5.1
Nearest Neighbour Algorithm for Hyperspectral Pixel Classification

```
        Input: k, X^L = list of training samples, x = a query pixel,
               distance metric d = d(a, b)
Output: y Class label for query pixel
k-NN(k, X^L, x)
{
        Y = 0, d = 0, ld = list of training distances
        FOR every i in X^L
            Ld_i = d(x, x_{t=i}), where x_{t=i} ∈ X^L

        Ld = Sort ascending¹ (Ld) #sort the list of distances
        y = max Σ_{i=0}^{k} δ(y, Ld_i)#take majority of the labels
            y ∈ Y

        where δ (y, kLd_i) = 1 if y = y of kLd_i and 0 if y ≠ y of kLd_i
            RETURN y
        }
```

Note:
1. This a simple implementation for better understanding of the algorithm. The sorting of the complete list may not be necessary, only top K distances can be remembered using an efficient data structure such as a min-heap.

Classification of Urban Land Use and Land Cover

problem. Nevertheless, it is possible because the laboratory references are prepared very carefully. The SAM algorithm is given in the table of Algorithm 5.2.

The success of SAM also depends upon the accurate atmospheric correction procedure. This is because the reference spectra collected using field spectrometers or laboratory spectrometers generally are devoid of any atmospheric effects (other than the noise in water absorption bands). Given the maturity of image-based atmospheric correction methods and advancement in Radiative Transfer Codes (RTC), atmospheric effects are effectively removed, enabling the effectiveness of SAM.

In Algorithm 5.2 the distance metrics $d(a, b)$, SAM (a, b) are the distance measures and similarity measures, respectively. There are many such measures which satisfy the axioms of the distance measure. Any such distance or similarity measure can be used in these algorithms. The distance and similarity have an inverse relation. That is, the distance between the similar vectors is a small number. As the similarity increases the distance decreases. If the similarity is normalized between 0 to 1 where 1 indicates the perfect match, the distance can be calculated by subtracting similarity from 1. For example, 1-SAM (a, b) would provide the distance between a and b. To define the distances formally we use the following notations:

Let $L = \{r_{r=1}, r_2, r_3 \cdots r_R, \}$ be the library of R references. "r" subscript indicates the index of the reference library. Each member of the library $r_{r=1}, r_2, r_3 \cdots r_R \in L$ are the reference spectra such as concrete, bitumen, metal roof and so on. Each member $r_1, r_2, \cdots r_R$ is represented as a vector of reflectance values r_d of each band as $r_r = \{r_{d=1}, r_2, r_3 \cdots r_D\}$ where D equals the number of bands. Thus, the spectral vector is denoted by a bold small letter and the reflectance values of the spectral vector are indicated with the same small letter with d superscript for band or wavelength index.

ALGORITHM 5.2
Pixel Classification Using Spectral Angle Mapper (SAM)

```
Input: X ∈ R(np) x d = list of pixels in the image; np = m x n;
L = list of reference pixels (R x d) where R = number of references;
   distance metric d = SAM(a, b)
Output: lC = list of class labels for corresponding pixels in the
   input list, y = class label of a query pixel

Create empty lC (npX1)

classSAM(x, r)
{
FOR every x_i in X
    FOR every r, in L
        SAM (x_i, r_r )
    y_i= Class (MIN of {SAM(x_i,r) for all rs})
    SET lC_i=y_i; assign i^th element of lC to y_i
}
RETURN lC
```

The target pixel vector is denoted by $x_t = \{x_{d=1}, x_2, x_3 \ldots x_D\}$. The target spectrum x_t can be from spaceborne or airborne hyperspectral image, UAU, or field spectrometer. With these notations, we can express the distances used in k-NN or SAM-like algorithms.

5.3.1 Spectral Angel or Cosine Similarity

SAM is one of the primary techniques that uses a spectral library to classify the pixels in an unsupervised manner. The goal is to classify a pixel by comparing the spectrum of that pixel with the spectrum of the known material. If the pixel spectrum matches with a particular reference spectrum of the known material in the library, we can assign that pixel to that particular material class. The "matching" or "closeness" of the two spectra is decided using the angle between the two spectral vectors with D dimensions where D is the number of bands in a spectrum, hence the name. Formally expressed as:

$$\frac{x_t \cdot r_r}{\|x_t\| \|r_r\|} \qquad \text{Equation 5.8}$$

Figure 5.4 illustrates the geometric representation of the SAM. A spectral angle mapper measures the angle between a reference spectrum and a target spectrum in D-dimensional space, where D is equal to the number of bands. The figure shows a reference spectrum and two-pixel vectors 1 and 2 in 3 Bands space. The pixel vector

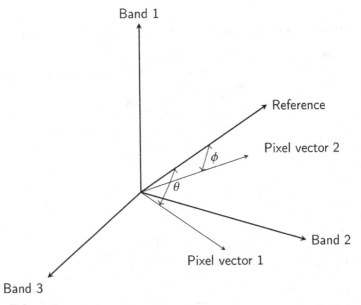

FIGURE 5.4 Geometric representation of spectral angel mapper or cosine similarity.

2 is closer to the reference spectrum than the pixel vector 1. SAM compares this angle and assigns the label of reference spectrum to the pixel vector 2.

Spectral Angle Mapper (SAM) has been a spectral similarity measurement and classification technique of choice for HS data (Goetz et al., 1985; Kruse et al., 1993; Yuhas et al., 2001). Various recent urban mapping studies using hyperspectral data deploy SAM for measuring similarity between reference spectra and target spectra and subsequent classification (Hepner et al., 1998; Chen & Hepner, 2001; Divya et al., 2014). Hepner et al. (1998) used SAM for classifying AVIRIS images to enhance the radar data analysis.

Chen and Hepner (2001) used SAM for imaging spectroscopy studies of Park City, Utah, USA. They attempted to analyse the fitness of the field spectra of urban materials in discriminating AVIRIS data. SAM was used for spectral matching with unmixing analysis. They found spectral variability increases uncertainty in the analysis of similar types of material. Hence, they suggest hybrid spectral analyses to resolve the spectral ambiguity, instead of a single analysis technique.

Divya, et al. (2014) compared a visual interpretation of Linear Imaging Self-Scanning Sensor (LISS) images covering urban and semi-urban areas of Valsad, Gujarat, India, to the classification results of Hyperion. A Hyperion image was corrected for atmospheric effects using Fast Line-of-sight Atmospheric Analysis of Spectral Hypercubes (FLAASH) and dimensionality was reduced using a combination of genetic algorithm and minimum noise fraction. SAM was used to identify six classes, namely built-up, built-up sparse, crop, fallow, scrubland, and water. They did not find any significant difference for the classes, but vegetation. Their conclusion about the discrimination ability of Hyperion data for urban classes is weak, as the standard illumination data for the on-screen interpretation of LISS image is not given. Further, the classification level is coarse and hypothetically, hyperspectral data may not provide any advantages any way at that level. Classes should have been broken down further to correct assessment and to exploit the strength of HS data.

Since its initial application in Spectral Image Processing System (SIPS) developed by Kruse et al. (1993), SAM has been used extensively for target detection and it provides a benchmark performance for other methods of spectral similarity. The distinct advantages of the method are its simplicity and computational efficiency. Further, the technique is amplitude invariant – the angle between the two vectors does not change with reflectance values amplitude of a particular band. The amplitude-invariant nature of the SAM is a disadvantage as well: In cases where two spectral curves are very similar in shape with a difference in amplitude values, SAM does not differentiate those materials. The Normalized Spectral Similarity Score (NS^3) method developed by Nidamanuri and Zbell (2011) is motivated to overcome this drawback.

5.3.2 Normalized Spectral Similarity Score (NS3)

The shape of the spectrum of some of the materials is very similar to each other with only differences in amplitude values. For example, cement concrete road and bitumen concrete road. It is not possible to discriminate such materials using SAM. Thus, including amplitude information in similarity measurement is required to discriminate between such targets. The NS^3 algorithm developed by Nidamanuri and

Zbell (2011) incorporated amplitude information in a similarity measurement of two spectra. The formulation is given below:
The spectral amplitude difference is calculated as:

$$A = \left[\frac{1}{D-1} \sum_{i=1}^{D} [x_d - r_d]^2 \right]^{0.5} \qquad \text{Equation 5.9}$$

where, N = number of interval and r_r and x_t are reference and target spectrum.
A values are further normalized using the following equation:

$$\hat{A} = \frac{A - A_{min}}{A_{max} - A_{min}} \qquad \text{Equation 5.10}$$

Further, this normalized spectral difference is incorporated into the formula to give a final match score as:

$$B = \sqrt{\hat{A}^2 + (1 - \cos\theta)^2} \qquad \text{Equation 5.11}$$

Value of B range from 0 to $\sqrt{2}$ where 0 indicates a perfect match. The value B – that is the similarity score between two spectra – is sufficient to detect the target material. However, it is further refined to give an empirical discriminative function as:

$$f = 0.6B + 0.4\Delta B \qquad \text{Equation 5.12}$$

where, $\Delta B = \frac{B_2 - B_1}{B_2}$ and B_1 is the highest similarity score for a given reference and target spectra, and B_2 = 2nd highest similarity score for the reference and target spectra.

Nidamanuri and Zbell (2011) investigated the proposed method by comparing it with SAM and CCSM methods. A spectral library was created using a field spectrometer. Historical HyMap imagery was used for validating the classification of different crop types. Though the results show improvement in overall accuracies, SAM and NS[3] show similar trends for most of the crops.

5.3.3 Hamming Distance

In their early review of hyperspectral data characteristics and analysis techniques, Goetz et al. (1985) suggested Hamming distance as a similarity measure between reference and target spectrum. The algorithm using binary encoding in such a manner provides a fast method for searching a hyperspectral image for a target material.

The reference spectrum and target spectrum are coded in binary format first and then Hamming distance between binary-coded spectra is calculated to see how close they are to each other. Binary coding is performed by a simple operation: reflectance values in each spectrum above its average reflectance are coded as 1 and below average are coded as 0. Formally, the scalar mean over all the values of pixel vectors is given by:

$$\mu_t = \frac{1}{D}\sum_{d=0}^{d=D} x_d \qquad \text{Equation 5.13}$$

and

$$\mu_r = \frac{1}{D}\sum_{d=0}^{d=D} r_d \qquad \text{Equation 5.14}$$

where, μ_t and μ_r are the target mean and reference mean respectively. The next step is the binary encoding of the spectra with the following steps:

$$\hat{x}_d = h(x_d - \mu_t) \qquad \text{Equation 5.15}$$

$$\hat{r}_d = h(r_d - \mu_r) \qquad \text{Equation 5.16}$$

where,

$$h(x) = \begin{cases} 0, & x < 0 \\ 1, & x \geq 0 \end{cases} \qquad \text{Equation 5.17}$$

is a binary encoding function.

The Hamming distance between the two spectra is given by:

$$d^H = \sum_{d=0}^{D} (\hat{x}_d \oplus \hat{r}_d) \qquad \text{Equation 5.18}$$

where $\oplus$ is a logical exclusive OR operator.

5.3.4 CROSS CORRELOGRAM SPECTRAL MATCHING (CCSM)

A cross correlogram is calculated using the ordinary linear correlation between the target spectrum and the reference spectrum at different positions on the x-axis. To achieve this, the reference spectrum is shifted on the x-axis and at each shifted position correlation between the two spectra is calculated. The cross correlation (ccr) for m match position is given by:

$$ccr_m = \frac{n\sum r_t x_t - \sum x_r \sum r_t}{\sqrt{\left[n\sum x_r^2 - (\sum r_r)^2\right]\left[n\sum x_t^2 - (x_t)^2\right]}} \qquad \text{Equation 5.19}$$

where, r_r and x_t are reference and target spectrum respectively, and n is the number of overlapping positions. By convention, the negative shift indicates that the reference spectrum is shifted towards lower wavelengths.

Meer and Bakker (1997) used a cross correlogram to measure similarity between the target spectrum and the reference spectrum of kaolinite. They showed that the addition of 5% noise in the target spectrum did not affect the skewness of the correlogram. Subsequent to these studies, Meer (2000) extended the CCSM by removal of the continuum and tested it on an AVIRIS target spectrum of kaolinite. They found CCSM with continuum removal to be more sensitive to subtle spectral features and noise. They suggested the use of an appropriate wavelength interval covering single absorption instead of selecting multiple absorptions.

Senthil Kumar et al. (2010) combined CCSM with the Variable Interval Spectral Average (VISA) method and tested its performance on two datasets: one a laboratory simulated image and the other an AVIRIS sensor 92AC3C for a test site called Indian Pine in north-western Indiana (soybean, wheat, grass, etc.). VISA and SAM provided almost similar results. Mixed pixel performance for VISA gave the poorest classification accuracy. However, the VISA and CCSM method applied together improved the accuracy of the classification of mixed pixels than their individual counterparts.

5.4 MAXIMUM LIKELIHOOD CLASSIFIER

We can use conditional probability as a discriminant function for classifying a pixel. More specifically, we can use the probability of a particular class given a pixel. A pixel is assigned to the class which is most probable given a pixel vector. Formally,

$$x \in y_{k=i} \text{ if } p\left(y_{k=i}/x\right) > p\left(y_{k=j}/x\right) \text{ for all } j \neq i \qquad \text{Equation 5.20}$$

We can use Bayes' theorem to calculate $p\left(y/x\right)$:

$$p\left(y/x\right) = \frac{p(y) p\left(x/y\right)}{\sum_{k=0}^{K} p(y_k) p\left(x/y_k\right)} \qquad \text{Equation 5.21}$$

where $p(y)$ is a prior probability, $p\left(x/y\right)$ is a likelihood of x, that is, the chances of observing the particular values of x within a class y_k, and the denominator is a constant for normalizing $p\left(y/x\right)$ between 0 to 1. It is the probability of x in the general population. In other words, it is the probability of observing x in any class. The

Classification of Urban Land Use and Land Cover

probabilities $p(y)$ and $p\left(x/y\right)$ are estimated from the training data. In discrete cases, they are simply counts of some values of x in **x** for all y.

If we assume that the x_d/y_k for all the values of d from 0 to D are independent of each other, then these probabilities can be multiplied to calculate the likelihood. For example, $x_{d=0}$ and $x_{d=1}$ are independent given a class $y_{k=1}$. This is a Naïve Bayes' classifier. The naïve assumption is the assumption about the conditional independence of the x values. In reality, the x values would be correlated. However, given the class, we can assume them to be independent.

If the denominator is ignored, as it is the same for all the y values (Richards & Jia, Supervised classification techniques, 2006), the discrimination functions becomes

$$p\left(y/x\right) = p(y) p\left(x/y\right) \qquad \text{Equation 5.22}$$

More specifically:

$$x \in y_{k=i} \text{ if } p(x) p\left(x/y_{k=i}\right) > p(x) p\left(x/y_{k=j}\right) \text{ for all } j \neq i \qquad \text{Equation 5.23}$$

The posterior probability is a product of the prior probability, and the likelihood is an English description of Equation 5.22. The likelihood term $p\left(x/y_k\right)$ for multivariate normal distribution is given by:

$$\left(2\pi^{-D/2}\right) \left|\Sigma_k\right|^{-1/2} e^{\left\{(x-\mu_k)^t \left|\Sigma_k\right|^{-1} (x-\mu_k)\right\}} \qquad \text{Equation 5.24}$$

where, μ is the mean vector and the covariance matrix for class k. If we take a natural log and remove the common terms, the discriminant function becomes:

$$\ln p\left(y/x\right) = \ln p(y) - \frac{1}{2} \ln\left|\Sigma_k\right| - \frac{1}{2}(x-\mu_k)^t \Sigma_k^{-1}(x-\mu_k) \qquad \text{Equation 5.25}$$

On assuming every class is equal, likely we can ignore $\ln p(y)$ as well, simplifying the equation further:

$$\ln p\left(y/x\right) = -\ln\left|\Sigma_k\right| - (x-\mu_k)^t \left|\Sigma_k\right|^{-1}(x-\mu_k) \qquad \text{Equation 5.26}$$

ALGORITHM 5.3
Maximum Likelihood Classification

```
Input: X^L = list of training samples, x = a query pixel
Output: y Class label for query pixel
ML-train(X^L)
For every k=0 to K
}
Calculate μ_k, and Σ_k #mean vector and covariance
Return μ_k, Σ_k
}
ML-deploy(μ_k, Σ_k, x)
{
        p
        FOR k=0 to K
            d_k = -ln|Σ_k|-(x - μ_k)^t | Σ_k|^-1(x - μ_k)
            d[k] = d_k #assign k^th element of p to f_k
        y = index of {max{p}}
        return y
        }
```

The equation is a maximum likelihood classifier. The algorithm is given in Algorithm 5.3.

The number of training samples required to calculate the statistics with a high degree of confidence is ~10 to 100 times the number of bands D (Richards & Jia, "Supervised classification techniques", 2006). This is a large number for hyperspectral data. For example, 100 bands entail at least 1000 pixels for training. Getting 1000 training samples in the image is a challenge for urban areas as the regions are not homogenous. This necessitates the reduction of dimensions by some means in case sufficient training data is not available. Ideally, the machine learning models should be trained on the training samples equal to or greater than $10D$ to $100D$. However, in case enough training samples are not available, the models can be trained on as low as $2D$ to $3D$ training samples. The training samples 2 to 3 times of the bands provide 95% of the accuracy achieved with the training samples 30 times of the bands (Van Niel et al., 2005). These lower limits on the number of training samples should be treated as a general guideline rather than a strict rule. Another associated problem with the small number of training samples is the possibility of a singular covariance matrix (discussed in detail in Section 3.3.2, "The Bhattacharyya distance analysis for urban classes"). The covariance matrix may be singular because of the high dimensionality of the space if the number of samples is less.

Maximum Likelihood (ML) has been one of the most used classifiers for urban LU/LC classification for both medium and high spectral resolution remote sensing data (Weng, 2012). Platt and Goetz (2004) use ML to classify urban fringe areas in Fort Collins, Colorado, USA. They used a modified level 2 USGS classification schema (Anderson et al., 1976). They observed "modest but real" improvement in classification accuracies by hyperspectral (HS) data over multispectral data. Analysis of results indicated the improvement because of a large number of bands rather than signal to noise ratio. Tan and Wang (May, 2007) used an Advance Spaceborne Thermal Emission and Reflection Radio meter (ASTER) and Chris-Proba data for urban mapping of Beijing city, China.

Platt and Goetz (2004) pointed out limited research on the mapping of urban land use land cover using hyperspectral data. They conducted a study to compare factors affecting the accuracy of LU/LC classification using AVIRIS data and then compared it with a synthetic Landsat ETM+ image created using the same AVIRIS image. The additional focus of the study was to see if the improvement in accuracy provided by the hyperspectral data was because of the number of bands or SNR ratio. They used a modified level II USGS classification schema. A total of eight classes were considered for the accuracy assessment: *residential, commercial/industrial, water, irrigated cropland, fallow, dry, rangeland, grassland*, and *irrigated urban*. ML classifier was used for classification. The results showed a 5% improvement in overall accuracies for hyperspectral data over multispectral data. Further, User's Accuracy showed more improvements than Producer's Accuracy indicating reduced false positives.

Research by Platt and Goetz (2004) was one of the important studies that established a few outcomes: HS data provided better accuracies than multispectral data, the accuracy improvement was because of more number of bands rather than SNR, and HS increased User's Accuracy. The conclusion that more bands provide better accuracy needs further investigation as the authors use dimensionality reduction. The quality of the features with and without dimensionality reduction should have been compared as well.

Tan and Wang (May, 2007) too pointed out that there is limited research in mapping LU/LC of urban areas using hyperspectral data. They compared LU/LC classification results of Beijing city using ASTER and Chris-Proba images. They did not correct hyperspectral images for atmospheric effects and used an ML classifier for segmenting images into nine urban classes as per the Anderson classification schema. Their (Tan & Wang, May, 2007) study showed that hyperspectral data provided better classification accuracy than multispectral data. They found hyperon data to be comparable with high-resolution data combined with multispectral data (LISS III).

Heiden et al. (2007) employed features derived from spectral signatures and used an ML classifier to detect impervious surfaces. The difference in their approach was two-fold: secondary features from a spectral curve were derived instead of using reflectance values of each band, and each material was detected using a separate classifier (employing spectral features in different wavelength ranges for different materials). For example, two of all the spectral features considered for polyethylene and roof tiles are "increase" and "absorption" of 486–880 and 1130–1259 for polyethylene and 515–2402 and 454–622 for roof tiles, respectively. "Brightness" was the additional feature added to facilitate the detection of spectrally bland dark surfaces such as asphalt. They created a spectral library of urban materials using HyMap imagery for Dresden City (1999, 2000, 2003), and Potsdam City (1999), Germany. Four types of features namely: absorption, reflectance peak, increase and decrease of reflectance, and the continuity of spectral curve were extracted from the training data from a spectral library (1999, 2000 images). The results were compared with grey value (or reflectance values) features as suggested by Herold et al. (2004). Spectral features showed better results than the grey values.

The approach by Heiden et al. (2007) is a logical extension of absorption features. However, the approach needs further investigation. Their current result comparison (between spectral features and grey value features) used different sets of wavelength ranges for extracting spectral features and grey values – grey values are always extracted from the wavelengths 445 nm, 576 nm, 638 nm, 759 nm, 1100 nm, 1316 nm, and 1989 nm. The true comparison would be if the grey values were derived from the same wavelength range corresponding to the range of spectral features.

In addition to ML, other supervised or semi-supervised machine learning algorithms have been used for urban HS data classification. Researchers have used Artificial Neural Network (ANN) (Zurada, 1992), Support Vector Machine (SVM) (Suykens & Vandewalle, 1999; Melgani & Bruzzone, 2004; Pal & Mather, 2005), Decision Tree (DT) classifiers (Pal & Mather, 2003) with reasonable success for various urban studies. Ridd et al. (April 1992) used ANN to classify AVIRIS and SPOT data to study urban morphology and VIS configuration of Salt Lake City, Utah, USA. Herold et al. (2002) computed landscape metrics for studying an urban land use structure of Santa Barbara, California, USA, using IKONOS images. Benediktsson et al. (2005) devised extended morphological kernels for analysing the Pavia dataset. Fauvel et al. (2008) further integrated SVM with extended morphological kernels for analysing urban images of Washington DC, USA. Myint et al. (2011) compared pixel and object-based approaches for urban planning.

Another alternative to hard classification is to calculate the fractions of various land covers within the pixel by the spectral unmixing analysis. Wu and Murray (2003) performed a spectral unmixing analysis to calculate VIS fractions in the metropolitan area of Ohio, USA, for ecological studies. Yang et al. (2003) developed a regression model between fractions of urban classes and Landsat band values for urban mapping of western Georgia, USA. Weng et al. (2008), one of the notable hyperspectral studies of urban areas, used Principal Components Analysis (PCA) for dimensionality reduction and then an unmixing analysis to calculate VIS fractions. In their work, hyperspectral data provided more accurate results as compared with multispectral data – especially in the case of low-reflectance surfaces such as asphalt roads. They attributed an improvement in accuracy to the large number of bands in the HS data.

The techniques covered in Sections 5.2, "Classification with urban indices", and 5.3, "Nearest neighbour", are more suitable for detecting a target material or a land cover. Most of them are instance-based learning techniques and use good references of the target materials to complete the task. Other classifiers, covered in Section 5.4, "Maximum likelihood classifier", can be used for both land cover and land use classification. In the next section, we provide a working example of land use classification. More specifically we showcase how the configuration of land covers over a group of pixels or a single pixel can be leveraged to identify land use. To this end, we illustrate how VIS fractions can be used effectively to identify land use classes. The VIS land covers are links between human activities and their impact on the environment.

Classification of Urban Land Use and Land Cover

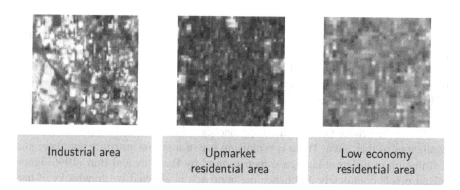

FIGURE 5.5 VIS fraction of different land use classes in Pune city.

Socio-economic activities change the land covers. Changing the land cover changes the physical response of the area with respect to the temperature, water permeability, and so on. VIS and socio-economic activities are tightly intertwined so much so that VIS fractions in a given pixel/s provide enough indication of its economic status.

This is illustrated in Figure 5.5. The figure shows three regions extracted from the EO-1 Hyperion image of Pune city (USGS, 2013-1), namely up-market residential, industrial and low-economy residential (slum). Each region has distinct proportions of the basic VIS classes. The up-market residential area has a lot of open spaces, good vegetation cover, and buildings. That is displayed by a purple colour in FCC indicating a mixture of vegetation (red) and built-up (blue) classes. In contrast, the low-economy residential zone is dominated by buildings, has fewer open spaces, and is almost devoid of vegetation. A new industrial area is marked by large buildings and open spaces with little vegetation.

We focus on identifying economic zones in the city. Identification of economic zones within the city is important for estimating urban consumption such as water, electricity and so on and emissions such as air, water, and soil pollutants; for example, CO_2. The vulnerability of the given area to various natural and artificial hazards is found to be correlated to the economic status of that area (Cutter et al., 2003). This is intuitive as, generally speaking, up-market areas have adequate infrastructure for rescue and recovery efforts as compared to slums and low-economy residential zones.

5.5 ECONOMIC ZONE CLASSIFICATION

It is important to identify good proxy features for a theme such as the economic zones of the city. The first-order spectral or spatial features are not useful. Economic zones entail a collection of various land covers and/or urban objects having a strong association with each other. They together function as a single cohesive unit. They may show the dominance of some of the land covers; for example, the up-market residential zone may have more open areas and vegetation. As postulated by Ridd (1995), different zones within the city show different proportions of the basic VIS land covers.

The proportions of some of the basic properties extracted from the image of the urban areas are good candidates for proxies. Landscape metrics are extensively studied for understanding ecology. Irrespective of the comprehensive work on the landscape metrics for understanding land use classes, important units like patches, background matrixes and corridors are not defined systematically for urban areas yet (McGarigal, 2014). Therefore, landscape metrics are not effective to identify economic zones in their present form. Besides, the majority of the studies use high-resolution imagery and hence they choose object-based approaches, which are complex and face certain challenges of their own. There is promising use of a spatial structure template; joint distribution of different landscape metrics used by Barnsley and Barr (1996) ignore mixed pixels completely and hence the approach is affected by drawbacks of hard classification of a pixel, especially at medium resolution. The approach by Rashed and Weeks (2003) used unmixing analysis and the VIS fractions as the proxy feature but does not consider joint distribution over the VIS to be a proxy signature for a given land use. Further, fractions of the VIS are converted to hard classes using a majority rule. Other similar studies (Phinn et al., 2002; Wu & Murray, 2003) are limited to multispectral data and use VIS for modelling urban growth patterns as per the VIS model (Ridd, 1995).

Deshpande et al. (2015) proposed the use of the fractions or abundance ratios of VIS classes at a broader level within a single pixel or a window of H^P rows and W^P columns (where, H^P and W^P >=1, referred to as a patch henceforth) as a proxy to the economic status. Instead of using a fraction of a single land cover class as a feature independently, they used the complete distribution over the VIS classes to define an economic class. Fractions of the VIS classes in a given region indicate the economic health of the area, broadly speaking. For example, the up-market residential areas exhibit a good road network, ample vegetation and open spaces. On the contrary, the slums exhibit a poor road network, little vegetation and open areas. In addition, instead of using hard classification, they calculate the VIS fractions directly using unmixing analysis.

Another important aspect of the work was the verification of the extracted endmembers and their agreement with field VIS spectra. The verification of the endmembers by field or lab reference spectra is an important aspect of an unmixing study and is ignored often. The agreement of the endmembers with the VIS material spectra helps in identifying the mixture of materials in the mixed pixel. A Simplex Volume Maximization (SiVM) algorithm (Thurau et al., 2010) was explored effectively. The study provides a comprehensive quality assessment of the endmembers extracted by SiVM for the field applications.

We provide a detailed account of the study as an example for land use classification using proxies. As discussed earlier, these features can be used with any classifier. The CNN learns the features automatically, however the features learned are spatial features such as edges of shapes, etc. or regions in the image. These features do not consider the spatial configuration of the learned features, which is essential in a higher-order class such as an economic zone. Most of the CNN architectures are bag of features classifiers[3] but capsule network. The use of capsule network in analysis of hyperspectral data and for land use land cover classification is limited at present.

Classification of Urban Land Use and Land Cover

5.5.1 Study Area

Pune city is a cultural, educational, and manufacturing hub of Maharashtra, Western India and has nearly doubled its size during the last decade (from ~3.76 million to ~5.05 million) (Office of the Registrar General & Census Commissioner, India, 2011-1). Pune is the seventh-most populous city in India and the second largest in the state of Maharashtra, India. Pune (PMC) and Pimpri-Chinchwad (PCMC) are the two municipal corporations in Pune District and together occupy an area of ~900.0 km^2. Pune is located 560 m above sea level on the western margin of the Deccan plateau. Pune has a hot, semi-arid climate close to tropical wet and dry with average temperatures ranging from 20 to 28 °C (Nalawade, 2015).

The uncontrolled growth of Pune[4] has adversely affected the urban environment and the quality of life, in general. Impervious surfaces have increased by ~50% during the period from 2001 to 2014, resulting in ~5% loss of natural drainage in Pune city. An estimated increase in runoff in PMC and PCMC administrative boundaries over the decade is in the order of 87% (Yadav & Deshpande, 2016). Pune experiences 2°–3° higher ambient air temperatures than its immediate surroundings on average and 5°–6° in the worst cases (More et al., 2015). The region is now affected by increased water runoff, decreased natural drainage streams and reduced groundwater table (MPCB, 2016).

5.5.2 Hyperspectral Data

EO-1 Hyperion (USGS, 2011-2) records the data in 242 bands with a 30 m spatial resolution and ~10 nm (FWHM) spectral resolution over 355–2577 nm. Only 198 bands of 242 are calibrated (7–57 and 77–224) (Beck, 2003). Further removal of water vapour-affected bands results in 141 useful bands. The central wavelengths of Hyperion bands are separated by ~10 nm and the SRF of each band overlaps by ~12% with the SRF of each single preceding and following bands. The Gaussian spectral response function with a central wavelength and FWHM as per the Hyperion documentation provide a very close approximation to the observed spectral responses of Hyperion bands (Figure 5.6).

Two Hyperion images on path 147 and row 47 with scene centres 18.5020 N, 73.8151 E and 18.5020 N, 73.7457 E were acquired (USGS, 2013-1; USGS, 2013-2). Coverage of ~25 km^2 is required for the entire Pune city, Maharashtra, India, and that is covered by three Hyperion images, each 7.7 wide and ~100 km in length. The first image is 102 km, and the second image is 107 km long. We focused on the western fringes of the city as the area has seen tremendous urbanization in the recent past because of emerging information technology hubs (Deshpande et al., 2013).

Overall, we removed channels 1–8, 58–81, 98–101, 119–134, 164–187, and 218–242 (Deshpande et al., 2013) from further processing. The further processing of digital images requires the conversion of Digital Numbers (DNs) to radiance values for selected Hyperion bands. Selected bands within the range 1–70 were divided by 40 and 71–242 were divided by 80 to convert the DNs to radiance (W/m^2 SR μm) (Beck, 2003). The April 22 image, which is the main image for experiments shows four distinct regions:

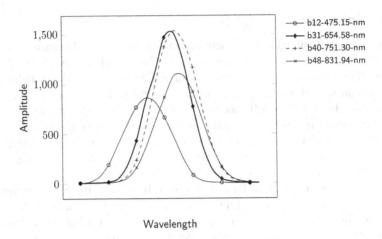

FIGURE 5.6 Sample Hyperion spectral response functions (ftp.eoc.csiro.au).

1. Area 1, rows 550–1050 (AGR) – dominated by agricultural and open areas (plains), with a small town near the city,
2. Area 2, 1050–1550 (IND) – dominated by industrial development, large open areas and large buildings with common industrial roof covers (coated iron sheets). It shows varied patterns – old industrial areas show the development of industrial blocks within the vicinity, and new industrial set-ups show ribbon development,
3. Area 3, rows 1550–2050 (RES) – dominated by urban materials (mixed pixels of open areas and vegetation, and built-up and vegetation), urban trees are distributed over the entire region with the exception of large areas of tree cover in reserved parks and the campus of an old educational institute, with a small fraction of open areas of different sizes distributed over the region,
4. Area 4, rows 2050–2550 (OPN) – predominantly open (pure pixels of open areas covered with dry grass, mixed pixels of open and vegetation) – hill ranges of the Sahyadri, shrubs and deciduous tree cover, lower hill slopes are covered with dry grass, a little built-up area in the form of small villages in river catchments along with small agricultural fields (Figure 5.7).

The large areas within these four zones were identified. The land use classes were verified by high-resolution imagery as well as ground field trips. The reference data used is the same as in Chapter 3, we reproduce the land use-related references for convenience (Table 5.3 and Table 5.4).

5.5.3 Semi-supervised Learning Background

A semi-supervised methodology using a cluster-and-label strategy was used. The cluster-label strategy takes advantage of the internal structure of VIS fractions for a particular land use class (Zhu, 2005). The method can be extended further to

Classification of Urban Land Use and Land Cover 211

FIGURE 5.7 Partitions of the EO-1-Hyperion image of Pune city.

TABLE 5.3
Landcover References from the Hyperion Image Used in the Study

Code	Class	Description
GGPA	Green grass (GG)	Marsh shrubs near the lake (Pashaan)
IRCR	Indutrial roof (IR)	Industrial area at Chakan road
IRKI	Indutrial roof (IR)	Industrial roof of Kirloskar factory, near Mula-Mutha Sangam
IRMR	Indutrial roof (IR)	Industrial area at MIDC road
IRNR	Indutrial roof (IR)	Industrial area at Nashik Road
IRNR2	Indutrial roof (IR)	Industrial area at Nashik Road 2
PLCI	Plain (PL)	Chakan Indrayani open area/plain
PLMU	Plain (PL)	Mutha plains
PLS12	Plain (PL)	Sector 12 Chakan industrial area
PLS12'	Plain (PL)	Sector 12 right half, Chakan industrial area
PLS5	Plain (PL)	Sector 5, Chakan industrial area
REDN	Residential (RE)	Datta Nagar
REHN	Residential (RE)	Hanumaan Nagar
RENS2	Residential (RE)	New Sangavi 2
REOS	Residential (RE)	Old Snagavi
REPG	Residential (RE)	Sinhagad medical institute, Narhe road
RERN	Residential (RE)	Raam Nagar
RUFC	Residential (RE)	Ferguson College road
RULC	Residential (RE)	Law College road
RUMO	Residential (RE)	Model colony
RUMO2	Residential (RE)	Model colony 2
RUNS2	Residential (RE)	Nava Sahyadri
TRBS	Tree (TR)	Trees near Baner STP
TRCH	Tree (TR)	Chattushrungi
TRLP	Tree (TR)	Leftbank Pashaan
TRLP2	Tree (TR)	Leftbank Pashaan 2-near bypass road
TRRP	Tree (TR)	Rightbank Pashan
TRSN	Tree (TR)	Shastri Nagar military farm
TRSN2	Tree (TR)	Shastrinagar military farm along the river
TRUN	Tree (TR)	University trees
TRUN2	Tree (TR)	University trees – North side
TRVG	Tree (TR)	Chittaranjan Vatika Garden

Notes:
* First two letters of the code indicate broader classes such as **RE**sidential, **I**ndustrial **R**oof and so on as given in column "Class". Next two letters of the code are the abbreviation of a specific location. For example, KH stands for **Kh**dakwasla, a particular locality of the study area. In case the location is indicated by two words then first letter of each word is taken as a part of code. For example, NS stands for **N**ew **S**angavi.

TABLE 5.4
Row and Column Coordinates of the Reference Sites Used in the Study

Code	R1	R2	C1	C2	Number of pixels
GGPA	1976	1979	28	31	16
IRCR	1182	1250	67	133	4623
IRKI	1787	1792	195	197	18
IRMR	1494	1548	105	163	3245
IRNR	1496	1517	40	65	572
IRNR2	1313	1342	94	127	1020
PLCI	1362	1371	71	76	60
PLMU	2205	2208	113	117	20
PLS12	1471	1479	138	146	81
PLS12'	1476	1490	114	121	120
PLS5	1508	1523	87	95	144
REDN	1990	2003	123	133	154
REHN	2002	2010	173	179	63
RENS2	1744	1764	87	106	420
REOS	1757	1769	89	99	143
REPG	2224	2233	229	238	100
RERN	1577	1606	188	218	930
RUFC	1970	2008	190	236	1833
RULC	1977	2007	198	225	868
RUMO	1927	1940	208	222	210
RUMO2	1913	1941	178	243	1914
RUNS2	1981	1998	203	223	378
TRBS	1847	1850	48	50	12
TRCH	1934	1937	196	199	16
TRLP	1974	1978	5	11	35
TRLP2	1976	1983	18	21	32
TRRP	1975	1981	53	62	70
TRSN	1690	1700	54	60	77
TRSN2	1700	1702	49	60	36
TRUN	1857	1862	166	169	24
TRUN2	1846	1851	165	166	12
TRVG	1926	1930	212	216	25

incorporate simultaneous optimization of the objective function (Demiriz et al., 1999) or augmented labels using the Hungarian method (Albalate et al., 2010). Another alternative to the two-stage cluster-and-label process is to use the K class centroid vectors of different economic zones as K cluster seeds. The disadvantage of the seeding approach is that the distributions over VIS types of some unseen land use patterns remain undiscovered. In other words, the clusters formed by this method would be limited to the K cluster prototypes/centroids. Hence in case there is an additional zone not represented well by the prototypes/centroids then it will be merged with one of the *K* clusters. This is not a desired state generally as we would like to discover more land use patterns and their corresponding VIS proportions.

The feature modelling is more critical in the design of the semi-supervised algorithm than the management of unlabelled training data. The designed features should help the clustering step to form a cohesive cluster so that the internal structure of the data for useful inferences is revealed (Zhu, 2005; Chapelle et al., 2006). The efforts by Deshpande et al. (2015) were focused on assessing the effectiveness of the distribution over VIS classes in a pixel or a patch as a proxy definition for the land use pattern.

The summary of the proposed methodology is as follows:

1. **Unmixing:** Extract fractions of VIS or VIS types in the pixels of an image using any unmixing algorithm. The unmixing model linear or intimate does not make any difference to the overall process. Next to endmember extraction, determine the identity of the endmembers using reference VIS signatures identified using supervision.
2. **Clustering:** Aggregate the VIS fractions for a patch using an average over the VIS fractions for each pixel within the patch. The aggregation step is not mandatory. Depending upon the resolution of the image and the configuration of the study area, a single pixel can represent a land use pattern. Use k-means clustering to cluster the pixels or patches into *k* clusters using the VIS fractions as features.
3. **Training:**[5] Extract the centroid VIS fraction vector for a group of pixels of the same class identified using supervision. These extracted VIS vectors are considered to be the definition of the economic class. The centroid vector resembles the topic generated by topic modelling (Blei et al., 2003); the economic zone is the visual topic, and the VIS or VIS types are the visual words.
4. **Cluster labelling:** Use the Spectral Angle Mapper (SAM) to label the individual pixels or the pixels in a patch. Use the reference VIS vectors created in the training phase (see Figure 5.8).

The detailed procedure followed is described below:

5.5.3.1 Unmixing

The processing begins with the endmember extraction and unmixing analysis. First, radiance values are converted to relative reflectance values using internal area average relative reflectance (Deshpande, Inamdar, & Vin, 2017) and then performed the unmixing analysis for each image section separately. We determined the identity of the extracted endmembers using the spectral angle between the endmembers and the reference signatures. We then used $e = 3$ endmembers corresponding to VIS classes to calculate the respective fractions in each pixel or patch. Various experiments were performed to extract different numbers of endmembers e; for $8>e>3$ as well.

There are two main approaches to model mixing: linear and intimate (Bioucas-Dias et al., 2012). In this study, we consider mixed pixels to be a linear mixture of some pure pixels within the image – especially the VIS. The fractions can be

Classification of Urban Land Use and Land Cover

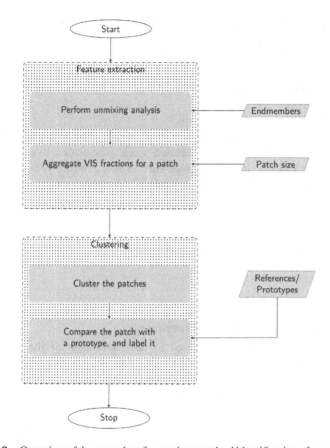

FIGURE 5.8 Overview of the procedure for semi-supervised identification of economic zones.

calculated by any suitable unmixing algorithm. The SiVM algorithm is chosen for its computational efficiency and its scalability (Chan et al., 2011; Thurau et al., 2010). It is O(kn) – where n is the number of pixels. A Windows version of python matrix factorization (PyMF) was used (Thurau et al., 2012). Unlike N-FINDR (Winter, 1999), VCA (Nascimento & Bioucas-Dias, 2005), and SVMAX (Chan et al., 2011) algorithms, the SiVM uses distance geometry for calculating the volume of the simplex. Next, Clustering and Training are two parallel steps (see Figure 5.8).

5.5.3.2 Clustering

A patch was considered for aggregation of the fractions. The step is required only if the operation is performed on a window size of m and $n > 1$. The fractions of the VIS are calculated for the patch by a simple average over all the pixels within the patch. The pixels or patches are then clustered using k-means using VIS fractions as features. The patch of 1×1 or 2×2 is considered in the present study. The parameters m and n are configurable and can be used further for understanding the effect of window size on the results. The benchmark patch size is decided based on the resolution and average small plot size of development in the study area. In the existing study, one pixel of 30 m × 30 m corresponds to ~200% of the smallest plot size (~20 m × 20 m) of urban development – 2 × 2 blocks hence cover ~4 buildings and hence a sufficient area for identifying a meaningful economic zone label to a patch. Experiments with different patch sizes were conducted where m=n=1 to 9 for studying optimal patch size for detecting economic zones.

5.5.3.3 Training

The centroid vector of each economic zone was calculated by taking a simple average of VIS fractions within the group of labelled pixels. The centroid vector models the distribution over VIS fractions for a given class. These centroid vectors are used for labelling the clusters generated by an unsupervised learning step.

Cluster labelling: Finally, labelling of cluster members that is pixels or a patch is performed. Cluster members are assigned a class using a spectral angle between the class centroid and the cluster centroid. All the cluster members within a given cluster are assigned to the class that is having the smallest angle between class centroids and the cluster centroid.

5.5.4 Experimental Setup

5.5.4.1 Image Partitions for the Experiments

The specific partitions were considered to take advantage of the different land cover configurations within the scene to verify whether the extracted endmembers truly match pure VIS pixels within the respective regions. For example, one of the endmembers extracted from the RES section should match the "concrete" signature, and one of the endmembers from the AGR section should match the "crop" signature, and so on. In the present work, we are interested in defining economic zones using distribution over VIS and do not consider a particular VIS type. Only 3–4 of the endmembers extracted from the whole image may or may not match the VIS type in a particular part of the image. For example, a vegetation endmember extracted from the whole image may represent a "tree" signature, whereas the subsection vegetation would be better represented by a "crop" signature. In such cases, the VIS distribution of pixels or groups of pixels in a particular part of the image would not be consistent with the VIS distribution as calculated using whole image endmembers. Further, most of the urban extent is covered by a stretch of ~15 km (~500 rows). Thus, although the endmembers can be extracted from the whole image, in principle, the existing process provides a few practical advantages:

 a. consistency and simplicity of interpretation of the fractions in a pixel or a patch in an image part and verification of the endmembers,
 b. computational efficiency of unmixing analysis, and
 c. optimal processing of the image for economic zone identification for an urban area.

Classification of Urban Land Use and Land Cover 217

TABLE 5.5
Site Codes of VIS Prototypes for Economic Zones

V	I	S	Location code
High-economy residential zone			
0.35	0.60	0.05	RULC
0.47	0.39	0.14	RUMO2
0.32	0.60	0.08	RUFC
Low-economy residential zone			
0.02	0.84	0.14	RERN
0.01	0.82	0.17	REDN
0.04	0.84	0.12	REHN
Industrial zone			
0.08	0.38	0.54	IRCR
0.09	0.36	0.55	IRMR
0.10	0.37	0.53	IRNR2

5.5.4.2 Data

We provide below the site codes used for various experiments and an evaluation of their results. Table 5.5 provides the site codes used for prototypes for cluster labelling and the respective VIS fractions vector. The average signature of the sites for each class was considered for cluster labelling. The average prototype VIS signature of each class is matched with the cluster centroid generated in the clustering step.

5.5.4.3 Comparison of the Results with Supervised Classification

The two image parts namely IND and RES using the proposed algorithm were classified. The accuracy was calculated for semi-supervised classification using a standard confusion matrix (Congalton, 1991). In addition, the User's accuracy and the Producer's accuracy of the classification results were calculated. Further, the results of the proposed algorithm were compared to the results by various supervised algorithms such as k-NN (Altman, 1992), Support Vector Machines (SVM) (Suykens & Vandewalle, 1999), ML (Strahler, 1980; Harris, 1985) and SAM (Goetz et al., 1985). The classification by these algorithms was performed using reflectance values of 141 bands as features. We compared the Producer's accuracy and the User's accuracy of all the classifiers with the proposed algorithm.

The data set in Table 5.6 was used for evaluation. The data as indicated by the second column of the table that is set A, represents the most practical worst-case scenario for image classification using remotely sensed data. The data set B was used for five-fold cross-validation. The average of five-fold cross-validation is calculated for a comparative assessment of classification results. The data size B ensures a training sample size of 2 to 3 times the dimensions in each kth validation step, that is, 141 in this case. The training samples 2 to 3 times the dimensions provide 95% of the accuracy achieved with the training samples 30 times the dimension (Van Niel et al.,

TABLE 5.6
Data Set Used for Evaluation

Land use class	Training / testing Site codes (Set A)	Data set for 5-fold cross-validation (Set B)
Vegetation	TRLP / TRUN	GGPA, TRBS, TRCH, TRRP, TRSN, TRSN2, TRUN2, TRVG (total 355 pixels)
Common open areas or Plains	PLS12 / PLS5	PLCI, PLMU, PLS12 PLS12^2, PLS5 (total 425)
Low-economy residential	REPG / RENS2	REPG, RENS2, REOS (total 663 pixels)
Up-market residential	RUNS2 / RUNS	RUMO, RUNS2 (total 588 pixels)
Industrial	IRKI / IRNR	IRNR (total 572 pixels)

2005). The general rule of choosing training samples 10 times the dimension is not feasible in this situation. Further, the class balance is also maintained in dataset B.

The patch size, the number of clusters, and the number of endmembers used to define the economic class play a critical role in the proposed method. We used the Silhouette coefficient (Rousseeuw, 1987) to study the effect of the patch size, the number of classes, and the number of endmembers in cluster quality. We calculated the average Silhouette coefficient for clusters n = 2, 3, 4, 5, 7, 9 for patch sizes p = 1, 2, 3, 4, 5, 7, 9 for the two sub-images IND and RES for endmembers e = 4, 5, 8. The Silhouette coefficient combines cohesion and separation of an object and varies from −1 to +1. Negative values are undesirable as they indicate that the average distance to all other objects in the same cluster is greater than the minimum of the average distance to points in another cluster.

5.5.5 Results of Land Use Analysis Using Proxy Features

Fractions of the VIS classes in a given region indicate the economic health of the area, broadly speaking. For example, the up-market residential areas exhibit a good road network, ample vegetation and open spaces. On the contrary, the slums exhibit a poor road network, little vegetation, and open areas (Figure 5.5). In the present work, we performed an analysis of high-income residential, low-economy (slums and mid-economy dense residential area) zones and industrial zones.

5.5.5.1 Quality of the Endmembers Extracted Using the SiVM

General observations: As we increase e (where, e = number of endmembers), more than one endmember of the same class is extracted, for example, 2–3 types of industrial roofs in an industrial area, 2–3 plain endmembers in an open area. Similar behaviour is observed even in the case of lower e values such as 3–4, where the image is dominated by a single material. Consequently, some of the obvious endmembers are extracted in the later stages. For example, the open areas with dry grass cover (plains)

Classification of Urban Land Use and Land Cover

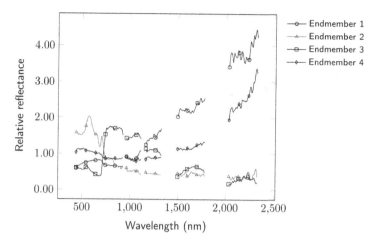

FIGURE 5.9 Extracted endmembers for e=4 for IND.

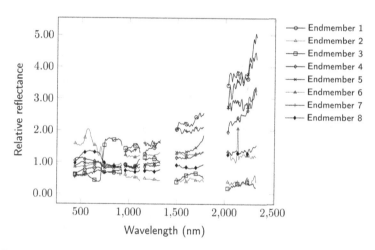

FIGURE 5.10 Extracted endmembers for e=8 for IND.

or exposed bare soil because of the site clearing activities are extracted for e>4 in an industrial area (Figure 5.9 to Figure 5.12).

Section AGR: As expected, considering the dominant classes, vegetation, plain, and industrial roof, are extracted as the endmembers when e=4. The town covers only a small portion of the image part and hence only limited pixels would exhibit concrete, as one of the mixture components hence residential roof (concrete) might not have been identified as an endmember. On the other hand, there are chances of large constructions with the industrial roof and hence the result. Irrespective of impervious surface type, most of the pixels in such areas can be considered as a mixture of vegetation, plain (soil), and impervious surfaces (industrial roofs/concrete).

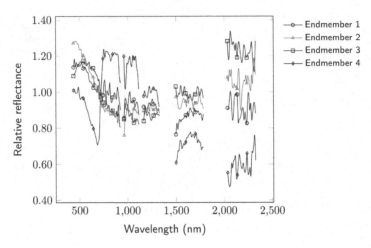

FIGURE 5.11 Extracted endmembers for e=4 for and RES.

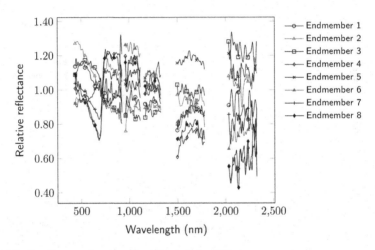

FIGURE 5.12 Extracted endmembers for e=8 for RES.

Section IND: This area is dominated by old and upcoming industrial buildings of large size. In addition, the area is characterized by large open areas, and large plots cleared for new construction. As a result, the endmembers extracted correspond to industrial roofs, open areas (plain), and vegetation. Water is also extracted as an endmember correctly (Figure 5.9 to Figure 5.12).

Section RES: This section comprises an up-market residential area, commercial buildings, and a medium- to low-economy residential area adjacent to an industrial zone. The first four endmembers extracted resemble concrete (parking lot concrete), residential area (concrete roof), bare soil, and vegetation (trees). Subsequently, plain (open areas), and different vegetation signatures are extracted as endmembers. Considering the variety of vegetation and its spatial distribution, it is natural to have

TABLE 5.7
Spectral Angle between Endmembers and References

Image	Vegetation	Plain	Bare soil	Industrial roofs	Residential roofs
550 – Agr	5.12	3.62		5.73	
1050 – Indus	3.62	5.73[a]	7.69[b]	14.53	
1550 – Res	6.27		5.12		3.62 / 5.73
2550 – Open	6.27	3.62			3.62[b]

Notes:
[a] The plain with domination of dry grass, [b] Indicate the endmembers are extracted at the later stage $k>5$

TABLE 5.8
Distribution over VIS Fractions for Prototype Economic Zones

High-economy residential zone				Low-economy residential zone				Industrial zone			
V	I	S	Location code	V	I	S	Location code	V	I	S	Location code
0.35	0.60	0.05	RULC	0.02	0.84	0.14	RERN	0.08	0.38	0.54	IRCR
0.47	0.39	0.14	RUMO2	0.01	0.82	0.17	REDN	0.09	0.36	0.55	IRMR
0.32	0.60	0.08	RUFC	0.04	0.84	0.12	REHN	0.10	0.37	0.53	IRNR2

pure pixels of different vegetation types like a thick canopy and vs shrubs (Figure 5.9 to Figure 5.12).

Section OPN: Endmembers extracted reflect the nature of dominant land covers and landforms. Trees and different signatures of plains (open areas) with a different degree of dry grass cover are extracted. The signature of concrete is also one of the endmembers for e=8. Table 5.7 shows spectral angles between extracted endmembers, by the SiVM algorithm, and references.

5.5.5.2 Distribution over VIS Fractions in Different Economic Zones

We discuss in detail VIS distribution for IND, RES:

Low economic zones: Low-economy zones (Table 5.8) show little vegetation cover and a few open spaces. The model is reflected in a distribution over VIS fractions. Three different low-economy zones explored show a more or less similar pattern of VIS distribution. In the case of the slums, impervious surfaces, at times, comprise Galvanized Iron (GI) roofs and are identified so (Deshpande, Inamdar, & Vin, 2017). The use of cost-effective roof covers such as GI sheets and asbestos sheets is a common practice for low-economy houses in urban India.

High-economy zone: All three different high-economy zones we studied show a very similar distribution over the VIS fractions (Table 5.8). The vegetation fraction increases, and the built-up area decreases as compared to low-economy zones. The open areas (playgrounds, other open spaces) fraction is comparatively lower than expected. The thick vegetation canopy might have been covering part of the open spaces and hence reduced fractions. This VIS distribution pattern needs to be verified with different study areas.

Industrial zone: Industrial zones (Table 5.8) also show a very distinct pattern. The zone is dominated by large open areas and little vegetation. The impervious surface is dominated by industrial roof covers, concrete (roof, parking lots etc.), and cleared sites for new construction. Large portions of open areas are the result of different spatio-temporal factors. The industrial zones are typically developed away from the existing residential zones, and most of the development in such demarked zones is managed in phases.

5.5.5.3 Optimal Patch Size and Number of Endmembers

For the image part IND that is predominantly industrial, the smaller numbers of clusters such as 1 to 3 provide cohesive clusters. This is shown for all the experiments for different numbers of endmembers e=4, 5, 8. Similarly, patch sizes of 2 × 2 and 3 × 3 provide better results as compared with larger patch sizes. The RES image shows more numbers of mixed pixels than IND and hence it affects the results slightly. The patch size of 1 × 1 or 2 × 2 produces more cohesive clusters as compared to larger patch sizes. With an increase in the number of endmembers used to define economic zones, the optimum patch size increases. In a scene where pixels are mixed (such as RES), optimal patch sizes for e=4, 5 are 1 and 2, whereas for e=8 optimal patch size is 7, 9.

The number of optimal classes that can be detected does not change much and remain in the range of 2–4. This makes sense, as it would be difficult to define more economic zones (or other similar land use classes) using distribution over more than ~4 unique materials. For all patch sizes, that is from 1 × 1 to 9 × 9 for all the images, the Silhouette coefficient is higher for a cluster number 2 and 3 and then drops by ~22%, ~18%, ~24% for e = 4, 5, 8 endmembers, respectively. Figure 5.13 shows the silhouette coefficient analysis performed for different endmembers and different patch sizes.

5.5.5.4 Comparison with Supervised Classification Results
Data Set A
The clusters using the VIS fractions as a feature form cohesive clusters. Each cluster indicates a particular socio-economic zone, as postulated. The most common classifier used for hyperspectral data SAM shows a high User's accuracy and poor Producer's accuracy for the industrial zones (100% vs 52%, Table 5.9). The industrial zone comprises large open areas and industrial sheds with metal roofs and hence it is very difficult to get the representative reference signature for the class for SAM classification. Contrary to that, both the residential zones, Low-economy and High-economy, show reasonable Producer's accuracy – 89% and 82%, respectively.

Classification of Urban Land Use and Land Cover

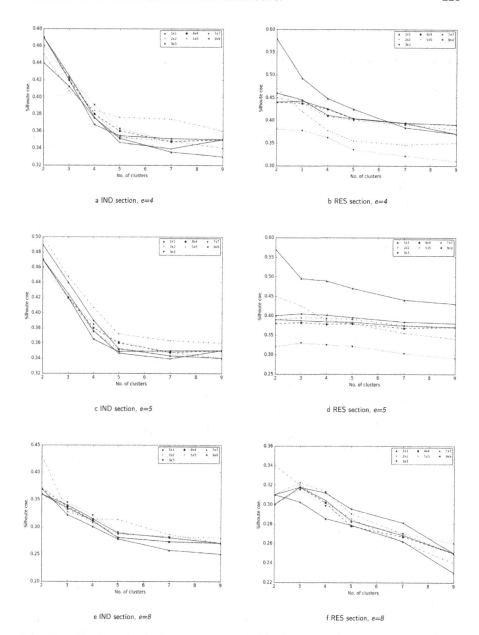

FIGURE 5.13 Variation in the Silhouette coefficient for different values of patch sizes for the IND and RES part.

TABLE 5.9
Confusion Matrix for Supervised Classification Using SAM

Classes

		Plain	Vegetation	Industrial	Low-economy residential	High-economy residential	Total	Producer's accuracy
References	Plain	101	0	0	43	0	**144**	70
	Vegetation	0	24	0	0	0	**24**	100
	Industrial	2	0	297	238	35	**572**	52
	Low-economy residential	37	0	1	372	10	**420**	89
	High-economy residential	0	58	0	11	309	**378**	82
	Total	**140**	**82**	**298**	**664**	**354**		
	User's accuracy	72	29	100	56	87	Overall accuracy	**72**

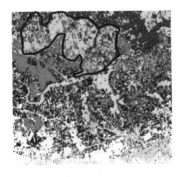
a Medium-economy residential area

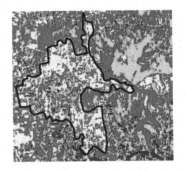
b Industrial area

FIGURE 5.14 Different economic zones identified using VIS fractions as features for RES and IND sections of the images.

Compared to SAM classification results, the classes using VIS fractions as features show a remarkable improvement: a jump of ~20% in overall accuracies is observed. The User's accuracy and the Producer's accuracy have been improved by a large margin (–9, Table 5.10). The average User's accuracy increased by ~11% (from 69% to 80%) and the Producer's accuracy increased by ~14% (from 79% to 93%) for all the classes. The User's accuracy and the Producer's accuracy of just three land use classes namely Industrial, Low-economy residential, and High-economy residential increased by ~18%; 81% to 99% and 74% to 92%, respectively. This is a significant improvement in the Producer's and the User's accuracies provided by the proposed method over standard classifiers such as the SAM. Moreover, the proposed method has reduced false positives by substantial margins. Figure 5.14 shows images of

TABLE 5.10
Confusion Matrix for Classes Identified Using VIS Fractions as a Feature

Classes

		Plain	Vegetation	Industrial	Low-economy residential	High-economy residential	Total	Producer's accuracy
References	Plain	130	0	0	014	0	**144**	90
	Vegetation	0	24	0	0	0	**24**	100
	Industrial	41	4	527	0	0	**572**	92
	Low-economy residential	42	0	0	378	0	**420**	90
	High-economy residential	0	28	0	0	350	**378**	95
	Total	213	**056**	527	392	350		
	User's accuracy	61	44	100	96	100	Overall accuracy	**92**

clusters formed using the VIS distribution for the sample residential (RES) and the industrial (IND) sections. It displays clustering results using a single-pixel window for better visualization. Each cluster indicates a socio-economic zone such as a low-economy zone or an industrial zone, and so on. Each cluster represents the common classes of vegetation and open areas also; this depends upon the land cover configuration of the area. Finding the correct number of clusters requires some experimentation. At times, a class is broken down into many clusters; for example, common open areas.

Thus, the analysis attests to the premise of defining economic zones with the urban area as a distribution over VIS classes. Further, it provides strong evidence for the patch size selected for this study. The lower number of endmembers producing better results for economic zone classification is encouraging as it indicated that VIS fractions are sufficient for basic economic zone classification.

Other classifiers namely k-NN, ML, and SVM produced similar results. The average overall accuracy of classification is ~66% and is poor as compared to a semi-supervised approach using VIS fractions. A reduced false-positive rate for all land use classes by the proposed method is reflected in the results by all other classifiers as well, and not just SAM (Table 5.11; a and b).

Data Set B – Five-fold Cross-Validation Results
The five-fold cross-validation results are consistent with the dataset A results. The only difference is the better performance of SVM classifiers. Table 5.12 and Table 5.13 show confusion matrixes for one of the datasets created in five-fold. We have chosen the subset which provides the best results by SAM. SAM shows a high User's accuracy and poor Producer's accuracy for the industrial zones (100% vs 62%, Table 5.12). Contrary to that, both the residential zones, low-economy and high-economy, show reasonable Producer's accuracy – 92% and 84%, respectively.

TABLE 5.11
Comparison of Proposed Method with Other Classifiers

	a) User's Accuracy						b) Producer's Accuracy					
UA	C1	C2	C3	C4	C5	OA	PA	C1	C2	C3	C4	C5	OA*
k-NN	42	51	100	48	80	66	k-NN	19	100	40	87	93	66
ML	00	41	97	48	78	63	ML	00	100	45	92	85	63
SVM	69	62	99	49	73	66	SVM	42	100	41	80	95	66
SAM	72	29	100	56	87	72	SAM	70	100	52	89	82	72
VISf[1]	61	44	100	96	100	92	VISf	90	100	92	90	95	92

* OA = Overall Accuracy, [1]VISf = Approach adopted in this study, C1 to C5 – Vegetation, Plain, low-economy residential, high-economy residential, industrial respectively.

TABLE 5.12
Confusion Matrix for Supervised Classification Using SAM-Five-Fold

	Classes	Plain	Vegetation	Industrial	Low-economy residential	High-economy residential	Total	Producer's accuracy
References	Plain	62	0	0	9	0	71	87
	Vegetation	0	82	3	0	0	85	96
	Industrial	0	10	123	0	0	133	92
	Low-economy residential	12	0	7	99	0	118	84
	High-economy residential	0	6	35	3	71	115	62
	Total	74	98	168	111	71	437	
	User's accuracy	84	84	73	89	100	Overall accuracy	84

Compared to SAM classification results, the VISf shows a remarkable improvement: a jump of 18% in overall accuracies is observed. The User's accuracy and the Producer's accuracy have been improved by a large margin (Table 5.12, Table 5.13). The average User's accuracy increased by 8% (from 86% to 95%) and the Producer's accuracy increased by 10% (from 82% to 92%) for all the classes. The User's accuracy and the Producer's accuracy of VISf show consistent improvement over an average of all the class accuracies for all the classes, except for the low-economy residential class.

Overall accuracy also improves by a substantial margin; the average improvement in average overall accuracy is 12%. The improvement over commonly used

TABLE 5.13
Confusion Matrix for Classes Identified Using VIS Fractions as a Feature-Five-Fold

Classes

		Plain	Vegetation	Industrial	Low-economy residential	High-economy residential	Total	Producer's accuracy
References	Plain	38	0	0	0	0	38	100
	Vegetation	0	17	0	6	0	23	74
	Industrial	0	0	161	0	4	165	98
	Low-economy residential	0	0	0	130	0	130	100
	High-economy residential	0	0	0	0	166	166	100
	Total	38	17	161	136	170	512	
	User's accuracy	100	100	100	96	98	Overall accuracy	98

TABLE 5.14
Comparison of Proposed Method with Other Classifiers-Five-Fold Validation

a) User's Accuracy							b) Producer's Accuracy						
UA	C1	C2	C3	C4	C5	OA	PA	C1	C2	C3	C4	C5	OA*
k-NN	92	97	81	85	88	86	k-NN	82	94	88	91	75	86
ML	78	94	64	78	83	74	ML	88	64	89	76	56	74
SVM	93	97	89	89	99	91	SVM	87	97	96	93	83	91
SAM	76	80	65	86	100	76	SAM	80	75	84	87	52	76
Avg.	85	92	75	84	93	82	Avg.	84	83	89	87	67	82
VISf[1]	95	99	79	93	95	94	VISf	100	88	94	93	93	94

Notes:
* OA = Overall Accuracy, 1VISf = Approach adopted in this study, C1 to C5 – Vegetation, Plain, low-economy residential, high-economy residential, industrial respectively

classifiers for hyperspectral data such as SAM and ML is 20%. SVM proves to be the second-best classifier as compared to VISf. However, VISf shows a 3% improvement over SVM as well. The overall improvement in the VISf is because of all-round performance; both Producer's and User's accuracies show consistent improvement for all the classes thus indicating increased true positives and reduced false positives as compared to most of the classifiers (Table 5.14).

5.5.5.5 Analysis of Improved Results

An in-depth analysis of all the parameters of the proposed method indicates three main reasons for improved results:

a. use of proxy features instead of spectral features,
b. use of distribution over VIS fractions as features, and
c. use of a patch.

Primary spectral features are not discriminating features for a land use class such as an economic zone. Though they contain information about the land cover, primary spectral features do not contain information about a higher-level class such as an economic zone. The spatial features and/or proxy features from spectral features are required to define an economic zone or other land use pattern. The present method captures the spatial configuration of land covers over a single pixel or a patch by calculating VIS fractions within them resulting in a better definition for land use class.

VIS or VIS-type fractions for each class provide discriminative vectors for clustering and classification. As can be seen from the study, vectors of the VIS fractions – within a single pixel or a patch – for different economic zones are very distinct (Table 5.7, Table 5.8). VIS fraction vectors provide high average similarity for a cluster and high dissimilarity between different cluster centres. The distribution over VIS fractions of two economic zones is very distinct. Thus, the improvement in results is because of descriptive and discriminative features.

A land use class such as an economic zone may or may not be assigned to a single pixel directly. Whether a single pixel represents a land use class such as an economic zone or a group of pixels is required for the same depends upon the resolution of the imagery and configuration of land covers in a given urban unit. In this case, cluster quality showed improvement for a patch size of 2 × 2, indicating the need for a group of pixels to define an economic zone for the study area.

5.5.5.6 Land Use Analysis Using Proxy Features

The research shows the effectiveness of the distribution of different land covers as a proxy for socio-economic zones. Distribution over VIS fractions in a given pixel is unique for a given economic zone (Table 5.8). The average spectral angle between mean VIS vectors of three economic classes, namely industrial, high-economy residential and low-economy residential, is above ~40° indicating a very good separation of classes. The unique VIS signature for a given economic class can be further verified by a topic discovery model (Blei et al., 2003). The spatial organization of pixels (Barnsley & Barr, 1996) can also be further incorporated to generalize the approach. Some of the specific observations are:

a. Compared to SAM classification results, the classes using VIS fractions as features show a remarkable improvement: a jump of ~20% in overall accuracies is observed. The User's accuracy and the Producer's accuracy have been improved by a large margin (Table 5.12, Table 5.13). The average User's accuracy increased by ~11% (from 69% to 80%) and the Producer's accuracy increased by ~14% (from 79% to 93%) for all the classes. The User's accuracy and the Producer's accuracy of just three land use classes namely Industrial, Low-economy residential, and High-economy residential increased by ~18%; 81% to 99% and 74% to 92%, respectively. The average overall accuracy of classification is ~66% and is poor as compared to a semi-supervised approach using VIS fractions. A reduced false-positive rate for all land use classes by the proposed method is reflected in results by all other classifiers as well, and not just SAM (Table 5.14; a and b).

b. The endmembers extracted by the SiVM algorithm show a high degree of resemblance with respective VIS classes: for e=3–4, signatures of vegetation, impervious surface, and soil are extracted. The observed average spectral angle between endmembers and references is ~5°. The process is thus very helpful in extracting signature distribution of the VIS fractions for different socio-economic attributes using hyperspectral data. Some class confusion exists between bright pixels of concrete, industrial roofs, and bare soil at times.

c. As we increase e, more than one endmembers of the same class are extracted, for example, 2–3 types of industrial roofs in an industrial area and 2–3 plain endmembers in an open area. As expected, considering the dominant classes, vegetation, plain, and industrial roof, are extracted as the endmembers when e=4.

d. The Silhouette coefficient is higher for cluster numbers 2 and 3 and then drops by ~22%, ~18%, ~24% for e = 4, 5, 8 endmembers respectively. The clusters using the VIS fractions as a feature form cohesive clusters. Each cluster indicates a particular socio-economic zone, as postulated.

e. The validity for the VIS distribution needs to be confirmed by further comprehensive studies of other cities in India and abroad. A similar pattern may or may not exist in other cities of the world. However, it is highly unlikely that the VIS distribution of different economic zones to be very similar to each other. Slum and low-economy residential, high-economy residential classes appear to have a universal signature: studies of Johannesburg (Verma et al., 2016), Nairobi, and a few other Indian cities show similar VIS configurations for various economic zones within the cities.

f. The VIS distribution might result in confusion for some economic zones. For example, new upmarket residential development on the fringes of the city would have large open areas, large building block sizes, and no vegetation, and hence it might be confused with an industrial zone. If endmembers of VIS types also were extracted correctly, for example, GI sheet roofs vs concrete roofs, this method would produce consistent results in such a situation as well.

5.6 ADVANCED MACHINE LEARNING METHODS SUCH AS DEEP LEARNING

An artificial neural network is one of the most researched classifiers in recent times. This is because its modern variants provide the best accuracy on many image-processing tasks, including hyperspectral image classification (Li et al., 2019;

Ghamisi et al., 2017). A neural network is a certain arrangement of connected units called neurons. Each neuron performs two functions – first, it aggregates the inputs by weighting them with different weights and then it squashes the weighted summation by a function such as sigmoid. The functions that convert the inputs to values between 0 to 1 are also known as squashing functions. The second step, that is the squashing step, converts the weighted summation from the first step to values between 0 to 1. A network-specific squashing function can be designed as well.

Figure 5.15 shows a single neuron of the second layer and shows how the weighted summation is squashed by a squashing function to produce an output. x are the input values and w are the weights of the connections. σ is the squashing function used. The superscript indicates the layer number, and the subscript index is a band number. Weights are numbered according to the input layer and output layer. For example, $w_{1,2}$ is the weight of the connection between the 2nd node of layer 0 to the 1st node of layer 2. That is input from the second node of the first layer is weighted by $w_{1,2}$ while performing summation at node 1 of layer 1. B is a bias term. The computation with matrix notation is illustrated.

The simplest network has three layers of nodes. The input nodes equal to the number of inputs are present in the first layer. The second layer often contains hidden nodes more than the number of the input layer. The last layer contains the nodes equal to the output. If it is a binary classifier, the output layer will have only one node, else equal to the number of classes. In the case of multiple output nodes, the squashing function for these nodes is logistic regression, all the nodes in all the layers are connected to each other. This is a description of a simple neural network also

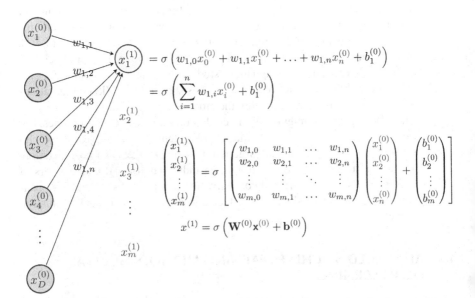

FIGURE 5.15 Single neuron computation in a layer.

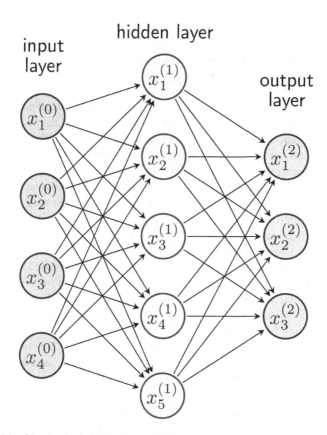

FIGURE 5.16 Simple single hidden layer MLP.

known as a multi-layer perceptron (MLP). Nodes of the network are referred to as neurons and a group of neurons performing some tasks are referred to as a perceptron.

Figure 5.16 shows a simple neural network (MLP) as explained earlier. The input nodes in the case of pixel-based classification are equal to the number of bands. They represent band values. Each node is weighted by the weight for each connection between a node in the input layer and a node in the hidden layer. The total number of weights between the input layer and the hidden layer would be 20 in this case. The same structure is repeated between the hidden layer and the output layer. The number of weights between the hidden layer and the output layer is 15. The output nodes represent the values after squashing. The number of nodes in the output layer is equal to the classes; for example, in this case, 3. In some of the cases, the output layer can be considered as a transformed input feature (especially when the output is not a class label) vector as well and a neural network can be seen as the model for transforming the input vector into some arbitrary feature vector.

The weights of the network are learned using the correct labels of the training samples, to be more specific, using the difference between the correct answer and the network answer. They are learned to minimize the error between the human labels of the training data and the labels given by the network. It is done by an algorithm known as a backpropagation algorithm. The learning begins with the initialization of the weights with arbitrary values. Then the network output is compared with the correct label of the training sample. The rate of change of weights with respect to the error (gradient) is calculated and then it is used to adjust the weights by propagating them in the backward direction through network layers. In a forward pass, the updated weights generate the label for a new instance of the sample and the process continues till some convergence criteria are met, such as the error between the network answer and the correct answer is negligible, or the difference between the network output and the class label for previous iteration and the present iteration is negligible.

There are many advantages of neural networks. An artificial neural network can be viewed as a universal function approximator. The neural network with sufficient complexity can approximate any mapping function between features and labels. That is, if there exists a mapping function between the features of the input data or just the input and its labels and has a sufficient number of hidden units then the neural network approximates that mapping function (Hornik et al., 1989). This implies that the connection weights or simply weights of the network can be learned such that the network approximates the mapping function if it exists. In other words, the finite sequence of a linear combination of input and squashing can approximate any input class label mapping function if it exists.

Convolutional neural networks (CNN) are a special class of networks useful in image processing, for a specific reason – they get rid of the need for handcrafted features in learning. A common architecture of a CNN has two types of layers; one performing convolution that learns filters and creates feature maps, and the other type of layers which are fully connected. The fully connected layers are also called dense layers. The convolution layers are not fully connected layers, they are locally connected layers. That is, the nodes within the scope of a convolution window are connected to a single node in the next layer. Thus, each cell in the convolution window has a one-to-one relation with the single node in the next layer corresponding to the central cell of the convolution window. Such an operation is followed by a pooling layer, a specific type of convolution, which reduces the size of the input image. The pooling layer, for example maxpooling, outputs the maximum value from the input receptive field and maps it to the next layer. Any number of convolution and pooling blocks can be arranged one after another for performing a classification task. For example, VGG16,[6] one of the well-known network architectures for labelling the image, uses 16 to 19 layers (Simonyan & Zisserman, 2014). The filters are learned during the process automatically and hence the features (feature maps). This is one of the biggest claimed advantages of CNN – it learns the features for image classification automatically. Thus, handcrafted features and explicit functions to extract them from the images are not required.

5.6.1 Understanding 1D, 2D, and 3D Convolutions

A 1D convolution is much simpler in operation than its cousins. A kernel is a simple row vector, and it spans over a pixel vector of D^f dimensions. Thus, the size of the filter would be $1xD^f$. The single filter placed over the receptive field creates a single

Classification of Urban Land Use and Land Cover

value for the central dimension in the next layer corresponding to the central dimension in the receptive field. That is, if the index of the receptive field is 2 and the filter is centred over 2 then it creates a transformed value for band 2 considering the scope of the receptive field of the filter. By moving over the entire D, the filter creates a truncated copy of the pixel vector if the padding is not used. If the padding is used it creates a copy of the pixel with a different set of D features for a pixel. Each filter creates a copy of the pixel. These copies can be stacked together to create a large pixel with $F*D$ features where F is the number of filters. Another possibility is a long list of newly created F pixels with D dimensions. The 1D convolution process is illustrated in Figure 5.17. It is an illustration of a snapshot during the convolution process when the 1D filter is placed on the starting position of the 1-dimensional pixel vector. The pixel vector with each band is marked with a grey cell. The vector is convolved by a 1x3 filter. At a particular timestep of convolution, each filter applied on the receptive field shown by local collection created a single output value.

2D convolution is slightly different from 1D convolution. The size of the filter spans H and W and D. However, the convolution operation does not operate in a depth (D) direction. The convolution operation runs across H and W for all the D of the filter. A 2D convolutional filter operating over a 3D space creates a 2D output that is H and W without any depth, with the right padding (Figure 5.18). A 3D convolutional kernel operating over a 3D space creates 3D output, a stack of 2D outputs where each output corresponds to a position across the depth axes.

Ji et al. (2012) introduced 3D convolution. The task was to recognize the human actions from video data at the airport, collected for surveillance purposes. The video

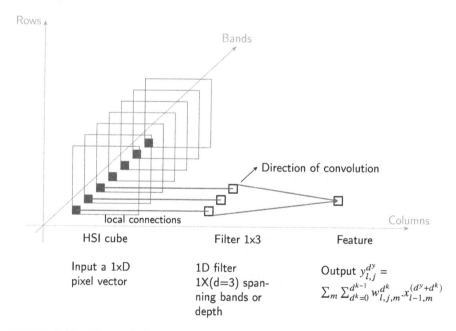

FIGURE 5.17 1D convolution over a hyperspectral pixel vector of D dimensions.

frames (a still image in the video sequence) at different times were stacked together along the depth. Thus, the input dataset was a 3D volume with $H \times W \times D$ dimensions where H and W were the height and width of the image/video and the D was temporal information. 3D convolution was proposed for extracting information from both the dimensions, spatial and temporal, simultaneously.

3D convolution is similar to 2D convolution in principle. Instead of moving over H and W, a 3D convolution filter moves over the depth as well. In this case of action recognition, it was a temporal sequence of human movements (Ji et al., 2012). In the case of the hyperspectral data, it is the spectral dimension. The 3D convolution filter is locally connected to the 3D volume of the input data cube. The depth extent of the filter is much smaller than the total depth. For every local volume connection, a 3D filter creates a single value which is mapped to a single cell of the output (Figure 5.19). By completing all the movements in HWD directions a 3D volume of the same size is generated (when input is padded appropriately). The number of kernels creates convoluted copies of the input layer equal to the number of kernels. For example, the input 33 feature maps were grouped into 5 {7,7,7,6,5} and convolved by 2 3D filters of 7x7x3 created 23 feature maps in the next layer (as 2 were truncated in each group totalling 10; 33−10) (Ji et al., 2012).

Table 5.15 summarizes the types of convolutions with their input and output. The right amount of padding is assumed. The superscript f indicates the dimensions for the filter. For example, D^f is a depth dimension of the filter.

One more special type of convolution, applicable in the case of a hyperspectral image, is a depth separable convolution (Chollet & others, Keras, 2015; Chollet, 2016). It is a combination of two 2D convolutions with a different focus and hence different filter sizes. A filter spanning H^f and W^f, say 3x3, and convolving across

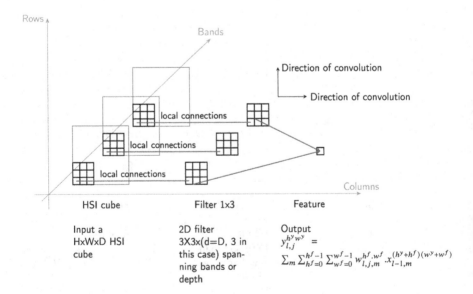

FIGURE 5.18 2D convolution over a hyperspectral cube of HxWxD.

Classification of Urban Land Use and Land Cover

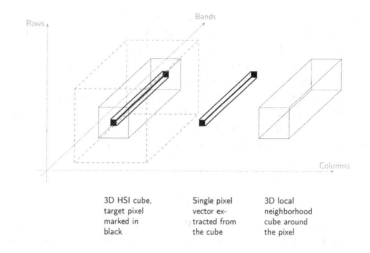

3D HSI cube, target pixel marked in black

Single pixel vector extracted from the cube

3D local neighborhood cube around the pixel

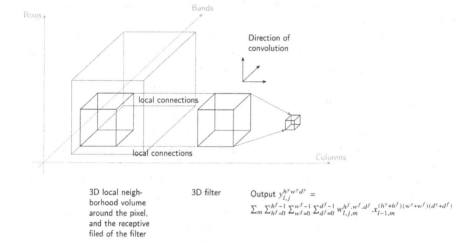

3D local neighborhood volume around the pixel, and the receptive filed of the filter

3D filter

Output $y_{l,j}^{h^y w^y d^y} = \sum_m \sum_{h^f=0}^{h^f-1} \sum_{w^f=0}^{w^f-1} \sum_{d^f=0}^{d^f-1} w_{l,j,m}^{h^f,w^f,d^f} \cdot x_{l-1,m}^{(h^y+h^f)(w^y+w^f)(d^y+d^f)}$

FIGURE 5.19 3D convolution over a hyperspectral cube of H×W×D.

TABLE 5.15
1D 2D and 3D Convolution Operation

Input	Filter size	Spans	Output
Pixel vector, $H \times W \times D$,	$1 \times D^f \mid D^f \neq 0$	D	$1 \times D$
Image $H \times W \times D$,	$H^f \times W^f \mid 0 < H^f < H, 0 < W^f < W, D^f = D$	$H \times W \times D, H \times W \times D$	$H \times W \times 1$
Image $H \times W \times D$	$H^f \times W^f \times D^f \mid 0 < H^f < H, 0 < W^f < W, D^f = D$	$H \times W \times D$	$H \times W \times D$

H and W direction for each band/channel is applied. This is also referred to as a depthwise convolution. One feature map created by each filter is stacked together in the next step. The resulting volume is convolved by $1x1xD$ filters to create feature maps equal to the number of filters. This is also referred to as a pointwise convolution. First filters exploit spatial correlations and then second spectral correlations. This type of convolution is equivalent to learning spatial and spectral features for hyperspectral data separately (Chollet, 2021). This type of convolution decouples spectral and spatial information in the hyperspectral image and thus does not exploit the joint spectral-spatial correlations. The sequence of operations, that is, depthwise and pointwise convolutions can be interchanged.

5.6.2 Deep Learning for Hyperspectral Images

Of these two broad types of networks, MLP and CNN both are useful in classifying hyperspectral data. MLP takes band values of a spectrum as an input and hence is suitable for classifying a spectrum from imaging as well as non-imaging devices. CNN is useful in the classification of a hyperspectral image as it learns the spatial features automatically.[7] However, the classification of a hyperspectral image with CNN is not straightforward as compared to a multispectral or panchromatic image. This is because a pixel of a hyperspectral image is rich in spectral information which is vital for classification. Extracting useful information from these large numbers of bands demands meaningful modifications in the convolution process, in addition to the careful overall design.

Figure 5.20 shows the initial input extracted from the hyperspectral image which is processed by the neural network processing. As we have discussed earlier, the focus of image classification is to assign a class label to a single pixel. The entire image is not an input to the classification process; it is a single pixel alone or along with the context pixels, that is, the local neighbourhood of the pixels. These two extracted subsets from the image, that is, the spectral features and the spatial features are the input to the neural network. The convolution operates on them as a set of features for the pixel to be labelled. The nature of the features, that is, if they are spectral or spatial, determines the nature of convolutions and how they are combined.

Once the constraints and the capabilities of each set of features are understood, the sequence of the convolution operation and the feature fusion operations can be designed properly. There are three to four basic strategies followed for classifying a hyperspectral image using neural networks (Li et al., 2019; Chen et al., 2014; Chen et al., 2015; Chen et al., 2016):

a. **Spectral-only MLP:** This is the simplest application of neural network to the hyperspectral image classification. The MLP is defined to take the band values from a pixel (or a reading from a non-imaging device as applicable) as input, and the network is built over it to output the class probabilities. These were one of the earliest applications of neural networks to hyperspectral data classification, or even multispectral image classification.

Classification of Urban Land Use and Land Cover

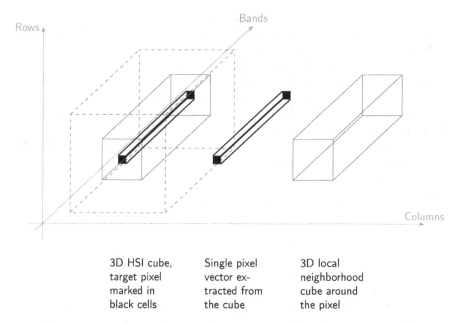

| 3D HSI cube, target pixel marked in black cells | Single pixel vector extracted from the cube | 3D local neighborhood cube around the pixel |

FIGURE 5.20 Inputs to neural network processing of hyperspectral image.

b. **Convolutional spectral-only deep features:** The difference in this way is to extract the deep features using 1D convolution over a spectral vector of D bands, instead of the fully connected layers. Once the deep features are extracted, dense layers follow these layers and produce the output.

c. **Primary feature fusion followed by a neural network:** In these types of architectures, primary spectral and spatial features are extracted and stacked together followed by MLP. To be more specific, the band values from a target pixel and the band values from the local neighbourhood of the target pixel are stacked together. Later, MLP or deep layers and dense layer combination is used to produce a final output. Though the primary features are extracted from the local neighbourhood of the target pixel, they are limited to band values only.

d. **Convolutional deep features fusion followed by dense neural network:** In these types of networks, spectral features are convolved with 1D convolution, the features within the local neighbourhood of the pixels are convolved with 2D or 3D convolution, and then the deep features extracted from both the paths are staked together followed by MLP. This is by far one of the well-established architectures for hyperspectral image classification (Yang et al., 2017).

e. **Integrated:** These types of networks perform joint spectral-spatial convolution using 3D convolution over the local neighbourhood of the target pixel. The convolution filter size needs to be designed carefully to extract joint spectral-spatial features from a hyperspectral image. These are conceptually simple yet computationally challenging networks.

Each one of these architectures is a progression of basic deep learning principles with increasing convolution scope such as 1D, 2D, and 3D, and spectral or spatial dimensions of the hyperspectral cube and so on. Building a neural network for hyperspectral image classification requires selecting a few network parameters. They are convolution scope and input such as pixel or local neighbourhood; stage of feature fusion such as before or after deep learning; and type of connections such as dense network or locally connected network and so on. We explain individual steps for extracting deep features from a pixel vector and a local neighbourhood cube around a target pixel. The combination of these steps with specific choices of the network parameters will result in a network for a specific task in hand (Figure 5.21 and Figure 5.22).

5.6.3 Spectral, Spatial, and Spectral-Spatial Neural Networks

Architectures that extract spectral features consider a single pixel vector and build the network for creating the output class label/s. The general steps are as follows.

1. Extract a target pixel vector as per the row-column index or the list index of the hyperspectral image.
2. Consider the band values as an input and build the first layer. The number of nodes in the first layer would be equal to the number of bands. The input values can be reduced by any standard methods for dimensionality reduction such as PCA on the original image with all the bands and then only a few transformed bands can be considered as the input.
3. If MLP/deep network is considered then, build one or more hidden layers as per the design. The number of layers would be large if a deep network is considered. Each hidden layer is fully connected to its previous and next layers. The activation functions such as sigmoid or any other type can be considered.
4. The last layer is a logistic regression layer. It will have a number of nodes equal to the number of classes. Each node value is normalized such that activations of all the output nodes sum to one. Thus, each node activation indicates the probability of the input belonging to the class node represents.
5. The built network is trained using the labelled samples. The model constitutes the network topology and the learned weights. The query pixel is given as an input to the model (trained neural network) to get the output, which is class label probability or a real value output in case the network was trained on real value output.
6. In the case of limited training samples, the hidden layers with a reduced number of nodes are deployed and again increasing number of nodes are deployed. The final nodes of such networks are designed to output the input sample itself. The network till the shortest size of the hidden layer (and including it) is called an encoder, and the rest of the network from the shortest hidden layer to the output layer is called a decoder network. Network learning is in two stages. The encoder is trained for reconstructing the original samples. The encoder is taken out of the network and is added with the logistic regression

layer to build a new classifier, which is fine-tuned to give the output class labels using the training data.

7. To build the spectral convolutional network, the steps remain the same at a higher level. However, the initial layers up to fully connected layers are replaced with convolutional layers. To build the 1D convolutional layers, consider the 1D filter size and the number of filters in each layer. The general strategy of the deep network design is to learn local features or fine granular features in initial layers and coarser or global features in later layers. Additionally, the number of feature maps generated in the layers should increase with the depth of the network, that is, the layer numbered with a large index value has more filters or feature maps. Once the network for deep features is built, the remaining layers remain the same as fully connected layers or dense layers.

Neural networks extracting spatial features use 2D CNN for a local neighbourhood of the target pixel to learn the features. A clarification is required here for spatial features. The deep learning hyperspectral classification literature uses the "spatial features" term for considering the local neighbourhood context of the target pixel. The term usage does not consider explicitly if the features learned are spectral or spatial (where the spatial features are edges, shapes, texture and so on). We have used the term in the same sense to avoid confusion. However, it should be noted that consideration of local neighbourhood does not entail spatial features. For example, stacking of flattened neighbourhood pixels with the target pixel vector or individually is still a long vector of band values of the pixels as per the scope. To extract the spatial features, CNN or a similar treatment is required.

1. If we consider the pixels in the local neighbourhood of the target pixel and flatten them and add them to the pixel vector, the network becomes what is known as spectral-spatial. However, it is to be noted that just stacking together flattened pixel vectors for the target pixel and the local neighbourhood does not make it extract spatial features, which are learned in common CNN over a local neighbourhood cube. The deep or standard neural network is built to take the vector created in the earlier step. The vector is a vector of band values as per the spatial scope. For example, let's say we consider a local window of 3x3, then the combined vector would be *1xD + 9xD = 10D* vector. This *10D* vector would be used for further processing of any kind.
2. The other pathway is to consider the cube of local neighbourhood pixels within a window of, say *(HxW)^{f}xD* where *(HW)^f* is much smaller than *H* and *W*, and build a CNN to take that cube as an input the way RGB images are processed by CNN. The number of bands of hyperspectral images is too large to process this efficiently for every pixel and hence some kind of dimensionality reduction is performed on the original image cube. The convolution steps are in principle similar to a 1D convolution network with differences in filter size and convolution scope. The dimensionally reduced image cube is taken as input. The cube around the target pixel is extracted as mentioned earlier and

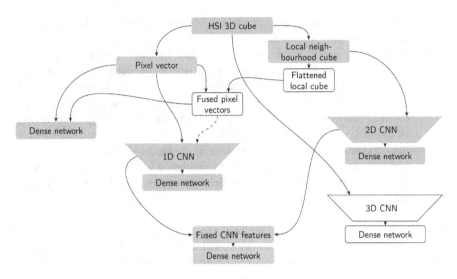

FIGURE 5.21 Various neural network pathways for building a hyperspectral image classifier.

then processed first by a convolutional network and then a dense network. The convolution filter spans in H and W dimensions.
3. One more alternative is to use 3D convolution instead of 2D. This type of convolution extracts joint spectral and spatial features from the input local cube in the convolutional network step.
4. The training and deployment operations remain the same as in the spectral step.

5.7 PERFORMANCE EVALUATION OF VARIOUS CLASSIFIERS ON BENCHMARK DATASETS

We have seen so far multiple deep-learning approaches for classifying a hyperspectral image. We have seen some of the common conventional classifiers as well. In this section, we evaluate the performance of the key classifiers on the benchmark datasets. The datasets are described earlier in Chapter 2. The conventional classifiers are implemented using a sklearn library. The training data was split into training data and testing data using a 80:20 split. The spectral-spatial deep model is the same as that implemented by Chen et al. (2015). The architecture is described in the table in detail (Table 5.16). For the remaining deep models, the architectural details are provided in the appendix. All the experiments were run on a personal laptop with the following specification:

CPU: Intel(R) Core (TM) i5-10310U CPU @ 1.70GHz 2.21 GHz
RAM: 16.0 GB (15.8 GB usable)
GPU: Intel(R) UHD graphics

The purpose of the experiments is not to select a favourite or best-performing classifier or recommend one. The purpose is to understand the variations in the results of

Classification of Urban Land Use and Land Cover

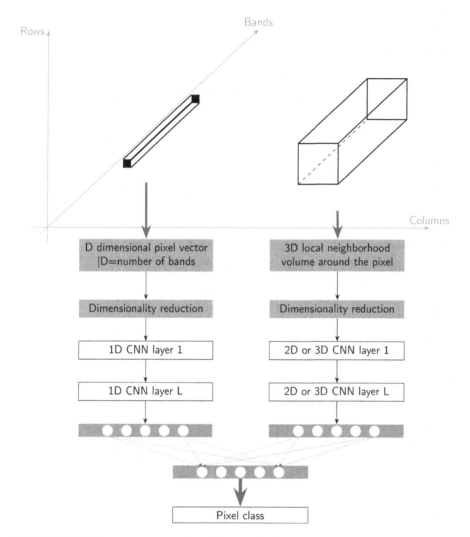

FIGURE 5.22 Most common neural architecture for hyperspectral image classification.

various classifiers and their general limits, trade-offs and so on. An additional focus is to understand the suitable conditions for each spectral, spatial, and spectral-spatial approach to work well. This is essential for building an appropriate model under the given circumstances.

As can be seen from the results, classification accuracies by various classifiers show that there are many good performers (Table 5.17 and Table 5.18). Spectral-spatial network which is specifically designed for hyperspectral image classification is performing well. It provides best or near-best results for some of the datasets. At the same time, conventional classifiers such as SVM, k-NN are providing good accuracies.

TABLE 5.16
Spectral-spatial Classifier

Layer	Kernel size	Step size	Feature maps	Drop out	Parameters	Connection
Spectral Network						
Input 1			200		0	
Dense			512		102912	
Dense			1024		525312	
Dense			1024	0.2	1049600	
Dense			2048		2099200	
Dense			2048	0.2	4196352	
Dense			4096		8392704	
Dense			4096		16781312	
Spectral-Dense			1000		4097000	
Spatial Network						
Input 2			(5, 5, 200)		0	
Convolution	3	1	(5, 5, 64)		115264	
Convolution	3	1	(5, 5, 64)		36928	
Convolution	3	1	(5, 5, 128)		73856	
Convolution	3	1	(5, 5, 128)		147584	
Convolution	3	1	(5, 5, 256)		295168	
Convolution	3	1	(5, 5, 256)		590080	
Convolution	3	1	(5, 5, 256)		590080	
Convolution	3	1	(5, 5, 512)		1180160	
Convolution	3	1	(5, 5, 512)		2359808	
Convolution	3	1	(5, 5, 512)		2359808	
Convolution	3	1	(5, 5, 512)		2359808	
Convolution	3	1	(5, 5, 512)		2359808	
Convolution	3	1	(5, 5, 512)		2359808	
Maxpooling	2	2	(2, 2, 512)		0	
Flatten			2048		0	
Dense			4096		8392704	
Dense			4096		16781312	
Spatial-Dense			1000		4097000	
Combined Network						
Concatenate			2000		0	Spectral-Dense, Spatial-Dense
Dense			1000		2001000	
Dense			16		16016	

Source: Chen, Zhao, & Jia, 2015.

TABLE 5.17
Classification Accuracies by Various Classifiers

Models	Pavia University	Pavia Centre	Indian Pines	Urban	Jasper Ridge	Washington DC	Botswana
(Fully connected NN/MLP) FNN	65%	82%	49%	95%	88%	69%	70%
1D CNN	75%	93%	40%	90%	92%	62%	64%
VGG	77%	96%	51%	97%	97%	68%	70%
Spectral-spatial CNN	90%	97%	59%	96%	97%	70%	82%
KNN	89%	98%	72%	95%	97%	67%	73%
Naive Bayes	68%	88%	49%	68%	92%	57%	80%
Decision Tree	88%	97%	72%	93%	95%	60%	69%
SVM	90%	98%	77%	96%	97%	67%	73%
Random Forest	90%	97%	74%	96%	96%	60%	72%
SAM	56%	83%	42%	60%	78%	38%	71%
NS3	59%	80%	40%	60%	79%	49%	67%

TABLE 5.18
Computational Performance of a Deep Model on Various Data Sets

Performance metric	Pavia University	Pavia Centre	Indian Pines	Urban	Jaspre Ridge	Washington DC	Botswana
Data size (MB)	33.1	123	5.67	21.8	2.84	143	75.2
Number of bands	103	102	200	162	198	191	145
Classes	9	9	16	6	4	7	14
Pixels	42776	148152	10249	94249	10000	392960	3248
Max RAM utilization in (GB)	7.1	8	7.3	7.6	7	8.9	5.4
Max CPU utilization	42%	51%	42%	40%	35%	48%	27%
Max GPU utilization	8%	6%	3%	4%	7%	8%	3%

5.8 EXPLAINING CONVOLUTIONAL SPECTRAL FEATURES

One of the advantages of a convolutional neural network (CNN) is that it learns the features and classification model simultaneously. In the case of image classification or segmentation, the learned features are spatial features such as edges, arcs, convex shapes-contours of the objects and the object parts and so on and are interpretable. There is an inherent hierarchy present in these features. The feature maps at the lower (closer to input) layers show the primitive image features such as edges with different orientations. The feature maps at the different higher levels show the higher-level features such as arcs, shapes, and object parts respectively. Thus, the features show the parts or whole of the primal sketch progressively and are consistent with the visual semantics. Often, the spatial dimensions (rows and columns) of the image are convolved to build the hierarchical features. The convolution over the spectral domain is not performed for a visible or a multispectral image. 3D filter, say *3x3xD* with a depth equal to the number of bands of the image is commonly used. This works for the visible or the multispectral image as the spectral features are not dominant in them. However, the common architectures for the visible or multispectral images that perform convolution over the spatial dimensions are not useful in the classification of a hyperspectral image. The spectral information is critical for the classification of the hyperspectral image.

As can be seen in the literature, the convolution is performed over a D-dimensional pixel vector where D = number of bands. The pixel vector is stacked vertically or horizontally, and the convolution is performed over the length of the vector. Additionally, the spatial features are stacked up together with the spectral features and the joined vectors are processed further. Thus, it is supposed to learn the deep features in the spectral domain. However, the features extracted by the spectral convolution in this manner are difficult to interpret in the spectral domain. For example, spatial convolution learns the features that represent edges, arcs, shapes and so on. In a similar manner, the spectral features learned by the spectral convolution should not be just specific wavelengths/bands, but diagnostic absorptions or reflections as of the spectral signature at a given position, slope or convexity or concavity of the spectral shape and so on. None of the approaches used in the past literature (barring a few) attempt to address this problem. They do not focus on the spectral features and their interpretation like the interpretation of the spatial features in a spatial domain. The convolution over a D-dimensional pixel vector is difficult to interpret as it appears to learn discontinuities in the pixel vector for a given receptive field. Visualizing the same is, again, a challenging task as it is a single row or column vector. Building the hierarchy of these spectral features like the spatial features is difficult. Furthermore, most of the research uses a dataset like Pavia (Kunkel, et al., 1988), which is having very high spatial resolution. Most of the accuracy improvement for such a high-resolution image is because of the spatial features. Whether the vector-based spectral convolution is useful in medium-to-coarse resolution imagery needs to be investigated further (as the spatial features are not dominant as compared to spectral features, in the images). The systemic mechanism of the spectral and the spatial features learning for the imagery with the dominant spectral features such as hyperspectral is required.

In contrast to that, the alternative mechanisms that can extract interpretable spectral features are:

a. a 1D convolution by filters dedicated to the different wavelength segments, possibly each one without sharing the weights. For example, the filter extended over the entire range of the spectrum (or fully connected layers with neurons equal to the number of bands) in the first convolutional layer (a single filter) would learn the significance of the different wavelengths and so on.

b. a 2D convolution over the spectral shape, that is, a graph of the spectrum in XY space, or a line diagram of the spectral signature.

The latter is more intuitive for multiple reasons. When the experts analyse the spectrum of a target material, they observe the shape of the spectrum at various wavelengths, that is, the diagnostic absorption, the convexity of the curve, the slope at different positions or over a particular wavelength range, and so on. Furthermore, when there are no specific diagnostic features for the target materials, the expert comprehends and compares the entire spectrum shape for discriminating the materials. The shape of the parts and/or the complete spectrum are intuitively useful. The shape features such as arcs and arc segments formed by joining multiple arcs show a hierarchy of the features as well. Thus, the representation would further aid in architecting the robust capsule network (Sabour et al., 2017).

5.8.1 Capsule Net

Deshpande et al. (2021) transform a hyperspectral pixel vector into a two-dimensional spectral shape and then perform the convolution over the image of the graph. The intuition is to learn the spectral features as represented by the shape of a spectrum or in other words the features which a spectroscopy expert uses to interpret the spectrum. Thus, now the filters would learn edges, arcs, arcs segments and the other shape features of the spectrum as applicable. The architecture was trained using cross-entropy loss. The preliminary experiments compared the results with the standard pixel vector representation and the capsule network. No spatial contextual information of the pixel was used in these experiments.

5.8.1.1 Feature Representation

The main difference between the proposed architecture and the architecture described so far is the alternative representation of the D-dimensional pixel vector. The pixel vector is converted to an image of spectral signature plotted in wavelength and reflectance space. So instead of the pixel vector, the new image of the pixel signature plot is input to the CNN. As the spectral signature is converted to the shape, it can be decomposed into hierarchical features learned at the different convolution layers at different levels. This is like recognizing a handwritten digit using its image. Now, because of this transformation, the deep architecture would be able to learn the features that are consistent with the spectroscopic interpretation. The features learned by deep architectures would be similar to the spectral signature features as domain experts seek while identifying the material corresponding to the signature. In the implementation (Deshpande et al., 2021) the signature was converted to a 128x128 image. The choice of image size was determined by the number of bands in the dataset. The parameter is configurable and can be changed according to the number of bands in the working dataset. Any size that represents the graph without loss of information is suitable. The other alternative for the image format would be showing the signature plot by two regions above and below, instead of a line plot image.

5.8.1.2 Architecture

The two architectures compared with each other are one-dimensional waveform architecture (Dai et al., 2017) and one by the CapsuleNet (Sabour et al., 2017). We describe the 1D architecture first.

The architecture comprises three blocks of convolution followed by one block of a fully connected layer. Each block of the convolution consists of three convolution layers and each convolution layer is followed by a batch normalization layer. The last convolution layer of each block is also followed by a dropout layer to prevent overfitting. The dropout rate of 40% is maintained. The first two convolution layers in each of the blocks are having a filter size of 3x3 and the last convolution layer of each block has a filter size of 5x5. The filter uses strides of three and five respectively. Max pooling was not used as it loses important spectral information. Instead, the convolution layer with strides three and five for sub-sampling is used. Each convolution layer of all the convolution blocks consists of 32, 64, and 128 feature maps respectively. The fully connected block of the network consists of two fully connected layers. The first fully connected dense layer consists of 256 units whereas the second fully connected layer, depending upon the number of classes considered, has nine units in it (Table 5.19). The second proposed architecture is the CapsuleNet modified with the reconstruction loss. This is designed to further enhance the performance and study

TABLE 5.19
Network Architecture for P1 and P2

Layer	Filter size	Step size	Feature maps	Dropout
Image			128x128	
Convolutional	3x3	1	126x126x32	
Normalization				
Convolutional	3x3	1	124x124x32	
Normalization				
Convolutional	3x3	3	40x40x32	0.4
Normalization				
Convolutional	3x3	1	38x126x32	
Normalization				
Convolutional	3x3	1	36x126x32	
Normalization				
Convolutional	3x3	3	11x126x32	0.4
Normalization				
Convolutional	3x3	1	9x126x32	
Normalization				
Convolutional	3x3	2	7x126x32	
Normalization				
Convolutional	3x3	5	1x1x128	0.4
Normalization				
Linear			256	0.4
Normalization				
Linear			9	

loss functions for spectral classification. The rest of the network is similar to the Caspsulenet (Sabour et al., 2017).

Rectified Linear Unit (ReLU) activation function was used for all the layers in the network and Softmax in the last layer. Categorical cross-entropy compares the prediction distribution with the true distribution, where the true class is represented as a one-hot encoded vector, and the closer the predictions are to that vector, the lower the loss. Stochastic Gradient Descent (SGD) with a learning rate of 0.01 and momentum of 0.5 was used. SGD with momentum tends to reach better optima and has better generalization than adaptive optimizers (Wilson et al., 2017).

5.8.1.3 Experimentation Details

The dataset used was the well-known Pavia University dataset (Kunkel et al., 1988). The image has 103 bands covering a 430 nm to 860 nm wavelength range and has a spatial resolution of 1.3 m. The image contains the nine classes namely asphalt, meadows, gravel, trees, painted metal sheets, bare soil, bitumen, self-blocking bricks, and shadows. As the study was to explore the semantics of the spectral deep features, only spectral features of the pixels were used for classification, by ignoring the spatial information of the pixel/s completely.

Multiple classification experiments using the proposed architectures were performed. The architectures compared were: P1 is the architecture as described in Section 5.8.1.2, "Architecture". The architecture P2 is P1 with the reconstruction loss (Table 5.19), P3 is the standard CapsuleNet (Sabour et al., 2017), and P4 is the CaspsuleNet with reconstruction loss (Table 5.20). For each of the architectures, the pixel vector and the spectral shape as proposed were provided as the input for comparative assessment. Five-fold cross-validation was performed. A stratified split was used to ensure that the proportions of the classes in the training and the validation datasets were maintained.

The accuracy improvements for the spectral shape input are evident in Table 5.20 and Table 5.21. The transformed pixel vector provided added advantage over the pixel vector in all the architectures. The improvement is substantial for all the architectures. Furthermore, if we compare the traditional layered architecture with the CapsuleNet's for the spectral features, the CapsuleNet's performance is enhanced. This enhancement can clearly be attributed to semantically consistent spectral features learned. Furthermore, preliminary experiments on loss functions suggest that the two-way optimization, that is, the Softmax loss and the reconstruction loss help in improving

TABLE 5.20
Comparison of Classification Accuracies (in %)

	Pixel Vector	Spectrum Shape
1d-wav	41.00	NA
P1	45.00	79.56
P2	46.73	81.24
P3 cap	32.81	84.63
P4 cap	52.18	91.71

TABLE 5.21
Precision and Recall for CapsuleNet (P4) for One of the Folds for Spectrum Shape Input (in %)

Class	Precision	Recall	F1-score	Support
Asphalt	96.6	91.0	93.7	1421
Bare soil	82.8	98.0	89.8	1030
Bitumen	79.5	90.6	84.7	287
Gavel	82.1	87.0	84.5	432
Meadows	99.5	92.2	95.7	3951
Painted metal	96.6	100	98.3	284
Bricks	85.5	89.8	87.6	729
Shadows	92.8	100	96.2	218
Trees	90.6	98.1	94.2	648
Accuracy			93.0	9000
Weighted avg.	93.6	93.0	93.2	9000

accuracy. The computational efficiency is the added advantage as no spatial features for the pixel were considered.

5.8.1.4 Visualization and Interpretability of the Spectral Features

The learned spectral features were pictorially represented using the activations by various layers of proposed architectures. Figure 5.23 shows the activated regions of the spectral shape. The activations indicate that those parts were used for classifying the pixel correctly. Each individual graph shows the activated part of the spectral shape, in thick black, by some of the first-layer capsules. a) subfigure shows red edge, b) shows red absorptions and infrared reflectance, c) shows red absorption, and d) shows infrared reflection as activated features respectively. These are the well-known spectral features of vegetation. The activations indicate that the capsules learn the spectroscopic features of vegetation when a spectral shape is provided as the input, which is not possible with the one-dimensional pixel vector. Please note that the actual activations are enhanced pictorially for illustration purposes.

The proposed transformation of the hyperspectral pixel vector to the two-dimensional spectral shape enabled the learning of spectral features that are semantically interpretable. Deep architecture designed for processing this transformed signature learns the spectral features automatically. These features are the same as those which spectroscopic experts seek in the material for its identification. The spectral features that are learned in different layers of the convolution, reflected the hierarchy of spectral features as well. For example, some of the short arcs of spectral shape learned are common diagnostic features of the signatures (Figure 5.23). Further, if the shape of the signature over a specific wavelength range is important, it is activated appropriately. Furthermore, the lower-level primitives can be successively combined to form the spectral signature (just like spatial convolution creates a hierarchy of spatial features).

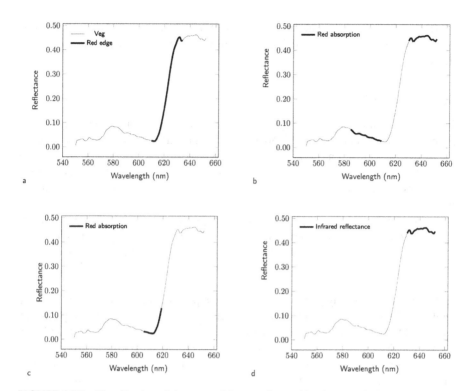

FIGURE 5.23 Visualization of the spectral features learned by the capsules.

The CapsuleNets are specifically designed to identify primitives and build semantically coherent associations between the primitives found in the two convolution layers. That is, the parts of the objects and objects are associated using dynamic routing. This behaviour is not spectrally reproduced by the CapsuleNets reported earlier as they use the pixel vector in deep processing (Arun et al., 2019). However, we can build a semantically sensible hierarchy of the spectral features because of the transformation of the one-dimensional vector to the two-dimensional spectral signature (Figure 5.23). These preliminary experiments indicate the potential of the shape features for extracting the spectral feature hierarchy using the CapsuleNets.

In addition to the work by Deshpande et al. (2021), recent work by Gao et al. (2018) attempts to use a 2-dimensional representation of the pixel vector. They reshaped a D-dimensional pixel vector into a 2-dimensional representation and used the 2-dimensional representation as an input to the 2D convolution, instead of 1D convolution over a D-dimensional pixel vector. The intention was to use the much richer representation of the pixel spectrum to achieve better accuracy. However, the transformed reshaped vector is merely an image unlike the one seen in earlier pixel vector transformation. The experts will not be able to interpret the representation similar to the spectral signature. The visual symbols of the spectral signature plot in wavelength and reflectance space are lost in the reshaping of the pixel vector.

These are a few initial efforts in the direction of learning the spectral features consistent with spectroscopy knowledge. More experimentation is required for investigating their utility.

5.9 SUMMARY

Image classification is one of the primary steps for hyperspectral image analysis. The purpose of image classification is to assign a single category label to each pixel in the image. Classification is used for identifying land use land cover of an area. Urban VIS land cover are hard targets for hyperspectral data analysis as they tend to have signatures similar to each other. Using complete information from the spectrum is helpful.

Unlike land cover, land use in an urban area needs to be inferred as it is not directly observed. Proxy features derived from primary features can be used for land use class effectively. Fractions of VIS within a single pixel or a group of pixels are unique enough to facilitate this. Different economic zones indicate unique VIS fractions vectors.

SAM is the mainstay of the classifiers for hyperspectral data as it can be easily applied to hyperspectral data from spectrometers and airborne/spaceborne platforms as well. Other classifiers found equally efficient are *k*-NN, Maximum likelihood, SVM, Random forest, and so on. Modern convolutional neural network (CNN) provides highly accurate classification results on benchmark datasets. The advantage of CNN over conventional classifiers is that it does not require handcrafted features. However, extracting the spectroscopic features that can be interpreted easily is still a challenging task.

NOTES

1. Some common steps such as conversion to radiance and atmospheric corrections are performed before classification. These pre-processing steps are incorporated into the classification process chain, usually.
2. Not be confused with discriminant function or analysis.
3. The CNN learns the features from the image in the first step and these features are used for the logistic regression classifier in the second step.
4. It is customary to use "Pune" for both PMC and PCMC urban regions. Historically, Pune's growth had urbanized the adjacent towns of Pimpri and Chinchwad to later form the independent Municipal Corporation of Pimpri-Chinchwad.
5. The word "training" is used for the process of determining a prototype vector. The model is entirely determined by an expert or represents a centroid VIS vector of a given training sample.
6. VGG16 is the architecture by Visual Geometry Group (VGG) Department of Engineering Science, University of Oxford.
7. We will see how well CNN learns the spectral features in Section 5.8, "Explaining convolutional spectral features".

WORKS CITED

Albalate, A., Suchindranath, A., Suendermann, D., & Minker, W. (2010). A semi-supervised cluster-and-label approach for utterance classification. *INTERSPEECH*.

Altman, N. S. (1992). An introduction to kernel and nearest-neighbor nonparametric regression. *The American Statistician, 46*(3), 175–185.

Anderson, J. R., Hardy, E. E., Roach, J. T., & Witmer, R. E. (1976). *Land use and land cover classification system for use with remote sensor data, geological survey professional paper 964*. Tech. rep., United States Government Printing Office: Washington.

Arnold, C. L., & Gibbons, C. J. (1996, Nov). Impervious surface coverage: The emergence of a key environmental indicator. *Journal of the American Planning Association, 62*(2), 243–258.

Arun, P., Buddhiraju, K., & Porwal, A. (2019). Capsulenet-Based spatial – spectral classifier for hyperspectral images. *Journal of Selected Topics in Applied Earth Observations and Remote Sensing, 12*(6).

As-syakur, A., Adnyana, I., Arthana, I., & Nuarsa, I. (2012). Enhanced Built-Up and Bareness Index (EBBI) for mapping built-up and bare land in an urban area. *Remote Sensing, 4*(10), 2957–2970. Retrieved from www.mdpi.com/2072-4292/4/10/2957

Barnsley, M. J., & Barr, S. L. (1996). Inferring urban land use from satellite sensor images using kernel-based spatial reclassification. *Photogrammetric Engineering & Remote Sensing, 62*(8), 949–958.

Beck, R. A. (2003, July). *EO-1 user guide v. 2.3*. Tech. rep., USGS Earth Resources Observation Systems Data Center (EDC), 47914 252nd Street, Sioux Falls, S.D., 57198–0001. Retrieved 10 19, 2022, from https://d9-wret.s3.us-west-2.amazonaws.com/assets/palladium/production/s3fs-public/atoms/files/EO1userguidev2pt320030715UC.pdf

Benediktsson, J. A., Palmason, J. A., & Sveinsson, J. R. (2005). Classification of hyperspectral data from urban areas based on extended morphological profiles. *IEEE Transactions on Geoscience and Remote Sensing, 43*(3), 480–491.

Bernabé, S., Lopez, S., Plaza, A., Sarmiento, R., & Rodríguez, P. (2011, 1). FPGA Design of an Automatic Target Generation Process for Hyperspectral Image Analysis. *Proceedings of the International Conference on Parallel and Distributed Systems – ICPADS*, 1010–1015.

Bertaud, A. (2004). *The spatial organization of cities: deliberate outcome or unforeseen consequence?* Retrieved October 10, 2014, from http://austinzoning.typepad.com/austincontrarian/files/WP-2004-01.pdf

Bioucas-Dias, J., Plaza, A., Dobigeon, N., Parente, M., Du, Q., Gader, P., & Chanussot, J. (2012, April). Hyperspectral unmixing overview: geometrical, statistical, and sparse regression-based approaches. Retrieved from http://arxiv.org/pdf/1202.6294v2.pdf

Blei, D. M., Ng, A. Y., & Jordan, M. I. (2003). Latent dirichlet allocation. *Journal of Machine Learning Research, 3*, 993–1022.

Bouzekri, S., Lasbet, A., & Lachehab, A. (2015, 1). A new spectral index for extraction of built-up area using Landsat-8 data. *Journal of the Indian Society of Remote Sensing, 43*.

Cadenasso, M. L., Pickett, S. T., & Schwarz, K. (2007, March). Spatial heterogeneity in urban ecosystems: reconceptualizing land cover and a framework for classification. *Frontiers in Ecology and the Environment, 5*(2), 80–88.

Chan, T.-H., Ma, W.-K., Ambikapathi, A., & Chi, C.-Y. (2011). A simplex volume maximization framework for hyperspectral endmember extraction. *IEEE Transaction on Geoscience and Remote Sensing, 49*(11), 4177–4193.

Chapelle, O., Schölkopf, B., & Zien, A. (Eds.). (2006). *Semi-supervised learning*. The MIT Press.

Chen, J., & Hepner, G. F. (2001). Investigation of imaging spectroscopy for discriminating urban land covers and surface materials. *AVIRIS Earth Science and Applications Workshop, Palo Alto, California. JPL Publication 02-1.* Jet Propulsion Laboratory: Pasadena, California. Retrieved from https://aviris.jpl.nasa.gov/proceedings/workshops/01_docs/2001Chen_web.pdf

Chen, Y., Jiang, H., Li, C., Jia, X., & Ghamisi, P. (2016). Deep feature extraction and classification of hyperspectral images based on convolutional neural networks. *IEEE Transactions on Geoscience and Remote Sensing, 54*(10), 6232–6251.

Chen, Y., Lin, Z., Zhao, X., Wang, G., & Gu, Y. (2014). Deep learning-based classification of hyperspectral data. *IEEE Journal of Selected topics in applied earth observations and remote sensing, 7*(6), 2094–2107.

Chen, Y., Zhao, X., & Jia, X. (2015). Spectral-spatial classification of hyperspectral data based on deep belief network. *IEEE Journal of Selected Topics in Applied Earth Observations and Remote Sensing, 8*(6), 2381–2392.

Chollet, F. (2016). Xception: Deep Learning with Depthwise Separable Convolutions. *CoRR, abs/1610.02357*. Retrieved from http://arxiv.org/abs/1610.02357

Chollet, F. (2021). *Deep learning with Python*. Simon and Schuster.

Chollet, F., & others. (2015). Keras.

Comber, A. J., Brunsdon, C. F., & Farmer, C. Q. (2012). Community detection in spatial networks: Inferring land use from a planar graph of land cover objects. *International Journal of Applied Earth Observation and Geoinformation, 18*, 274–282.

Congalton, R. G. (1991). A review of assessing the accuracy of classifications of remotely sensed data. *Remote Sensing of Environment, 37*, 35–46.

Cunningham, P., & Delany, S. (2020). k-Nearest Neighbour Classifiers: 2nd Edition (with Python examples). *CoRR, abs/2004.04523*. Retrieved from https://arxiv.org/abs/2004.04523

Cutter, S. L., Boruff, B. J., & Shirley, W. L. (2003). Social vulnerability to environmental hazards. *Social Science Quarterly, 84*(2), 242–261.

Dai, W., Dai, C., Qu, S., Li, J., & Das, S. (2017). Very deep convolutional neural networks for raw waveforms. *IEEE International Conference on Acoustics, Speech and Signal Processing*.

Demiriz, A., Bennett, K. P., & Embrechts, M. J. (1999). Semi-supervised clustering using genetic algorithms. *Proceedings of Artificial Neural Networks in Engineering*.

Deng, J., Dong, W., Socher, R., Li, L.-J., Li, K., & Fei-Fei, L. (2009). Imagenet: A large-scale hierarchical image database. *2009 IEEE conference on computer vision and pattern recognition*, 248–255.

Deshpande, S. S., Inamdar, A. B., & Vin, H. M. (2017). Urban land use land cover discrimination using image based reflectance calibration methods for hyperspectral data. *Photogrammatric Engineering and Remote Sensing, 83*(5), 365–376.

Deshpande, S., Inamdar, A., & Vin, H. (2013). Discrimination of Vegetation-Impervious Surface-Soil Classes in Urban Environment Using Hyperspectral Data. *Proceedings of Asian Conference on Remote Sensing (ACRS 2013), Bali, Indonesia, Oct. 20–24*. Bali.

Deshpande, S., Ray, K., Vin, H., & Inamdar, A. (2015). Semi-supervised extraction of economic zones in urban area using hyperspectral data. *Asian Conference on Remote Sensing*. Manila.

Deshpande, S., Sowmya, A., Yadav, P., Ladha, S., Verma, P., Vaiapury, K., ... Balamuralidhar, P. (2017). CogVis: attention-driven cognitive architecture for visual change detection. In A. Seffah, B. Penzenstadler, C. Alves, & X. Peng (Ed.), *Symposium on Applied Computing* (pp. 151–154). Marrakech, Morocco.

Deshpande, S., Thakur, R., & Balamuralidhar, P. (2021). Learning Deep Spectral Features for Hyperspectral Data Using Convolution Over Spectral Signature Shape. *2021 11th Workshop on Hyperspectral Imaging and Signal Processing: Evolution in Remote Sensing (WHISPERS)*, 1–5.

Deshpande, S., Thakur, R., & P., B. (2020). *Leveraging spatial structure with CapsuleNet for identification of the land use classes.* doi: 10.1117/12.2573980.

Divya, V. V., Ravi shankar, G., & Ravi shankar, T. (2014). Hyperspectral data for land use/land cover classification. *ISPRS Technical Commission VIII Symposium. sprsarchives-XL-8*, pp. 991–995. Hyderabad, India: The International Archives of the Photogrammetry, Remote Sensing and Spatial Information Sciences. Retrieved from www.int-arch-photogramm-remote-sens-spatial-inf-sci.net/XL-8/991/2014/isprsarchives-XL-8-991-2014.pdf

Ebert, A., Kerle, N., & Stein, A. (2009). Urban social vulnerability assessment with physical proxies and spatial metrics derived from air – and spaceborne imagery and GIS data. *Nat Hazards, 48*, 275–294. doi: 10.1007/s11069-008-9264-0

Egmont-Petersen, M., de Ridder, D., & Handels, H. (2002). Image processing with neural networks – a review. *Pattern recognition, 35*(10), 2279–2301.

Fauvel, M., Benediktsson, J. A., Chanussot, J., & Sveinsson, J. R. (2008). Spectral and spatial classification of hyperspectral data using SVMs and morphological profiles. *IEEE Transactions on Geoscience and Remote Sensing, 46*(11), 3804–3814.

Forman, R. T. (1995). Some general principles of landscape and regional ecology. *Landscape Ecology, 3*, 133–142.

Gamba, P. (2013, March). Human settlements: a global challenge for EO data processing and interpretation. *Proceedings of the IEEE, 101*(3), 570–581.

Gao, H., Lin, S., Yang, Y., Li, C., & Yang, M. (2018). Convolution neural network based on two-dimensional spectrum for hyperspectral image classification. *Journal of Sensors, 2018*.

Ghamisi, P., Plaza, J., Chen, Y., Li, J., & Plaza, A. (2017). Advanced spectral classifiers for hyperspectral images: A review. *IEEE Geoscience and Remote Sensing Magazine, 5*(1), 8–32.

Goetz, A. F., Vane, G., Solomon, J. E., & Rock, B. N. (1985, June). Imaging spectrometry for earth remote sensing. *Science, 228*(4704), 1147–1153.

Halle, W., Brieß, K., Schlicker, M., Skrbek, W., & Venus, H. (2002, 1). Autonomous onboard classification experiment for the satellite BIRD. Retrieved January 12, 2023, from www.researchgate.net/publication/242784134_Autonomous_Onboard_Classification_Experiment_for_the_Satellite_BIRD

Harris, R. (1985). Contextual classification post-processing of Landsat data using a probabilistic. *International Journal of Remote Sensing, 6*, 847–866.

Hashim, H., Abd Latif, Z., & Adnan, N. (2019). Urban vegetation classification with NDVI threshold value method with very high resolution (VHR) Pleiades imagery. *The International Archives of Photogrammetry, Remote Sensing and Spatial Information Sciences, 42*, 237–240.

Hastie, T., Tibshirani, R., & Friedman, J. (2008). Overview of supervised learning. In Second (Ed.), *Elements of statistical learning.* Springer: New York.

Heiden, U., Segl, K., Roessner, S., & Kaufmann, H. (2007, December). Determination of robust spectral features for identification of urban surface materials in hyperspectral remote sensing data. *Remote Sensing of Environment, 111*(4), 537–552.

Hepner, G. F., Houshmand, B., Kulikov, I., & Bryant, N. (1998, August). Investigation of the integration of AVIRIS and IFSAR for urban analysis. *Photogrammetric Engineering & Remote Sensing, 64*(8), 813–820.

Herold, M., Roberts, D. A., Gardner, M. E., & Dennison, P. E. (2004). Spectrometry for urban area remote sensing – development and analysis of a spectral library from 350 to 2400 nm. *Remote Sensing of Environment, 91*(34), 304–319. Retrieved from www.sciencedirect.com/science/article/pii/S0034425704000768

Herold, M., Scepan, J., & Clarke, K. C. (2002). The use of remote sensing and landscape metrics to describe structures and changes in urban land uses. *Environment and Planning A, 34*, 1443–1458.

Hornik, K., Stinchcombe, M., & White, H. (1989). Multilayer feedforward networks are universal approximators. *Neural Networks, 2*(5), 359–366. Retrieved from www.sciencedirect.com/science/article/pii/0893608089900208

Huete, A. (1988). A soil-adjusted vegetation index (SAVI). *Remote sensing of environment, 25*(3), 295–309.

Ji, S., Xu, W., Yang, M., & Yu, K. (2012). 3D convolutional neural networks for human action recognition. *IEEE transactions on pattern analysis and machine intelligence, 35*(1), 221–231.

Kruse, F. A., Lefkoff, A. B., Boardman, J. W., Barloon, K. B., & Goetz, A. F. (1993). The Spectral Image Processing System (SIPS) – Interactive visualization and analysis of imaging spectrometer data. *Remote Sensing of Environment, 44*, 145–163.

Kunkel, B., Blechinger, F., Lutz, R., Doerffer, R., Piepen, H. v., & Schroder, M. (1988). ROSIS (Reflective Optics System Imaging Spectrometer) – A candidate instrument for polar platform missions. In J. Seeley, & S. Bowyer (Eds.), *SPIE 0868 Optoelectronic technologies for remote sensing from space* (p. 8). SPEI. https://spie.org/Publications/Proceedings/Volume/0868?SSO=1

Li, F., Andreetto, M., Ranzato, M. A. & Perona, P. (2022, 4). Caltech 101.

Li, H., & Reynolds, J. F. (1993). A new contagion index to quantify spatial patterns of landscapes. *Landscape Ecology, 8*(3), 155–162.

Li, S., Song, W., Fang, L., Chen, Y., Ghamisi, P., & Benediktsson, J. (2019). Deep learning for hyperspectral image classification: An overview. *IEEE Transactions on Geoscience and Remote Sensing, 57*(9), 6690–6709.

Liu, C., Shao, Z., Chen, M., & Luo, H. (2013). MNDISI: a multi-source composition index for impervious surface area estimation at the individual city scale. *Remote Sensing Letters, 4*(8), 803–812. Retrieved from https://doi.org/10.1080/2150704X.2013.798710

Lopez, S., Vladimirova, T., Gonzalez, C., Resano, J., Mozos, D., & Plaza, A. (2013). The promise of reconfigurable computing for hyperspectral imaging onboard systems: A review and trends. *Proceedings of the IEEE, 101*(3), 698–722.

Lorenz, E., Halle, W., Fischer, C., Mettig, N., & Klein, D. (2017, 1). Recent results of the firebird mission. *ISPRS – International Archives of the Photogrammetry, Remote Sensing and Spatial Information Sciences, XLII-3/W2*, 105–111.

McFeeters, S. (1996). The use of the Normalized Difference Water Index (NDWI) in the delineation of open water features. *International journal of remote sensing, 17*(7), 1425–1432.

McGarigal, K. (2014, January 16). *FRAGSTATS help*. Retrieved October 10, 2014, from www.umass.edu/landeco/research/fragstats/documents/fragstats.help.4.2.pdf

Meer, F. V. (2000). Spectral curve shape matching with a continuum removed CCSM algorithm. *International Journal of Remote Sensing, 21*(16), 3179–3185.

Meer, F. V., & Bakker, W. (1997). CCSM: Cross Correlogram Spectral Matching. *International Journal of Remote Sensing, 18*(5), 1197–1201.

Melgani, F., & Bruzzone, L. (2004). Classification of hyperspectral remote sensing images with support vector machines. *IEEE Transactions on Geoscience and Remote Sensing, 42*(8), 1778–1790.

Mitchell, T. M. (1997). Instance-based learning. In *Machine learning*. McGraw-Hill Science/Engineering/Math.

Mitchell, T. M. (1997). Introduction. In *Machine learning*. McGraw-Hill Science/Engineering/Math.

More, R., Kale, N., Kataria, G., Yadav, P., & Deshpande, S. (2015). Understanding thermal fluxes in and around Pune city using USING remotely sensed data. *Asian Conference on Remote Sensing*. Manila, Philippines.

MPCB. (2016, February 11). *General features of Pune District*. Retrieved from http://mpcb.gov.in/relatedtopics/CHAPTER1.pdf

Myint, S. W., Gober, P., Brazel, A., Grossman-Clarke, S., & Weng, Q. (2011). Per-pixel vs. object-based classification of urban land cover extraction using high spatial resolution imagery. *Remote Sensing of Environment, 115*, 1145–1161.

Nalawade, S. B. (2015). *Geography of Pune urban area*. Retrieved from www.ranwa.org/punealive/pageog.htm

Nascimento, J. P., & Bioucas-Dias, J. (2005, April). Vertex component analysis: a fast algorithm to unmix hyperspectral data. *IEEE Transactions on Geoscience and Remote Sensing, 43*(4), 898–910.

Nidamanuri, R. R., & Zbell, B. (2011, March). Normalized Spectral Similarity Score (NS3) as An efficient spectral library searching method for hyperspectral image classification. *IEEE Journal of Selected Topics in Applied Earth Observations and Remote Sensing, 4*(1), 226–240.

O'Neill, R. V., Hunsaker, C., Timmins, S. P., Jackson, B. L., Jones, K. B., Riitters, K. H., & Wickham, J. D. (1996). Scale problems in reporting landscape pattern at the regional scale. *Landscape Ecology, 11*(3), 169–180.

Office of the Registrar General & Census Commissioner, India. (2011-1). *Provisional Population Totals Paper 2, Volume 2 of 2011: Maharashtra*. Retrieved June 2, 2014, from www.censusindia.gov.in/2011-prov-results/paper2-vol2/census2011_paper2.html

Pal, M., & Mather, P. M. (2003). An assessment of the effectiveness of decision tree methods for land cover classification. *Remote Sensing of Environment, 86*, 554–565.

Pal, M., & Mather, P. M. (2005, March). Support vector machines for classification in remote sensing. *International Journal of Remote Sensing, 26*(5), 1007–1011.

Phinn, S., Stanford, M., Scarth, P., Murray, A. T., & Shyy, P. T. (2002). Monitoring the composition of urban environments based on the vegetation – impervious surface – soil (VIS) model by subpixel analysis techniques. *International Journal of Remote Sensing, 23*(10), 4131–4153.

Platt, R. V., & Goetz, A. F. (2004, July). A comparison of AVIRIS and Landsat for land use classification at the urban fringe. *Photogrammetric Engineering & Remote Sensing, 70*(7), 813–819.

Rashed, T., & Weeks, J. (2003). Exploring the spatial association between the measures from satellite imagery and patterns of urban vulnerability to earthquake hazard. Retrieved October 10, 2014, from http://130.191.118.3/Research/Projects/IPC/publication/Rashed_Weeks_Regensburg_2003.pdf

Richards, J. A., & Jia, X. (2006). Multispectral transformations of image data. In *Remote Sensing Digital Image Analysis – an introduction, 4th ed.* (4 ed.). Berlin: Springer-Verlag.

Richards, J. A., & Jia, X. (2006). Supervised classification techniques. In *Remote Sensing Digital Image Analysis – an introduction, 4th ed.* Berlin: Springer-Verlag.

Ridd, M. (1995). Exploring a VIS (vegetation-impervious surface-soil) model for urban ecosystem analysis through remote sensing: comparative anatomy for cities. *International journal of remote sensing, 16*(12), 2165–2185.

Ridd, M. K., Ritter, N. D., Bryant, N. A., & Green, R. O. (April 1992). Neural network classification of AVIRIS data in the urban ecosystem. *Association of American Geographers Annual Meeting*. San Diego, California.

Rouse Jr, J., Haas, R., Schell, J., & Deering, D. (1973). *Monitoring the vernal advancement and retrogradation (green wave effect) of natural vegetation.*

Rouse Jr, J., Haas, R., Schell, J., & Deering, D. (1973). Paper a 20. *Third Earth Resources Technology Satellite-1 Symposium: The Proceedings of a Symposium Held by Goddard Space Flight Center at Washington, DC on, 351*, 309.

Rousseeuw, P. J. (1987). Silhouettes: a graphical aid to the interpretation and validation of cluster analysis. *Journal of Computational and Applied Mathematics, 20*, 53–65.

Russakovsky, O., Deng, J., Su, H., Krause, J., Satheesh, S., Ma, S., ... others. (2015). Imagenet large scale visual recognition challenge. *International journal of computer vision, 115*(3), 211–252.

Sabour, S., Frosst, N., & Hinton, G. E. (2017). Dynamic routing between capsules. *31st Conference on Neural Information Processing Systems (NIPS)*. Long Beach, California, USA.

Sameen, M., & Pradhan, B. (2016, 1). A novel built-up spectral index developed by using multiobjective particle-swarm-optimization technique. *IOP Conference Series: Earth and Environmental Science, 37*, 12006.

Samsudin, S., Shafri, H., & Hamedianfar, A. (2016, 1). Development of spectral indices for roofing material condition status detection using field spectroscopy and WorldView-3 data. *Journal of Applied Remote Sensing, 10*.

Senthil Kumar, A., Keerthia, V., Manjunatha, A. S., van der Werffb, H., & van der Meer, F. (2010). Hyperspectral image classification by a variable interval spectral average and spectral curve matching combined algorithm. *International Journal of Applied Earth Observation and Geoinformation, 12*, 261–269.

Simonyan, K., & Zisserman, A. (2014). Very deep convolutional networks for large-scale image recognition. *arXiv preprint arXiv:1409.1556*.

Strahler, A. H. (1980). The use of prior probabilities in maximum likelihood classification of remotely sensed data. *Remote Sensing of Environment, 10*(2), 135–163.

Suykens, J. A., & Vandewalle, J. (1999, June). Least squares support vector machine classifiers. *Neural Processing Letters, 9*(3), 293–300.

Tan, Q., & Wang, J. (May, 2007). Hyperspectral versus multispectral satellite data for urban land cover and land use mapping – Beijing, an evolving city. *ASPRS 2007 Annual Conference*. Tampa, Florida.

Thurau, C., Kersting, K., & Bauckhage, C. (2010). Yes we can – simplex volume maximization for descriptive web-scale matrix factorization. *CIKM '10 Proceedings of the 19th ACM international conference on Information and knowledge management* (pp. 1785–1788). Toronto, Ontario, Canada.

Thurau, C., Kersting, K., Wahabzada, M., & Bauckhage, C. (2012). Descriptive matrix factorization for sustainability: adopting the principle of opposites. *Data Mining and Knowledge Discovery, 24*(2), 325–354. Retrieved from www-kd.iai.uni-bonn.de/index.php

Turner, B. L., Skole, D., Sanderson, S., & Fischer, G. (1995). *Land Use and land-cover change science/research plan, IGBP Report No 35, HDP Report No.7*.

Turner, M. G., O'Neill, R. V., Gardner, R. H., & Milne, B. T. (1989). Effects of Changing Spatial Scale on the Analysis of Landscape Pattern. *Landscape Ecology, 3*(3/4), 153–162.

USGS. (2011-2). *Earth Observing 1 (EO-1)-Sensors*. Sioux Falls, SD, USA: U.S. Geological Survey. Retrieved 10 19, 2022, from www.usgs.gov/centers/eros/eo-1-sensors

USGS. (2013-1, April 22). EO1 Hyperion Scene: EO1H1470472013112110KZ_PF1_01, for Target Path 147, and Target Row 47. *EO1 Hyperion Scene: EO1H1470472013112110KZ_PF1_01, for Target Path 147, and Target Row 47*. Sioux Falls, SD, USA: U.S. Geological Survey. Retrieved from http://earthexplorer.usgs.gov

USGS. (2013-2, April 27). EO1 Hyperion Scene: EO1H1470472013117110PZ_PF2_01, for Target Path 147, and Target Row 47. Sioux Falls, SD, USA: U.S. Geological Survey. Retrieved from http://earthexplorer.usgs.gov

USGS. (2023). *What are the best Landsat spectral bands for use in my research?* Sioux Falls, SD, USA: U.S. Geological Survey. Retrieved January 12, 2023, from USGS: www.usgs.gov/faqs/what-are-best-landsat-spectral-bands-use-my-research

Van Niel, T. G., McVicar, T. R., & Datt, B. (2005). On the relationship between training sample size and data dimensionality: Monte Carlo analysis of broadband multi-temporal classification. *Remote Sensing of Environment, 98,* 468–480.

Veldkamp, A., & Lambin, E. F. (2001). Predicting land-use change. *Agriculture, Ecosystems and Environment, 85,* 1–6.

Verma, P., Yadav, P., Deshpande, S., Gubbi, J., & P, B. (2016). Urban growth studies for Johannesburg city using remotely sensed data. *Asian Conference on Remote Sensing.* Colombo, Sri Lanka.

Weng, Q. (2012). Remote sensing of impervious surfaces in the urban areas: requirements, methods, and trends. *Remote Sensing of Environment, 117*(0), 34–49. Retrieved from www.sciencedirect.com/science/article/pii/S0034425711002811

Weng, Q., Hu, X., & Lu, D. (2008, June). Extracting impervious surfaces from medium spatial resolution multispectral and hyperspectral imagery: a comparison. *International Journal of Remote Sensing, 29*(11), 3209–3232.

Wilson, A. C., Roelofs, R., Stern, M., Srebro, N., & Recht, B. (2017). The marginal value of adaptive gradient methods in machine learning. *31st International Conference on Neural Information Processing Systems.* Long Beach, California.

Winter, M. E. (1999). N-findr: An algorithm for fast autonomous spectral end-member determination in hyperspectral data. *SPIE Conf.Imaging Spectrometry V,* 266–275.

Wu, C., & Murray, A. T. (2003). Estimating impervious surface distribution by spectral mixture analysis. *Remote Sensing of Environment, 84,* 493–505.

Wu, J. (2004). Effects of changing scale on landscape pattern analysis: scaling relations. *Landscape Ecology, 19,* 125–138.

Xu, B., & Gong, P. (2007, August). Land-use/land-cover classification with multispectral and hyperspectral EO-1 data. *Photogrammetric Engineering & Remote Sensing, 73*(8), 955–965.

Xu, H. (2006). Modification of normalised difference water index (NDWI) to enhance open water features in remotely sensed imagery. *International journal of remote sensing, 27*(14), 3025–3033.

Xu, H. (2007, 1). Extraction of urban built-up land features from landsat imagery using a thematicoriented index combination technique. *Photogrammetric Engineering & Remote Sensing, 73,* 1381–1391.

Xu, H. (2008, 1). A new index for delineating built-up land features in satellite imagery. *International Journal of Remote Sensing, 29,* 4269–4276.

Xu, H. (2010, 1). Analysis of Impervious Surface and its Impact on Urban Heat Environment using the Normalized Difference Impervious Surface Index (NDISI). *Photogrammetric Engineering and Remote Sensing, 76,* 557–565.

Xue, J., & Su, B. (2017). Significant remote sensing vegetation indices: A review of developments and applications. *Journal of sensors, 2017.*

Yadav, P., & Deshpande, S. S. (2016). Assessment of anticipated runoff because of impervious surface increase in Pune Urban Catchments, India: a remote sensing approach. *proceedings of SPIE 10033, Eighth International Conference on Digital Image Processing (ICDIP 2016).*

Yang, J., Zhao, Y.-Q., & Chan, J.-W. (2017). Learning and transferring deep joint spectral – spatial features for hyperspectral classification. *IEEE Transactions on Geoscience and Remote Sensing, 55*(8), 4729–4742.

Yang, L., Xian, G., Klaver, J. M., & Deal, B. (2003, September). Urban land-cover Change detection through sub-Pixel imperviousness mapping using remotely sensed data. *Photogrammetric Engineering & Remote Sensing, 69*(9), 1003–1010.

Yuhas, R. H., Goetz, A. F., & Boardman, J. W. (2001). Discrimination among semi-arid landscape endmembers using the Spectral Angle Mapper (SAM) algorithm. *JPL, Summaries of the Third Annual JPL Airborne Geoscience Workshop, 1: AVIRIS Workshop*, pp. 147–149. Boulder.

Zha, Y., Gao, J., & Ni, S. (2003). Use of normalized difference built-up index in automatically mapping urban areas from TM imagery. *International journal of remote sensing, 24*(3), 583–594.

Zhu, X. (2005). *Semi-supervised learning literature survey.* Computer Sciences, University of Wisconsin-Madison. Wisconsin-Madison.

Zurada, J. M. (1992). *Introduction to artificial neural systems* (Vol. 8). West: St. Paul.

6 Appendix

This chapter will discuss the following topics:

- A summary of the literature on hyperspectral remote sensing in urban areas
- EO1-Hyperion irradiance values
- Help on the code available with the book

6.1 SUMMARY OF IMPORTANT WORK IN HYPERSPECTRAL REMOTE SENSING OF URBAN AREAS

Table 6.1 and Table 6.2 show the list of important works in hyperspectral remote sensing of urban areas. This is not a complete list. It provides a summary of some of the important work in hyperspectral remote sensing of urban areas. The new deep learning work is not included in this as of now.

TABLE 6.1
Previous Work in Hyperspectral Remote Sensing of Urban Areas

Papers	Objective/Domain	Sensor	Study area	Classification schema	Atmospheric correction	Classifier
Ridd, 1992	Urban ecology, urban morphology	AVIRIS, SPOT, TM	USA, Salt Lake, UT	VIS	NO	ANN
Harsanyi and Chnag, 1994	Mineral mapping, and dimensional reduction	AVIRIS	USA, Lunar Crater Volcanic Field, Northern Nye, NV	No schema,	NO, mention importance but do not correct	
Ridd, 1995	Urban ecology	AVIRIS, SPOT, TM	USA, Salt Lake, UT	VIS		
Arnold, 1996	Urban planning					
Barnsley and Barr, 1996	Land use classification			Spatial Reclassification Kernel		
Hoffbeck and Landgrebe, 1996	Mineral mapping	AVIRIS	USA, Cuprite, Nevada 1992		No. log residual normalization	ML
Hepner et al., 1998	Urban analysis	AVIRIS and InterFerometric (IF) SAR	USA, UCLA, Westwood		NO	SAM
Jensen and Cowen, 1999	Urban mapping, RS data suitability for urban LU/LC	Multispectral sensors				
Song, et al., 2001	When and how to correct atmospheric effects				NO	
Chen and Hepner, 2001	Spectral analysis	AVIRIS, ASD spectra,	USA, Park City Utah		ATREM	SAM

Reference	Topic	Data	Location		Atmospheric correction	Method
Pickett, et al., 2001	Urban ecology					
Phinn, et al., 2002	Urban VIS mapping, Suitable method for urban VIS mapping?	Landsat TM, SPOT, IRS	Australia, Brisbane City	VIS	NO	Unmixing, Unsupervised and then supervised
Herold, et al., 2002	Urban land use	IKONOS	USA, Santa Barbara CA		NO, caused problems in classification	Landscape metrics
Pal and Mather, 2003	Crops and semi-urban	DAIS 7915, Landsat ETM+.	Spain, La Mancha Alta; England, Littleport		NO	DT
Wu and Murray, 2003	Urban ecology, socio-ecological studies	ETM+, Columbus, Metropolitan area OH		VIS	NO, DN to radiance only	Unmixing, FC linear spectral mixture model
Yang, et al., 2003	Urban mapping	Landsat	USA, western Georgia	VIS	NO, because lack of data	Regression model
Thomas, et al., 2003	Urban mapping, runoff calculations	ADAR 5500, 1 m res	USA, Scottsdale, Arizona			
Platt and Goetz, 2004	Urban fringe LU/LC, MS vs. HS, accuracy improvements by SNR or HS?	AVIRIS/Synthetic Landsat image	USA, Colorado, Fort Collins, Horse tooth Reservoir	Anderson	HATCH	ML
Rogan and Chen, 2004	Planning perceptive, mapping	Review				
Melgani and Bruzzone, 2004	Vegetation, SVM and its effectiveness	AVIRIS,	USA, Indiana's Indian Pines 1992		NO	SVM
Benediktsson, et al., 2005	Integration of spectral and spatial features	Paiva data set,	Washington DC,		NO	Extended morphological profiles
Ham, et al., 2005	Non-urban	AVIRIS, Kennedy Space Centre, Hyperion	Florida, Hyperion Okavango Delta, Botswana		NO	RF–binary hierarchical classifier

(*Continued*)

TABLE 6.1 (Continued)
Previous Work in Hyperspectral Remote Sensing of Urban Areas

Papers	Objective/Domain	Sensor	Study area	Classification schema	Atmospheric correction	Classifier
Pal and Mather, 2005	Crops and semi-urban (rural)	DAIS 7915, Landsat ETM+	Spain, La Mancha Alta, England, Littleport		NO	SVM
Tan and Wang, 2007	Urban mapping, comparison of HS vs MS	ASTER/Chris-prob-a, Beijing	China, Beijing	Anderson	No	ML
Cadenasso, et al., 2007	Urban ecology			HERCULES main for urban ecology		
Xu and Gong, 2007	Urban mapping	Hyperion, ALI	USA, Fermont California	USGS 1 and 2	NO	Linear Discriminant Analysis
Fauvel, et al., 2008	Integration of spectral and spatial features	Paiva data set	USA, Washington DC		NO	SVM and Extended morphological kernel
Weng, 2008	Urban VIS mapping, comparison of ALI and Hyperspectral accuracies for VIS unmixing	Hyperion and Ali,	USA, Marion County, Indiana	VIS classes	NO	Unmixing, Linear spectral unmixing
Plaza, et al., 2009	Review of hyperspectral techniques					
Esch, et al., 2009	Urban mapping of impervious surfaces	Landsat 7 ETM+	Southern part of Germany	VIS	ATCOR-2/3	SV Regression
Mountrakis, et al., 2010						SVM

Appendix

Myint, et al., 2011	Urban mapping, Pixel vs. object-based approaches	Multispectral, QuickBird	USA, Phoenix, Arizona	NO	ML, Object segmentations
Gamba, 2013 Shafri, et al., 2012	Review–HS remote sensing of urban area				
Divya, et al., 2014	LU LC mapping/ visual interpretation vs Hyperion, vegetation, semi-urban area	Hyperion, Resourcesat-2 LISS III	India, Gujrat, Anand, Valsad	FLASH	SAM
Lin and Tsogt, 2015	When and how to correct atmospheric effects			NO	

TABLE 6.2
Some Important Conclusions (Column heading 1 – Lack of research in urban LU/LC using HS data; 2 – Urban area is complex; 3 – HS data provided more accurate results; 4 – Confusion between soil and built-up is observed)

Papers	1	2	3	4	Additional Remarks
Ridd, 1992		Yes			
Harsanyi and Chnag, 1994		Yes			
Ridd, 1995	Yes	Yes			Need to standardize urban building blocks for ecological studies
Arnold, 1996					No-point source pollutant as a major contributor, Adverse effects of impervious surfaces, Correlation between landscape/impervious surface and pollutant
Barnsley and Barr, 1996					High resolution is not synonymous to accuracy, in fact, cite articles where reduction is observed
Hoffbeck and Landgrebe, 1996			Yes		
Hepner, et al., 1998	Yes	Yes	Yes		
Jensen and Cowen, 1999		Yes			High resolution not required for Anderson level 1–3 classification, required for level 4 classification
Song, et al., 2001					Do not correct for single date image for multispectral data
Chen and Hepner, 2001	Yes	Yes	Yes,		HS data is required
Pickett, et al., 2001		Yes			
Phinn, et al., 2002				Yes	Provided indirect suggestion for VIS fractions as an indicator of land use class
Herold, et al., 2002		Yes			
Pal and Mather, 2003					Among {DT, ANN, ML}, DT is not found to be useful for high dimensional data
Wu and Murray, 2003					Provided indirect suggestion for mixture signatures for land uses such as low density residential, central business district
Yang, et al., 2003					
Thomas, et al., 2003					High resolution imagery is useful
Platt and Goetz, 2004	Yes		Yes		Modest but real advantages because of more number of bands than SNR
Rogan and Chen, 2004	Yes		Yes		Review article
Melgani and Bruzzone, 2004					SVMs are effective alternative with dimensionality reduction
Benediktsson, et al., 2005					
Ham, et al., 2005					RF-binary hierarchical classification is better
Pal and Mather, 2005					Comparison of SVM with ML and ANN; SVM is better
Tan and Wang, 2007	Yes		Yes		

TABLE 6.2 (Continued)
Some Important Conclusions (Column heading 1 – Lack of research in urban LU/LC using HS data; 2 – Urban area is complex; 3 – HS data provided more accurate results; 4 – Confusion between soil and built-up is observed)

Papers	1	2	3	4	Additional Remarks
Cadenasso, et al., 2007		Yes			
Xu and Gong, 2007	Yes		Yes	Yes	EO-1 results for vegetation outperformed ALI constantly
Fauvel, et al., 2008					Extension of 2003 work
Weng, 2008		Yes	Yes		EO-1 imagery provided more accurate results, especially in case of low reflectance surfaces such as asphalt roads. Accuracy is because of more number of bands.
Plaza, et al., 2009					Review article, Need for standardization, Urban mapping needs integrated use
Esch, et al., 2009					
Mountrakis, et al., 2010					Review article, SVM
Myint, et al., 2011				Yes	High resolution is more useful
Gamba, 2013	Yes	Yes	Yes		Yes–material detection, importance of HS data is recognized, VHR is not equal to improved results
Shafri, et al., 2012	Yes		Yes		Review article, Need for integrated spectral and spatial tech. fusion
Divya, et al., 2014			Yes	Yes	Study for vegetation types
Lin and Tsogt, 2015				Yes	Do not correct for single date image for multispectral data

6.2 HYPERION TOP OF THE ATMOSPHERE IRRADIANCE (WAVELENGTH IN NM; IRRADIANCE IN W M^{-2} µM^{-1})

Wavelength	Irradiance	Wavelength	Irradiance	Wavelength	Irradiance
355.59	949.37	559.09	1821.99	762.6	1235.37
365.76	1158.78	569.27	1841.92	772.78	1202.29
375.94	1061.25	579.45	1847.51	782.95	1194.08
386.11	955.12	589.62	1779.99	793.13	1143.6
396.29	970.87	599.8	1761.45	803.3	1128.16
406.46	1663.73	609.97	1740.8	813.48	1108.48
416.64	1722.92	620.15	1708.88	823.65	1068.5
426.82	1650.52	630.32	1672.09	833.83	1039.7
436.99	1714.9	640.5	1632.83	844	1023.84
447.17	1994.52	650.67	1591.92	851.92	964.6
457.34	2034.72	660.85	1557.66	854.18	938.96
467.52	1970.12	671.02	1525.41	862.01	982.06
477.69	2036.22	681.2	1470.93	864.35	949.97
487.87	1860.24	691.37	1450.37	872.1	954.03
498.04	1953.29	701.55	1393.18	874.53	949.74
508.22	1953.55	711.72	1372.75	882.19	931.81
518.39	1804.56	721.9	1235.63	884.7	929.54
528.57	1905.51	732.07	1266.13	892.28	923.35
538.74	1877.5	742.25	1279.02	894.88	917.32
548.92	1883.51	752.43	1265.22	902.36	894.62

Wavelength	Irradiance	Wavelength	Irradiance	Wavelength	Irradiance
1356	370.96	1558	265.73	1759	182.75
1366	365.57	1568	260.2	1769	180.09
1376	358.42	1578	251.62	1780	175.18
1386	355.18	1588	244.11	1790	173
1396	349.04	1598	247.83	1800	168.87
1406	342.1	1608	242.85	1810	165.19
1416	336	1618	238.15	1820	156.3
1426	325.94	1628	239.29	1830	159.01
1437	325.71	1638	227.38	1840	155.22
1447	318.27	1648	226.69	1850	152.62
1457	312.12	1659	225.48	1860	149.14
1467	308.08	1669	218.69	1870	141.63
1477	300.52	1679	209.07	1880	139.43
1487	292.27	1689	210.62	1891	139.22
1497	293.28	1699	206.98	1901	137.97
1507	282.14	1709	201.59	1911	136.73
1517	285.6	1719	198.09	1921	133.96
1527	280.41	1729	191.77	1931	130.29
1537	275.87	1739	184.02	1941	124.5
1548	271.97	1749	184.91	1951	124.75

Source: USGS, 2011-1.

Appendix

Wavelength	Irradiance	Wavelength	Irradiance	Wavelength	Irradiance
905.05	892.69	1006.8	710.54	1154	534.17
912.45	876.1	1013.3	714.26	1164	519.74
915.23	877.59	1016.9	703.56	1174	511.29
922.54	839.34	1023.4	698.69	1184	497.28
925.41	834.6	1027.1	695.1	1194	492.82
932.64	841.54	1033.4	682.41	1205	479.41
935.58	837.11	1037.3	676.9	1215	479.56
942.73	810.2	1043.5	669.61	1225	469.01
945.76	814.7	1047.5	661.9	1235	461.6
952.82	802.22	1053.6	657.86	1245	451
955.93	788.04	1057.6	649.64	1255	444.06
962.91	784.44	1063.7	643.48	1265	435.25
966.11	778.2	1073.8	623.13	1275	429.29
972.99	772.22	1083.9	603.89	1285	415.69
976.28	764.29	1094	582.63	1295	412.87
983.08	758.6	1104.1	579.58	1305	405.4
986.46	751.28	1114.1	571.8	1316	396.94
993.17	743.88	1124.2	562.3	1326	391.94
996.63	740.25	1134.3	551.4	1336	386.79
1003.3	721.76	1144	540.52	1346	380.65

Wavelength	Irradiance	Wavelength	Irradiance	Wavelength	Irradiance
1961	123.92	2163	84.64	2365	62.9
1971	121.95	2173	85.47	2375	61.68
1981	118.96	2183	84.49	2385	60
1991	117.78	2193	83.43	2395	59.94
2002	115.56	2203	81.62	2405	59.18
2012	114.52	2213	80.67	2415	57.38
2022	111.65	2224	79.32	2425	57.1
2032	109.21	2234	78.11	2435	56.25
2042	107.69	2244	76.69	2445	55.09
2052	106.13	2254	75.35	2456	54.02
2062	103.7	2264	74.15	2466	53.75
2072	102.42	2274	73.25	2476	52.78
2082	100.42	2284	71.67	2486	51.6
2092	98.27	2294	70.13	2496	51.44
2102	97.37	2304	69.52	2506	0
2113	95.44	2314	68.28	2516	0
2123	93.55	2324	66.39	2526	0
2133	92.35	2335	65.76	2536	0
2143	90.93	2345	65.23	2546	0
2153	89.37	2355	63.09	2556	0

6.3 THE BOOK CODE HELP

This book is accompanied with code which is shared at https://github.com/shail eshshankardeshpande/HSRS-2023-Deshpande. The code is written in Python language, and the code is mainly procedural. Each procedural step in the processing of hyperspectral data is logically organized into two to three basic Python files. For example, conversion of digital numbers to radiance, and then to normalized or atmospherically corrected reflectance values are organized sequentially in a single file. The code for classification or higher-order analysis such as optimal band analysis is organized in separate files. The code is provided without any warranty under a creative commons license. It runs on Google Colab and is tested. However, this help is not to be considered as a detailed instruction manual. The user is assumed to be friendly with Python language. The code should work on the shared example files and other data files. However, in case the code does not work or creates erroneous output on the data other than examples, the user needs to debug the code and fix it. All the other details are available under the help section of the repository.

The code is written for light functional use without any frills. In many places, the user is supposed to provide the working folder strings, or other string variables in the script. The error checking is minimal. The code uses bare bone numpy array processing and avoids any use of an advanced library unless and until it is absolutely required, for example, hdf reader. The communication is through csv files and reading them is handled mainly through numpy calls. The section below describes the scripts, their input and output, the additional context, and higher-level steps to run it. Please cite the following papers if you use the code (1 and 3 are must, 2 and 4 as appropriate):

1. **(Deshpande, Inamdar, & Vin, 2019)**
   Deshpande, S. S., Inamdar, A. B., & Vin, H. M. (2019). Spectral library and discrimination analysis of Indian urban materials. *Journal of the Indian Society of Remote Sensing, 47*, 867–877.

2. **(Deshpande, Yadav, Mutreja, & Balamuralidhar, 2019)**
   Deshpande, S., Yadav, P., Mutreja, G., & Balamuralidhar, P. (2019). Spectral Library of Indian Urban Materials–OGC Compatible Web Services "Tarang". *2019 10th Workshop on Hyperspectral Imaging and Signal Processing: Evolution in Remote Sensing (WHISPERS)*, (pp. 1–5). doi:10.1109/WHISPERS.2019.8921295

3. **(Deshpande, Inamdar, & Vin, 2017)**
   Deshpande, S. S., Inamdar, A. B., & Vin, H. M. (2017). Urban land use land cover discrimination using image based reflectance calibration methods for hyperspectral data. *Photogrammatric Engineering and Remote Sensing, 83*(5), 365–376.

4. **(Deshpande, Gupta, Yadav, Inamdar, & Vin, 2015)**
   Deshpande, S. S., Gupta, M., Yadav, P., Inamdar, A. B., & Vin, H. M. (2015). Spectral library search Tarang for Indian urban materials: OGC compatible web services for connected devices. *TACTiCs, TCS Technical Architect Conference*. Bangalore.

Appendix

6.3.1 CODE COMPONENTS

To access the code please follow the link. After reaching the repository the user will find the following components:

1. Data folder
2. Results folder
3. Scripts folder
4. Help
5. A README file, and
6. Requirements file.

Each folder also contains an additional README file wherever applicable.

1. Data folder: The Data folder contains all the raw data example files that are required as inputs to the scripts. It includes the image files and other supporting files required for running the scripts. The contents of the Data folder are described below.
    a. **asdlib** sub-folder is a folder containing signature files recorded by an ASD spectrometer for some of the urban samples discussed in Chapter 4. The sample signatures are used for optimal band analysis. They also can be used for any other purpose found suitable by the user.
    b. **EO1H1470472013112110KZ** is an EO1-Hyperion image used for many of the case studies (USGS, 2013-1). It is a Pune city image, a well-known city in western Maharashtra, India. The image was acquired on April 23, 2013. The image is of very good quality and can be used for hyperspectral data analysis. The ground truth is not available currently in the public domain. Please contact first the author in case you need further help.
    c. **6sinput** is a file required for atmospheric correction using 6sv radiometric transfer code. It is an input file taken as input by 6sv executable and it produces **xas, xbs, and xcs** coefficient files required for atmospheric correction (see Chapter 3 for details).
    d. **Bandcenters files** are the files for band centre wavelengths. A bandcenter csv file has three columns. The first column is for wavelength in nm, the second is for the band number, and the third column is for the inclusion or exclusion flag. 0 in this last column indicates the band is not considered for calculation and 1 indicates it is included. There are many bandcenter files created as required, most of them for Hyperion. For example, bandcenter-heiden file uses the bands significant for urban discrimination in Heiden et al. (Heiden, Segl, Roessner, & Kaufmann, 2007). asdbandcenters is a file similar to Hyperion bandcenters except it has band centre wavelengths for ASD spectrometer.
    e. **expr7_2_signaturesapr22_r3** this is a training or testing data file for SAM or any other classification exercise. The file has 5 columns. The first 4 columns indicate starting row, ending row, starting column, and ending column respectively. The row and column numbers are for the

Pune image in the Data folder. Values in the fifth column are class labels, and the last column indicates if the area bounded by the row-column coordinates is part of training or testing. For example, record "1448 1456 208 218 moshi 0" indicate that the area having class label "moshi" which is a stone quarry is not considered as a training area for the experiment. The record "2200 2204 118 123 mutha fp 1" indicates the area bounded by the coordinates according to the first four columns as a "mutha flood plain" (representative of an open area/soil) and is the training area for soil class. The "moshi" is bounded by rows 1448 to 1456 and 208 to 218 columns and son. The format of the training file or testing file remains the same. In the case of evaluation of classification results, the classes included (marked as 1 in the last column of the file) are used for calculating precision and recall. The class labels and the detailed description of labels can be found in the Chapters 3 and 4.

 f. **hirrad** is the file showing irradiance for Hyperion bands, it is the same file in the appendix (6.2, "Hyperion top of the atmosphere irradiance (Wavelength in nm; Irradiance in W m-2 μm^{-1})"). These values are required in atmospheric corrections as per the EO-1 equation in Chapter 3.

2. **Results** folder contains the sample results created by the scripts in the "Scripts" folder.
3. **Scripts** folder are the folders with Python scripts. They are explained one by one below briefly, and later in relatively more detail. The explanation is not in alphabetical order. We explain them in the order of image processing steps.

 a. **iarv3** stands for internal area relative reflectance. The name is because of the development legacy and contrary to what the name may suggest it does have code for flat field reflectance, simple Hyperion equation, log residual and so on.

 b. **sixscorr** is a file exactly similar to the **iarv3** file with the only difference last part where the pixel vector is divided by a reference spectrum. Instead, the code part is replaced by the correction factors multiplication calculated by 6sv executable. Those factors are stored in **xas, xbs, and xcs** files.

 c. **sam** is a small library of classifiers. It includes SAM implementation as explained in Chapter 5. In addition, it includes many of the conventional classifiers implemented in the "sklearn" Python library. It provides the services of training and testing for the consumer script. Many of the classifiers or wrapper function around that is designed for simplicity. They take a training data file as an input (as explained in the "**Data**" folder section), a list of pixel vectors, and a list of labels. It produces class labels for the query pixels and calculates the accuracies if the labels are provided.

 d. **crossvalscript-kfold** is a script file for evaluating the results by the classifiers in SAM by applying random data splits for training and testing data.

 e. **Bdistasdbatch, bdistv3, plotasdsignatures** these all files are a group of files required to perform the optimal band analysis discussed in Chapter 4. **Plotasdsignatures** is a script which reads the **asdlib** folders structure and creates an array of signatures from each folder of the working directory. For example, **Plotasdsignatures** will create an array of signatures from Cement C_Concrete folder in **asdlib** working directory, **bdistv3** is a script that calculates the Bhattacharyya distance given the signatures of two classes (which are arranged in the folders of **asdlib**). And the last script **Bdistasdbatch** is a wrapper function that takes class name string, working directory etc. as inputs and runs it for multiple folders. It calculates the Bhattacharyya distance between each class pair for a different number of bands and other algorithm parameters and aggregates the result (see "Most significant wavelengths for discrimination of urban classes" of Chapter 4 for more details of the algorithm). The code can be repurposed for image classes as well.

   f. **MLP, 1DCNN, VGG, SP2-CNN-Chen** are the deep learning model files written in Keras – a Python API for machine learning library TensorFlow. Keras is a popular programming interface for building and deploying deep learning models quickly. MLP is a simple fully connected network. **1DCNN** is a model treating the hyperspectral pixel as a D-dimensional row vector, and convolution is over spectral bands. VGG is a standard VGG implementation (Simonyan & Zisserman, 2014). **SP2-CNN-Chen** is a spectral-spatial convolutional neural network as implemented in (Chen, Zhao, & Jia, 2015). Please see Chapter 5 for more details on **SP2-CNN-Chen**. Other models are given in the Section 6.3.3, "Deep learning models".

4. The Help folder provides detailed instructions for running the script files. It also provides additional details to understand the code better. The help is organized according to the scripts.
5. The README file describes the files in detail.
6. Requirements mentions the versions of necessary libraries required to run. No installation is required. The correct versions of the libraries can be configured in Google Colab and the Colab will use the appropriate version to run the scripts.

You can find the README file in the **Script** folder for a more detailed description of each of the scripts. At present, the files are set with the variables to work with the data in the Data folder. The user needs to just run the script on Colab.

6.3.2 Main Calibration Script

We discuss the iarv3 script in detail here, which is one of the most important scripts.

1. The following imports are required to run the script: numpy, pyhdf.SD, pyhdf.SDC, matplotlib.pyplot imshow, pylab, matplotlib.image AxesImage, scipy.interpolate interp1d, scipy.ndimage zoom, createBandCenters. createBandCenters is a script that reads the band centre files and loads the band centre array in the calling program.

2. There are multiple supporting functions, for example, for smoothing the spectral signature for plotting, returning the list of pixels, labels, signature array for the given coordinates, average signature for the given coordinate, and so on.
3. The script begins by opening the Hyperion hdf file, say "EO1H1470472002342110PZ.L1R", into a 3-dimensional array.
4. At times processing the entire image is not required. It may not run successfully on low RAM (say 8GB and below) machines as well. In any of the cases, a copy of the image with the required number of rows is created and used.
5. The next step is removing the unwanted channels, dropping duplicate bands and so on. The reasons for dropping each band may be different. For example, a few initial bands and some of the last bands are not calibrated and hence they are removed.
6. Next, an array of scaling factors for calibrating the bands is created. VNIR, and SWIR bands have different scaling factors and hence the two separate arrays and created and then joined.
7. Next, each pixel of the image is divided by the new scaling factors. Direct dimension-wise array division would have been efficient. However, a simple loop is implemented as the size of the array in some of the early cases was going out of bounds.
8. Next, a simple dark object or improved dark object is applied to the image.
9. Next the normalization section is optional. This is an equal area of normalization applied to pixels for normalizing the illumination differences. For SAM, it does not make any difference. However, any method that considers the amplitude difference between two spectra, this method should not be applied as it shifts the entire spectrum upward or downward during the normalization.
10. Next, IAR, FAR, a simple Hyperion equation can be applied to the image processed so far to get the reflectance values.
11. The generated signatures can be analysed by using "plotsignatures" to plot the signatures.
12. IAR is applied by taking some over all the pixels in the image and the flatfield signature is loaded using the flatfield file which is similar to the signature file.

6.3.3 Deep Learning Models

Tables from 6.3 to 6.5 provide the details of models used in Chapter 5, Table 5.17. Please remember that these are generic models, the output shape needs to be changed as per the requirement of the classification experiment.

TABLE 6.3
MLP

Layer	Kernel Size	Step size	Feature maps	Dropout	Parameters
Input			28x28		0
Flatten			784		0
Dense			128	0.2	100486
Dense			10	-	1290

TABLE 6.4
1DCNN

Layer	Kernel Size	Step size	Feature maps	Dropout	Parameters
Convolution	3	1	(198, 64)		256
Convolution	3	1	(196, 64)	0.5	12352
MaxPooling	2		(98, 64)		0
Flatten			6272		0
Dense			100		627300
Dense			16		1616

TABLE 6.5
VGG

Layer	Kernel Size	Step size	Feature maps	Dropout	Parameters
Image			224x224x3		0
Convolution	3	1	224x224x64		1792
Convolution	3	1	224x224x64		36928
MaxPooling	3x3	1	112x112x64		0
Convolution	3	1	112x112x128		73858
Convolution	3	1	112x112x128		147584
MaxPooling	3x3	1	56x56x128		0
Convolution	3	1	56x56x256		295168
Convolution	3	1	56x56x256		590080
MaxPooling	3x3	1	28x28x256		0
Convolution	3	1	28x28x512		1180160
Convolution	3	1	28x28x512		2359808
Convolution	3	1	28x28x512		2359808
MaxPooling	3x3	1	14x14x512		0
Convolution	3		14x14x512		2359808
Convolution	3		14x14x512		2359808
Convolution	3		14x14x512		2359808
MaxPooling			7x7x512		0
Flatten			25088		0
Dense			4096		102764544
Dense			4096		16781312
Dense			1000		4097000

WORKS CITED

Chen, Y., Zhao, X., & Jia, X. (2015). Spectral-spatial classification of hyperspectral data based on deep belief network. *IEEE Journal of Selected Topics in Applied Earth Observations and Remote Sensing, 8*(6), 2381–2392.

Deshpande, S. S., Gupta, M., Yadav, P., Inamdar, A. B., & Vin, H. M. (2015). Spectral library search Tarang for Indian urban materials: OGC compatible web services for connected devices. *TACTiCs, TCS Technical Architect Conference.* Bangalore.

Deshpande, S. S., Inamdar, A. B., & Vin, H. M. (2017). Urban land use land cover discrimination using image based reflectance calibration methods for hyperspectral data. *Photogrammatric Engineering and Remote Sensing, 83*(5), 365–376.

Deshpande, S. S., Inamdar, A. B., & Vin, H. M. (2019). Spectral library and discrimination analysis of Indian urban materials. *Journal of the Indian Society of Remote Sensing, 47*, 867–877.

Deshpande, S., Yadav, P., Mutreja, G., & Balamuralidhar, P. (2019). Spectral Library of Indian Urban Materials–OGC Compatible Web Services "Tarang". *2019 10th Workshop on Hyperspectral Imaging and Signal Processing: Evolution in Remote Sensing (WHISPERS)*, (pp. 1–5). doi:10.1109/WHISPERS.2019.8921295

Heiden, U., Segl, K., Roessner, S., & Kaufmann, H. (2007, December). Determination of robust spectral features for identification of urban surface materials in hyperspectral remote sensing data. *Remote Sensing of Environment, 111*(4), 537–552.

Simonyan, K., & Zisserman, A. (2014). Very deep convolutional networks for large-scale image recognition. *arXiv preprint arXiv:1409.1556*.

USGS. (2011-1). *Earth Observing 1 (EO-1)-FAQs*. Retrieved December 16, 2015, from www.usgs.gov/media/files/hyperion-irradiance-band-information

USGS. (2013-1, April 22). EO1 Hyperion Scene: EO1H1470472013112110KZ_PF1_01, for Target Path 147, and Target Row 47. *EO1 Hyperion Scene: EO1H1470472013112110KZ_PF1_01, for Target Path 147, and Target Row 47*. Sioux Falls, SD USA: U. S. Geological Survey. Retrieved from http://earthexplorer.usgs.gov

Index

A

Absolute reflectance, 96, 129
Activation function, 238, 247
Additive component, 73, 87, 95, 98, 105
Additive effects *see* Additive component
Aerosol model, 77, 79, 80
  average continental, 79
  boundary layer, 79, 80
  continental, 79, 80
  dust, 79–81
  maritime, 79, 80
  oceanic, 79, 80, 122
  soluble aerosols, 79, 80
  soot, 79, 80
  stratosphere, 81
  troposphere, 79, 80
  urban, 79, 80
Again, 5, 8, 9, 11, 16, 34, 82, 98, 99, 129, 238, 244
Air pollution
  carbon dioxide, 13, 27, 31, 78
  particulate matter, 29, 30
  sulphur dioxide, 29
Airborne platform, 52, 183
  AVIRIS, 33, 56, 58, 61, 128, 130, 167, 199, 202, 205, 206, 260, 261
  AVIRIS-NG, 56, 58
  HyMap, 58, 71, 200, 205
  HyspIRI, 58, 131
  HyTES, 58, 71
Analytical spectral devices (ASD), 127, 129–130, 133
  FieldSpec, 3, 133
Apparent reflectance
Artificial colour, 147, 160, 161
Artificial neural network, 206, 229, 232, 252
Asphalt sample, 129
  pavement, 5
  road/bitumen, 167, 265, 108
  surface, 65, 165
ASTER, 126, 130, 205, 262
At surface reflectance, 66
Atmospheric absorption, 82, 97
  residual, 97, 99, 114, 260, 270
  water vapour, 78, 89, 100, 209
Atmospheric conditions, 83, 84, 88, 95, 99, 135
Atmospheric correction, 59, 66, 73, 74, 76–79, 81, 88, 95, 99, 119, 120, 197, 269
  image-based, 59, 73, 74, 77, 78, 82, 95, 97, 100, 105–107, 113–115, 119, 197

  physics-based, 73, 74, 76–81, 100, 113–115, 117, 119
Atmospheric effects, 6, 43, 74–76, 77, 94, 95, 100, 105, 197, 199, 205, 260, 263
Atmospheric model, 78, 84, 92, 93, 95, 99, 105
  Mid-Latitude Summer, 79
  Mid-Latitude Winter, 79
  Sub-Arctic Summer, 79
  Sub-Arctic Winter, 79
  tropical, 79, 209
  U.S. Standard, 79
Atmospheric scattering, 76, 83, 87, 88, 92
Atmospheric transmittance, 81, 88
Average spectrum, 97, 98
AVIRIS *see* Airborne platforms
AVIRIS-NG *see* Airborne platforms

B

Band centre, 45, 53, 84–86, 87, 91, 170, 193, 269, 271
Bandwidth, 50
Banyan tree, 104, 136, *see also* Tarang samples
Bare soil, 67, 94, 104, 107, 112, 115, 119, 148, 152, 167, 169, 179, 184, 189, 219–221, 229, 247, 248
Bayes' theorem, 202
Bhattacharyya distance, 144, 151–153, 171, 179, 204, 271
B-distance, 144–146, 151–155, *see also* Bhattacharyya distance
Bitumen composites, 144, 148, 149, 152, 154, 169
Bottom-up model, 29
Bright soil, 10, 126, 179
Brightness temperature, 8
Broad band, 43, 50, 53, 69, 191
Broadening, 44
Brown soil, 104, 136

C

Calibrated bands, 85
Calibration
  calibration performance, 109, 113
  calibration-SLEUTH model, 22
  reflectance calibration, 73–77, 95, 98, 106, 107, 111, 112, 114–119, 268
CDP 27, 28, 30–1
Cellular automata (CA), 22
Chromatic discrimination, 147, 161
Chromatic properties, 147, 157

275

Index

CIE, 147, 148, 157, 161
  CIE colour-matching function, 147
  CIE Lab space, 157, 161
  CIE 1964 Standard observer, 147, 148
Class label, 146, 176–180, 193, 196, 197, 204, 232, 236, 238, 270
Classification, 43, 59, 61, 105–108, 115, 117, 118, 166, 175–182, 188–191, 193, 195–197, 199–209, 211, 213, 215, 217, 219, 221–229, 231–233, 235–239, 241, 243–245, 247, 249–257, 260–262, 264, 268–270, 272, 273
  deep learning, 69, 175, 229, 236, 238, 239, 252, 254, 259, 271, 272
  k-NN, 175
  maximum likelihood (ML), 182, 204
  random forest, 67, 121, 243, 250
  SAM, 107, 164, 167, 197, 199, 200, 202, 217, 222, 224–229, 270
  semi-supervised, 206, 210, 214, 215, 217, 229, 250, 252
  spectral angle mapper, 67, 105, 164, 178, 181, 182, 196, 197, 199, 214, 258
  supervised classification, 43, 176, 203, 204, 217, 222, 224, 226, 255
  SVM, 23, 30, 67, 182, 191, 192, 206, 215, 217, 225–227, 241, 243, 250, 253, 261, 262, 264, 265
  unsupervised classification, 43, 176, *see also* Clustering
  URBAN INDICES, 175, 182, 188, 190, 193, 194, 206
  VIS classification 107, 118, 119, 184, 185
Clay minerals, 131
Climate change, 2, 17, 27, 37
Clustering, 59, 214, 215, 217, 225, 228, 252
Coarse resolution, 171, 181, 244
Colour, 50, 62, 64, 128, 133, 135–137, 143, 147, 148, 150, 152, 153, 155–157, 160, 161, 169, 171, 172, 187, 207
  colour composite, 59, 62
  false colour, 62
Colour conversion, 147
Colour matching function, 147
Colour sensation, 148
Colour vision, 147
Concrete, 2, 3, 44, 46, 64, 70, 94, 97, 98, 101, 104, 105, 107, 108, 110–113, 115–117, 119, 126, 129–130, 131, 135, 136, 141, 143, 148, 153, 157, 160, 161, 167, 169, 170, 173, 174, 188, 189–192, 195, 197, 199, 216, 219, 220–222, 229, 271
  degradation, 44, 189
  pavement, 98, 135, 148, 169
  surface, 44, 135, 167
Construction material, 129, 131, 136

Continuum, 45, 46, 67, 202, 254
Convolutional Kernel, 233
Convolutional layer, 239, 244
Convolutional neural network, 182, 232, 243, 250, 252, 254, 271
Convolutional spectral features, 243
Covariance-difference, 144
Covariance matrix, 144, 203, 204
Cross-validation, 217, 218, 225, 247

D

Dark object subtraction, 82, 87, 91, 94, 99
Dark pixel, 82
Database, 4, 121, 162, 163, 165, 252
Deegree 3 WPS, 164
Dense layers, 232, 237, 239
Depth separable convolution, 234
Diagnostic absorption, 45, 47, 48, 69, 108, 119, 141, 148, 155, 167, 169, 188, 190, 244, 245
Diagnostic band, 178
Diagnostic feature, 42, 67, 125, 146, 171, 245, 248
Diagnostic reflection, 150, 169
Digital elevation model (DEM), 6, 24
Digital number (DN), 9, 24, 65, 119, 209, 268
Digital spectral library, 126, 127, 172, 174
Dimensions, 41, 51, 52, 60, 179, 180, 183, 198, 204, 217, 234, 238, 244
Discriminant function, 182, 202, 203, 250
DISORT, 81
Drainage, 3–8, 209
  artificial drainage, 3, 5
  drainage carrying capacity, 3
  drainage density, 4, 37
  natural drainage, 3–7, 209
Drone, 52, 60, 183
Dry grass, 62, 63, 67, 101, 102, 104, 110, 112, 161, 210, 218, 221

E

ECOSTRESS, 130–132, 171
EDGAR, 27
Electromagnetic energy, 42, 44, 45, 47
Electromagnetic radiation, 42, 46, 48, 69, 71
Electromagnetic spectrum, 146
Emissions, 2, 13, 15, 17, 19, 21, 22, 27–32, 34, 207
  CH_4 30, 78
  CH_4-methane 78
  carbon dioxide (CO_2), 2, 13, 15, 17, 19–22, 27, 29–32, 55, 78, 207
  particulate matter, 2, 29, 30
  scope 1 emissions, 27, 28, 30–32
  scope 2 emissions, 28, 31

Index

sulphur dioxide (SO2), 29
Emission spectrum, 42
Empirical line method, 97–99
Enhanced thematic mapper (ETM), 8, 9, 190, 194, 195, 205, 261, 262

F

False positive, 205, 224, 227
FAR, 95–97, 106, 107, 111–116, 119, 174, 191, 195, 212, 237, 240, 245, 252, 256, 272
Farms, 62, 104
FCLS, 30, 166
Feature fusion, 236, 237
Feature map, 232, 234, 236, 239, 242, 243, 246, 272, 273
Feature selection, 145
Field measurements, 69, 98, 99, 105, 133–135
52o North WPS, 164, 171
Filter size, 234, 235, 237, 239, 246
Fine resolution, 48
FLAASH, 78, 81, 121, 199
Flat field, 95, 97, 98, 105, 107, 111–115, 123, 270
  concrete pavements, 98, 148, 169
  construction site, 98, 102
  dam faces, 98
  industrial roof, 65, 98, 101, 104, 107, 111, 113, 114, 117, 119, 210, 212, 219–222, 229
  playground, 63, 98, 101, 102, 104, 108, 110, 112, 113, 119, 136, 148, 169, 222
  Playa, 97, 98
  residential roof, 219, 221
  salt flats, 98
  sandy beaches, 98
Flat field relative reflectance (FAR), 95, 97
Flood, 2–6, 34, 36, 37, 176, 270
  flash flood, 2, 5
  urban flood, 3, 5, 6, 34, 36
  water logging, 6
Forest fire, 104, 190
Forward model, 29
Fully connected layers, 232, 239, 244
FWHM, 44, 45, 57, 58, 60, 69, 89, 91, 100, 133, 209

G

Gain, 5, 8, 9, 11, 16, 34, 82, 98, 99, 238, 244
Galvanized iron (GI), 150, 221
GDP, 25, 26
GER 1500, 133
GIS, 1, 6, 24, 37, 54, 57, 68, 173, 182, 209, 230, 238, 250, 253, 255
Global effects, 2
Global warming, 2, 27
GPC, 30, 36
Green grass, 62, 101, 102, 104, 185, 212
Gulmohar, 136, see also Tarang samples

H

Hamming distance, 200, 201
Handcrafted features, 182, 232, 250
Haze removal, 82
Heatwaves, 7
Hidden layer, 231, 238
Hidden Markov Model (HMM), 25-7
High resolution, 8, 53, 59, 253, 264, 265
Histogram method, 83, see also Dark object subtraction
Hue, 62, 67, 194, 254
HYDICE, 61
Hydrological cycle, 2–4
Hydrological structure, 4
Hydroxides, 132
HYMAP, 58, 71, 200, 205, see also Airborne platforms
Hyperion, 51, 59, 61, 62, 67, 86–89, 100, 101, 147, 166, 207, 209–212, 259, 261–263, 266, 269, 270, 272
Hyperspectral
  hyperspectral data, 3, 32–34, 41–43, 48, 49, 51, 52
  hyperspectral image, 42, 59, 65, 67, 176, 179, 181–183, 188
HyspIRI see Airborne platforms
HyTES see Airborne platforms

I

IAR, 95–98, 105–117, 119, 272
ICLEI, 30, 36
Image interpretation, 121
Imaging spectroscopy, 48, 49, 120, 171, 172, 199, 251
Impact of urbanization, 1, 3–5, 7, 9, 11
Impervious surface, 1–5, 7, 19, 21, 22, 24, 26, 31, 32, 44, 50, 64, 65, 72, 107, 110, 111, 113, 116, 119, 122, 123, 147, 157, 161, 166, 167, 176, 183, 187, 205, 209, 219, 221, 222, 229, 251, 254, 255, 257, 262, 264
  spatial distribution, 4, 17, 19, 20, 220
Improved dark object subtraction, 87, 91
Index-based methods, 189
Indicator framework, 12, 34, 38
Induced absorption, 46
Induced emission, 46, 47
Input layer, 230, 231, 234
Intensity, 8, 9, 41, 42, 44, 45, 76, 148
  UHI, 8, 9
Internal area relative reflectance, 270
Inverse models, 29
IPCC, 17, 30

J

JPL, 130

K

Keras, 234, 252, 271

L

Land use and land cover, 2, 8, 15, 20, 24, 32, 59, 109, 175, 177, 179, 181–183, 185, 187, 189, 191, 193, 195, 197, 199, 201, 203, 205, 207, 209, 211, 213, 215, 217, 219, 221, 223, 225, 227, 229, 231, 233, 235, 237, 239, 241, 243, 245, 247, 249, 251, 253, 255, 257
    LULC models, 17
    LU/LC, 2, 20, 183, 204, 205, 260, 261, 264, 265
    land cover, 2–9, 15, 17, 22, 43, 49, 100, 104, 105, 168, 176, 177, 182–191, 206
    land use, 2, 10, 20, 59, 100, 104, 105, 109, 113, 148, 155, 163, 175–177, 181–187, 205, 206, 207, 208, 218
Landsat, 9, 30–31, 51, 84, 87, 170
Leodoit-Wolf estimator, 144
Line profile, 44, 45
Local neighbourhood, 181, 236–239
Locally connected layers, 232
Log residuals, 105
Logistic regression, 26, 182, 230, 238
Logistic regression layer, 238

M

Machine learning, 20, 51, 67, 108, 175, 176, 180, 182, 204, 206, 229, 271
    ANN, 206, *see also* Artificial neural network
    CNN/(CNN), 182, 208, 232, 236, 239, 243, 250
    deep learning (DL), 69, 175, 229, 236, 238, 239, 252, 254, 259, 271, 272
    maximum likelihood (ML), 181, 182, 190, 191, 202, 204, 206, 250, 256
    Naive Bayes (NB), 53, 71, 88, 100, 181, 182, 189, 190, 243, 253, 254
    spectral angle mapper (SAM), 107, 164, 167, 197, 199, 200, 202, 217, 222, 224–229, 270
    SVM, 23, 30, 67, 182, 191, 192, 206, 215, 217, 225–227, 241, 243, 250, 253, 261, 262, 264, 265
Mango tree, 104, 136, *see also* Tarang samples
Markov chain, 26, 27
Mast tree, 136, *see also* Tarang samples
Maxpooling, 232, 242, 273
Mean-difference, 144
Medium resolution, 208
Metabolic parameter, 17, 19–21

Metal roofs, 104, 136, 148, 150, 152, 153, 155, 156, 169, 190–192
Minerals, 50, 131, 132
    arsenates, 132
    borates 10, 132
    carbonates, 132
    elements, 132
    halides, 132
    oxides, 132
    phosphates 132
    silicates, 132
    sulphates, 81, 132
    sulphides, 132
    tungstate, 132
Mixture signature, 101, 104, 107, 127, 264
Moderate resolution, 81, 182, 183
MODTRAN, 78, 81, 120, 122
Moore-Penrose pseudoinverse, 145
Mortality, 7
Multi-layer perceptron (MLP), 231
Multiplicative effects, 82
    absorption, 41–48, 50, 66, 67, 69, 73, 74, 77, 78, 81, 82, 91, 97, 100, 105, 108, 114, 116, 119, 128, 131, 141, 148, 149, 151, 155, 161, 167–169, 188, 190, 197, 202, 205, 206, 244, 245, 248
    scattering, 66, 73, 74, 76–79, 81–84, 86–88, 92, 94, 97, 119, 120, 122, 133
Multispec, 62, 145
Multispectral, 41–43, 47–51, 69, 74, 84, 87, 191, 236, 244

N

Narrow band, 48–50, 87, 191
NASA, 9, 52, 56, 58, 69, 71, 120, 122, 123, 128, 130, 172, 173, 251
Neural network, 37, 122, 176, 179, 182, 189, 206, 229–232, 236–240, 243, 250, 252–255, 271
Neuron, 230, 231, 244
Non-clay minerals, 131
NS3, 108, 119, 121, 164, 199, 200, 243, 255

O

Offset, 98, 99
OGC, 127, 162–164, 172, 173, 268, 273, 274
1D convolution, 232, 233, 237, 239, 244
Optimal bands, 125
Optimal flat field, 113
Optimal span, 145, 154–156
Optimal spectral range, 126
Output layer, 230, 231, 238
Overlap correction, 87, 88, 91–95, 120
Ozone, 78, 79

Index

P

Path radiance, 73, 76, 82–85, 87–95, 98, 99, 105
Peepal, 136, *see also* Tarang samples
Pixel vector, 41, 51, 59, 66, 97, 180, 182, 183, 198, 199, 201, 202, 232, 233, 235, 238, 239, 244, 245, 247, 248, 249, 270
Plains, 4, 5, 102, 104, 210, 212, 218, 221
Platform
  airborne, 52, 183
  remote sensing, 7, 37
  satellite, 8, 52
  sensing, 7, 32, 162
  spaceborne, 51, 52, 69, 250
Precision, 27, 50, 56, 166, 248, 270
Producer's accuracy, 105, 106, 110–117, 205, 217, 222, 224–229
Propagation factor, 84, 91, 92, *see also* Scaling factor
Prototype-based learning methods, 108
Proxies, 4, 19–21, 33, 185, 186, 208, 253
Proxy features, 19, 21, 30, 175, 183, 185, 207, 218, 228, 250
Pseudo inverse, 179
Pure pixels, 30, 101, 107, 109, 179, 210, 214, 221
Py6S, 78, 79, 123

Q

Quarry, 102, 104, 110–112, 114–116, 119, 161, 270

R

Radiance, 8, 9, 23, 43, 61, 65, 66, 69, 73, 74, 76–78, 81–85, 87–96, 98–100, 105, 113, 119, 122, 209, 214, 250, 259, 261, 266–268, 270
  at sensor radiance 43, 88
Radiative transfer code, 77, 81, 120, 197
Raintree, 104, 136, *see also* Tarang samples
Recall, 248, 270
Rectified Linear Unit (ReLU), 247
Reference pixel, 105, 107, 197
Reference sites, 98, 99, 103, 104, 213
Reflectance, 23, 30, 42–44, 50, 53, 61, 66, 67, 69, 70, 73–77, 81, 83, 85, 87–89, 91, 93–101, 103, 105–109, 111–123, 128, 129, 131, 141–146, 148, 150–152, 155–157, 162, 163, 167–169, 172, 190, 197, 199, 200, 205, 206, 214, 217, 248, 249, 252, 265, 268, 270, 272, 274
Reflectance calibration, 73–77, 79, 81, 83, 85, 87, 89, 91, 93, 95–99, 101, 103, 105–107, 109, 111, 113–117, 119–121, 123, 172, 252, 268, 274
Regression, 8, 26, 30–31, 67, 98, 99, 172, 177, 181, 182, 206, 230, 238, 250, 251, 261, 262
Regularized regression, 31
Relative reflectance, 95–97, 105, 114, 129, 214, 270
Relative scattering model, 83, 84
  clear, 84, 86, 87, 92, 93, 95, 105
  hazy 84, 86, 87, 93
  moderate, 84, 86, 87
  very clear, 84, 92, 95, 105
  very hazy, 84, 86, 87, 93
Resample, 127, 167
Residential
  area, 105, 107, 111, 112, 119, 148, 207, 208, 218, 220
  land use class, 111, 112, 229
  up-market residential area, 207, 208, 218, 220
  low-economy residential area, 220
Resilience, 11–14
RGB, 41, 42, 50, 148, 179, 180, 239
Runoff, 2–5, 7, 21, 37, 39, 209, 257, 261

S

SAM, 67, 105, 107, 164, 178, 182, 196, 197, 199, 200, 202, 214, 217, 222, 224–229, 258, 270, *see also* Spectral Angle Mapper
Sampling interval, 45, 46, 69, 133
Saturation, 62, 64, 67
Scaling factor, 84–87, 91–93, 95, 272
Scan line corrector (SLC), 6, 34, 38
Scattering model, 83, 84, 86, 87, 92
Scope 1 emissions, 27, 28, 30–2
Scope 2 emissions, 28, 30–31
SDG, 12, 14, 16, 17, 32, 34, 38
SDG goals, 12
  SDG indicator framework (*see* Indicator framework)
Separability 144, 151–153, 167, *see also* Bhattacharyya distance, B-distance
Significant span, 145, 146
Singular matrix, 144, 179, 204
6SV, 77–79, 88, 93, 94, 110, 113–117, 119, 123, 269, 270
SLEUTH, 22
Social economic environmental/ecological (SEE), 10–12, 14, 15, 19, 32
Softmax, 247
Solar elevation angle, 23–4, 66
Solar irradiance, 77, 88
Spaceborne platforms, 51, 52, 69, 250
  ASI, 53, 57
  CHIME, 55, 57, 70
  CHRIS, 53, 57, 70, 205, 262
  DESIS, 54, 57, 70

Index

DLR, 54, 55, 57, 70, 189
EnMAP, 54, 55, 57, 70
EO-1 Hyperion 51, 67, 76, 82, 83, 85–87, 91, 92, 118, 207, 209
HISUI, 54, 57, 69, 71
Hyperion, 51, 59, 61, 62, 67, 86–89, 100–102, 105, 107, 114, 118, 120, 123, 147, 161, 166, 174, 199, 207, 209–212, 256, 259, 261–263, 266, 269, 270, 272, 274
HySpecIQ, 56, 71
international space station (ISS), 54, 57
ISA, 53
METI, 54, 57
Orbital sidekick (OSK), 55
Pixxel, 55, 69, 72
PRISMA, 53, 57, 69, 70
PROBA-1, 1, 53, 57, 70
SHALOM, 53, 70
Span 145, 146, 154–159, 232–235, 240
Spatial features, 59, 68, 175, 177, 178, 181–183, 207, 208, 228, 237, 239, 240, 243, 244, 248, 257, 261, 262
Spatial resolution, 6, 9, 32, 37, 49, 51–58, 89, 100, 120, 121, 171, 182, 247, 255
Spatial signatures, 59, 64, 67
Spatio-temporal, 16, 25, 39, 187, 189, 222
Spectral characteristics, 44, 49, 50, 88, 141, 142, 166, 167
Spectral deconvolution, 89, 122
Spectral features, 67–69, 127, 131, 149, 157, 169, 172, 176, 181–183, 202, 205, 206, 228, 236–238, 243–245, 247–250, 252, 253, 274
Spectral indices, 174, 190, 256
   BSI, 192, 195
   NDBI, 190–192, 194
   NDVI, 9, 188, 189, 191, 194, 253
   RVI, 188
   SAVI, 123, 191, 192, 194, 254
Spectral library, 43, 121, 125–127, 130–132, 141, 142, 161, 164, 165, 168, 170–174, 198, 200, 205, 253, 255, 268, 273, 274
Spectral line, 44, 45, 171
Spectral matching, 43, 50, 52, 66, 67, 108, 141, 175, 183, 199, 201, 254
Spectral processing services, 163, 164
   match (*see* Spectral processing services)
   search (*see* Spectral processing services)
   view (*see* Spectral processing services)
Spectral resolution, 42, 44, 47, 49, 54–56, 74, 89, 90, 100, 107, 114, 119, 125–127, 147, 167, 171, 190, 204, 209
Spectral response, 43, 45, 69, 73, 88–90, 100, 101, 209, 210
Spectral signatures, 43–45, 49, 52, 59, 62, 66, 67, 69, 98, 119, 126, 129, 130, 132, 143, 144, 149–151, 166–169, 171, 179, 181, 190, 205, 245, 248, 249, 252
Spectral-spatial features, 175
Spectra Vista Corporation, 133
Spectrometer, 33, 38, 45, 53, 54, 56, 58, 70–72, 120–123, 127, 128–129, 130, 133, 135, 136, 141, 167, 192, 254, 269
   ASD, 60, 127–130, 133, 135, 142, 147, 160, 167, 171, 172, 260, 269, 270, 271
   Beckman, 127, 129, 130, 132
   Field, 69, 127, 167, 178, 192, 197
   laboratory, 127, 197
   Nicolet, 129, 130–132
   SVC, 133, 135, 142, 146, 151, 174
Spectroscopy, 44, 48, 49, 70–72, 120, 128, 171, 172, 174, 199, 245, 250, 251, 256
Spectrum, 32, 41, 69, 87, 88, 91, 96–98, 105, 125, 128–131, 141–144, 146, 147, 156, 163, 164, 167, 168, 198–202, 236, 244, 245, 247–250, 253, 272
Splib06a, 127, 172
Spontaneous emission, 46, 47
SRF, 69, 87–92, 95, 209
   overlap, 88, 92, 95
Standard deviation, 91, 141, 142, 146
Stochastic Gradient Descent (SGD), 247
Subabul, 104, 136, *see also* Tarang samples
Superspectral, 50
Supervised machine learning, 175, 176, 180, 206
Surface reflectance, 43, 66, 76, 120, 121
Surface runoff, 5
Surface temperature, 7–9
Sustainability, 1, 10–12, 14, 15, 32–34, 36, 256
SWIR, 56, 57, 60, 72, 141–144, 146–148, 150–152, 154–157, 167, 169, 170, 190, 195, 272

T

Tarang, 132, 133, 135–137, 142, 149–153, 161–166, 168, 171, 172, 268, 273, 274
Tarang samples, 135–137
   Acacia arabica, 136
   Albizia julibrissin 136
   Albizia saman, 136
   American Lawn Grass, 136
   Asbestos/Swastik, 136
   Brown soil, 104, 136
   cement concrete, 108, 135, 136, 141, 143, 148, 153, 169, 199
   Corrugated GI sheet, 136
   Delonix regia 136
   Ficus bengalensis, 136
   Fibre (Fbr), 136
   Mangalore tiles, 136, 141

Index

Mangifera indica, 136
metal roofs, 104, 136, 148, 150, 152, 153, 155, 156, 169, 190, 191, 192
  plain GI sheet, 136
  Polyalthia longifolia, 136
  Taiwan Lawn Grass, 136
  Terminalia catappa, 136
Target pixel, 67, 125, 181, 182, 198, 237, 238, 239
Temporal resolution, 53, 57, 58
Temporal variables, 25
Testing data, 240, 269
Thematic class, 101, 111, 152
Thematic map, 53, 68, 87, 189, 190
Thematic mapper (TM), 87, 190
3D convolution, 232–235, 237, 240, 254
TOA, 43, 76
Top of atmosphere, 43, 88
Top of atmosphere reflectance, 23
Training data, 30, 176, 178, 181, 203, 204, 205, 214, 232, 239, 240, 270
Training data size *see* Training samples
Training samples, 179, 193, 196, 204, 217, 218, 238
Tri stimulus, 147
True positive, 227
True reflectance, 43, 66, 74, 119
2D convolution, 233, 234, 245, 249

U

UAV, 52, 60, 181
Ultraspectral, 50
Unmixing, 59, 61, 125, 165, 166, 172, 173, 179, 185, 187, 190, 199, 206, 208, 214–216, 251, 261, 262
Unsupervised machine learning, 175, 176, *see also* Clustering
Urban catchment, 1, 3, 4, 6, 7, 39, 257
Urban growth, 19, 20, 22–26, 38, 39, 50, 175, 187, 208, 257
Urban growth sprawl, 23
Urban heat island, 7–9
Urban indices, 8, 175, 182, 188, 190, 193, 194, 206
Urbanization, 1–7, 9, 11, 15, 23, 24, 183, 190, 209
  local effect, 2
  global effect, 2
Urban materials, 44, 125, 126, 130, 132, 133, 142, 147, 148, 149, 151, 152, 154, 156, 161, 166, 167, 169, 170, 171, 172, 179, 185, 190, 191, 205, 210, 268, 273, 274
  asbestos roof sheets, 137
  aggregates, 135, 141, 149, 152, 153, 169, 230, 271
  cement composites, 143, 147, 148, 149, 150, 152, 153, 154, 155, 156, 169, 179
  clay tiles, 136, 141, 143, 150–152, 169
  colour-coated materials, 136, 143, 149–150
  concrete, 2, 3, 5, 7, 42, 44, 46, 64, 70, 94, 97, 98, 101, 104, 105, 107, 108, 110, 111, 112, 113, 115, 116, 117, 119, 125, 126, 129–130, 131, 135, 136, 141, 143, 148, 153, 157, 160, 161, 167, 169, 170, 173, 174, 188, 189, 190, 191, 192, 195, 197, 199, 216, 219, 220, 221, 222, 229, 271
  fibre reinforced plastic (FRP), 137
  pavement blocks, 135, 141, 143, 148, 153, 174
Urban metabolism, 1, 15, 16, 17, 18, 19, 20, 22, 29, 32, 34, 36
Urban vegetation, 133, 253
User interface, 129, 130
User's accuracy, 105, 106, 110–117, 119, 205, 217, 222, 224–228
USGS, 6, 9, 34, 38, 51–53, 62, 70, 72, 76–78, 88, 94, 100, 101, 107, 120, 122, 123, 127–130, 166, 170–175, 183, 184, 190, 195, 204, 205, 207, 209, 251, 256, 257, 262, 266, 269, 274
  classification, 184, 204, 205
  splib07a, 129–130
  spectral library, 173

V

VIS, 9, 19, 20, 22, 24, 25, 26, 30, 31, 37, 38, 41–44, 47, 50, 58, 59, 71, 72, 85, 93–95, 101, 105–109, 113–115, 117–121, 133, 142–149, 154–157, 161, 166, 167, 169, 173–176, 180, 183–185, 187, 195, 198, 199, 202–204, 206–208, 210, 213–218, 221, 222, 224–229, 244, 248–256, 258, 260–264, 272
Visual interpretation, 30, 41, 59, 155, 199
VNIR, 56, 57, 60, 61, 85, 141–148, 151–152, 154, 156, 167, 169, 272
Vulnerability, 1, 12, 14, 15, 187, 188, 207, 252, 253, 255

W

Water absorption, 44, 105, 168, 197
Water bodies, 14, 22, 62, 82, 85, 104, 105
Water hyacinth, 62, 111, 116
Watershed, 4–6, 38, 184
Water vapour, 78, 79, 100, 209
Wavelength range, 44–46, 58, 78, 88, 113, 128, 130–131, 133, 145, 146, 148, 154, 156, 193, 195, 205, 206, 245, 247, 248
Web engine, 163–165
Web interface, 162, 165

Web services, 161, 162, 164–166, 171, 172, 268, 273, 274
Weighted deconvolution, 91, 120
White plate reference, 133
WPS, 162–164, 171, 173
  deegree 3 WPS, 164

52o North WPS, 164, 171
ZOO WPS, 164

Z

ZOO WPS, 164

Printed in the United States
by Baker & Taylor Publisher Services